AF675713

Operational Art

This book describes and explains the theory of operational art – an intermediate field of theory and practice between tactics and strategy.

This book is one of relatively few works in the English language on the theory of operational art. It deals predominantly with the theory of operational art as applied to preparing, planning and conducting a high-intensity conventional war. In contrast to other works, the book is written in broad terms, rather than focusing on national military culture and experience, which allows more flexibility in adjusting to new evidence. The book explains the theory of operational art in all three physical dimensions (land, sea, and air/space) and describes/analyses various aspects of operational art in general terms, using historical examples to facilitate the understanding of the theory. The case studies focus on examples of major campaigns and operations in both the more recent and distant history of warfare, including the current Ukraine–Russia war.

This book will be of much interest to students of strategic studies, military history, defense studies, and International Relations, as well as military professionals.

Milan Vego is University Professor and Admiral R. K. Turner Professor of Operational Art at the Naval War College, Rhode Island, United States. He is the author of fifteen books, including *Operational Warfare At Sea* (2009, 2017). He is also the author of the textbook *Joint Operational Warfare* (2007, 2009), used at the Naval War College and several staff colleges in the United States and overseas.

Routledge Advances in Defence Studies

Series editors: Timothy Clack
University of Oxford, UK
Advisory Board: Tarak Barkawi
London School of Economics, UK
Richard Barrons
Global Strategy Forum, UK
Kari BingenTytler
Center for Strategic and International Studies, USA
Ori Brafman
University of California, Berkeley, USA
Tom Copinger-Symes
British Army, UK
Karen Gibsen
Purdue University, USA
David Gioe
West Point, USA
Robert Johnson
Oxford University, UK
Mara Karlin
John Hopkins University, USA
Tony King
Warwick University, UK
Benedict Kite
British Army, UK
Andrew Sharpe
Centre for Historical and Conflict Research, UK
Suzanne Raine
Cambridge University, UK

Routledge Advances in Defence Studies is a multi-disciplinary series examining innovations, disruptions, counter-culture histories, and unconventional approaches to understanding contemporary forms, challenges, logics, frameworks, and technologies of national defence. This is the first series explicitly dedicated to examining the impact of radical change on national security and the construction of theoretical and imagined disruptions to existing structures, practices, and behaviours in the defence community of practice. The purpose of this series is to establish a first-class intellectual home for conceptually challenging and empirically authoritative studies that offer insight, clarity, and sustained focus.

Understanding UK Defence Exports
The International Trade in Defence Capabilities
John Louth

Climate Change, Conflict, and (In)Security
Hot War
Edited by Timothy Clack, Ziya Meral, and Louise Selisny

Defence Offsets and the Global Arms Trade
Explaining Cross-National Variations
Jonata Anicetti

Governance and Intervention in Mali
Elusive Security
Susanna D. Wing

Great Power Cyber Competition
Competing and Winning in the Information Environment
Edited by David V. Gioe and Margaret W. Smith

Human Factors in Effective Counter-Terrorism
A Comparative Study
Richard Warnes

The Russia-Ukraine War
Towards Resilient Fighting Power
Viktoriya Fedorchak

Defence Planning for Small and Middle Powers
Rethinking Force Development in an Age of Disruption
Edited by Tim Sweijs, Saskia van Genugten and Frans Osinga

European Total Defence
Past, Present and Future
Edited by Gjermund Forfang Rongved

Operational Art
Theory and Practice
Milan Vego

For more information about this series, please visit: www.routledge.com/Routledge-Advances-in-Defence-Studies/book-series/RAIDS

Operational Art

Theory and Practice

Milan Vego

Routledge
Taylor & Francis Group
LONDON AND NEW YORK

First published 2025
by Routledge
4 Park Square, Milton Park, Abingdon, Oxon OX14 4RN

and by Routledge
605 Third Avenue, New York, NY 10158

Routledge is an imprint of the Taylor & Francis Group, an informa business

© 2025 Milan Vego

The right of Milan Vego to be identified as author of this work has been asserted in accordance with sections 77 and 78 of the Copyright, Designs and Patents Act 1988.

All rights reserved. No part of this book may be reprinted or reproduced or utilised in any form or by any electronic, mechanical, or other means, now known or hereafter invented, including photocopying and recording, or in any information storage or retrieval system, without permission in writing from the publishers.

Trademark notice: Product or corporate names may be trademarks or registered trademarks, and are used only for identification and explanation without intent to infringe.

British Library Cataloguing-in-Publication Data
A catalogue record for this book is available from the British Library

ISBN: 978-1-032-36769-9 (hbk)
ISBN: 978-1-032-36770-5 (pbk)
ISBN: 978-1-003-33368-5 (ebk)

DOI: 10.4324/9781003333685

Typeset in Times New Roman
by Apex CoVantage, LLC

Contents

Figures

Abbreviations

AA	antiaircraft
AAA	antiaircraft artillery
ABCCC	airborne command and control center
ABDA	American–British–Dutch–Australian (Command)
ACR	armored cavalry regiment
AFSOUTH	Allied Forces South
AOR	area of responsibility
ASCM	anti-ship cruise missile
ASM	air-to-surface missile
ATCM	air traffic control measures
ATO	air tasking order
BDA	battle damage assessment
BEF	British Expeditionary Force
BMD	ballistic missile defense
C2	command and control
CAOC	Combined Air Operations Center
CBI	China-Burma-India
CBRN	chemical, biological, radiological, nuclear
CCD	cover, concealment and deception
CCS	Combined Chiefs of Staff
CENTCOM	(U.S.) Central Command
CIA	Central Intelligence Agency
CINC	commander in chief
CINCPac	Commander in Chief Pacific Fleet
CJCS	Chairman of the Joint Chiefs of Staff
CJTF	Combined Joint Task Force
COMINCH	Commander in Chief, United States Fleet
COMSOPAC	Commander, South Pacific Area
DE	directed energy
EA	electronic attack
EMP	electromagnetic pulse
EMS	electromagnetic spectrum
EP	electronic protection

ES	electronic support
EW	electronic warfare
FEAF	Far East Air Force
FM	field manual
GCC	Gulf Cooperation Council
HNS	host nation support
HUMINT	human intelligence
I&W	indications and warning
IAF	Israeli Air Force
IDF	Israeli Defense Forces
IGHQ	Imperial General Headquarters
IJN	Imperial Japanese Navy
INFOSYS	information system
IRG	Iraqi Republican Guards
ITO	Iraqi Theater of Operations
IW	information warfare
IW-D	defensive information warfare
JCS	Joint Chiefs of Staff
JFACC	Joint Force Air Component Commander
JFCE	Joint Forces Command – East
JFLCC	Joint Force Land Component Commander
JFMCC	Joint Force Maritime Component Commander
JFSOCC	Joint Force Special Operations Commander
JIC	Joint Intelligence Center
JP	Joint Publication
JTF	Joint Task Force
KTO	Kuwaiti Theater of Operations
LACM	Land-Attack Cruise Missile
LCT	Landing Craft, Tank
MCM	mine countermeasures
MEB	Marine expeditionary brigade
MILDEC	military deception
MTB	motor torpedo boat
NAAF	Northwest African Air Forces
NASAF	Northwest African Strategic Air Force
NATAF	Northwest African Tactical Air Force
NEI	Netherlands East Indies
NSFS	naval surface fire support
OHL	Oberste Heeresleitung
OKH	Oberkommando des Heeres
OKL	Oberkommando der Luftwaffe
OKM	Oberkommando der Marine
OKW	Oberkommando der Wehrmacht
OMG	operational maneuver group
OOB	order of battle

OPSEC	operations security
POL	petroleum, oil and lubricants
POW	prisoner of war
PRC	People's Republic of China
PSYOP	psychological operation
RAAF	Royal Australian Air Force
RAF	Royal Air Force
RDF	radio direction finding
RKKA	Workers' and Peasants' Red Army (Soviet)
ROE	rules of engagement
ROK	Republic of Korea
SACEUR	Supreme Allied Commander Europe
SAM	surface-to-air missile
SAS	Special Air Service
SHAPE	Supreme Headquarters Allied Command Europe
SIGINT	signals intelligence
SOPAC	South Pacific Area
SP	self-propelled
SSN	nuclear-powered attack submarine
SSO	Special Security Organization (Iraqi)
SWPA	Southwest Pacific Area
TF	task force
TG	task group
TO&E	table of organization and equipment
TOAS	Teatro de Operaciones del Atlantico Sur
TOM	Teatro de Operaciones Malvinas
TV	*teatr voyennykh* (military theater)
UAV	unmanned surface vehicle
UN	United Nations
UNC	United Nations Command
USAAF	U.S. Army Air Forces
USFK	U.S. Forces Korea
USV	unmanned surface vehicle
UUV	unmanned underwater vehicle
WMD	weapons of mass destruction

Preface

This work is based on the revised/updated version of the *Joint Operational Warfare. Theory and Practice* published in 2007 (reprinted in 2009) and now titled *Theory of Operational Warfare*. As the new title implies, the focus is on the theory of operational art illustrated by selected historical examples. This work deals predominantly with the theory of operational art as applied for preparing, planning and conducting a high-intensity conventional war.

The term "operational art" (*operativnoye iskusstvo*) was for the first time used by the former Imperial Russia's General Aleksander A. Svechin in his book *Strategy* (*Strategiya*) published in 1926. In its original meaning, it pertained to what the Soviets/Russians called the study of "operations" (or major operations, in U.S./NATO terms). The operational art is an intermediate component of the military art or art of the war between strategy and tactics. It consists of theory and practice. The application of operational warfare is an art, while its study is based on scientific principles.

Since the end of the Napoleonic Wars in 1815, technological and societal changes have been the principal factors in the steady evolution of warfare at all levels, especially at the operational level. Operational art has played an irreplaceable and critical role in orchestrating tactical actions for the accomplishment of strategic and operational objectives through the planning, preparation and conduct of campaigns and major operations. The gap between tactics and strategy is too large to be bridged by physical combat only. While full mastery of tactics is, and will remain, fundamental for success in combat, there is no greater error than to assume that technology and tactics alone are sufficient for winning the nation's wars. The excessive focus on tactics of platforms and weapons/sensors reduces all fighting to simple targeting and shooting. In too many cases, tactical considerations have come to not only dominate operational art but also heavily influence policy and strategy. History also provides numerous examples of what happens when strategy and policy are neglected or, even worse, completely ignored. All three components of military art – strategy, tactics and operational art – must be in harmony; otherwise, success in war will be elusive at best, or simply unattainable.

The author's intent was, by using diverse and open sources, to provide a synthesis of the most important theoretical aspects of operational art. This work represents the author's interpretations of the selected literature on the subject. Readers should critically read and evaluate the author's views on the subject, read other sources and then make their own judgment.

The author's intent was not to present a set of firm and unchangeable rules or beliefs on operational art. Sound military theory should be general to allow enough flexibility to adjust to new empirical evidence. One should avoid making assertions without any proof. Experience shows that no new technologies, no matter how advanced, can replace operational art or other components of the art of war. In all eras, new technologies have always affected the character of war, but not its nature, as articulated by General Carl von Clausewitz. Any war is a clash of opposing wills, uncertainties and friction abound. There are invariably unforeseen events, often caused by pure chance or simple luck. Human actions and reactions in war cannot be predicted or measured with any degree of certainty. All too often, human behavior and actions can be irrational. All this makes the course and outcome of any war highly unpredictable and uncertain.

This work consists of ten chapters. Chapter 1, "Introduction," defines and describes the historical evolution of operational art, its importance and the main characteristics. Chapter 2, "The Objectives," explains and illustrates distinctions among military-/theater-strategic, operational and tactical objectives, respectively. Chapter 3, "Operational Factors," explains and analyzes the operationally/strategically significant elements of the factors of space, time and force – individually and then collectively.

Chapter 4, "Methods," defines and explains the main methods of combat forces' employment aimed to accomplish a strategic objective – campaigns and operational objectives – major operations. Chapter 5, "Elements," elaborates in some detail the main elements of major operations and campaigns, specifically, operational/strategic deployment, center of gravity, operational maneuver, point of culmination, strategic/operational deception and operational reserve.

Chapter 6, "Operational Command and Control," describes the purpose and key elements of operational command and control and related operational organization. Chapter 7, "Operational Support," describes and evaluates the key components of support for a major operation or campaign (operational intelligence, information warfare, fires, logistics and protection, respectively). Chapter 8, "Operational Design," explains and assesses the main elements of the design of major operations and campaigns: ultimate/intermediate objectives; identifying operational/strategic centers of gravity; geostrategic central/exterior position; interior/exterior lines of operation, operational/strategic direction (thrust); and the operational idea and its main elements.

Chapter 9, "Operational Thinking," defines and describes the importance, the main attributes and methods of acquiring operational thinking. The operational commanders and their staff must think broadly and far into the future instead of thinking more nearly and tactically. They need to know and understand not only military but also non-military aspects of a situation in a given theater. Chapter 10, "Tenets," describes in some detail the main requirements for successful performance of the operational commanders in a war. They are based on the historical experiences of many wars. The main tenets include acceptance by the operational commanders of the absolute primacy of policy and strategy, unwavering and firm focus on the objectives, obtaining and maintaining the freedom to act, exercising the imitative, using overwhelming power, high speed of action, taking a high and prudent risk, achieving surprise and creativeness in employing combat forces.

This work does not end with a separate chapter "Conclusion" because it deals with military theory. A sound military is never definitive but open-ended. New empirical evidence or reevaluation of past experiences should result in modifications, amendments or even drastic changes in certain aspects of military theory.

To facilitate the understanding of theory, historical examples have been liberally used throughout the work. The focus is on examples of campaigns and major operations in the past century. However, whenever possible, examples of more recent conflicts, such as the Gulf War I (1990–1991), the war in Iraq in 2003 (Operation Iraqi Freedom) and the most recent Ukraine–Russia War have been presented. The only purpose of these examples is to illustrate various theoretical points or assertions made in the text. Some historical examples provide more details; others, less. A more detailed discussion of various historical examples would obscure the theoretical point to be made. To understand the points being made, a broad knowledge of past wars, and especially past campaigns and major operations, is a prerequisite.

This work would not have been possible without the understanding and staunch support of John Porado, CAPT, U.S. Navy, Chairman of Joint Military Operations, U.S. Naval War College and his deputy Fred B. Horne, CAPT, U.S. Navy (Retired). I'm also greatly in debt to Andrew Humphrys, Senior Editor, Military Strategic and Security Studies at Routledge, for his support and great patience in dealing with my delays in submitting the manuscript. Many thanks to Dr Tim Clack, General Editor of the Routledge Advances in Defence Studies series, for his support and prompt approval of my book manuscript. Thanks also to Max Gethings, editorial assistant to Andrew Humphrys, for preparing my manuscript for publication. I am also greatly appreciative of all the work and patience by Mr. Promoth Jaikishan and his team at Apex CoVantage for their copyediting work and the production team at Routledge.

1 Introduction

Modern warfare has conclusively demonstrated that although successes can be achieved by force-on-force combat or attrition and relying primarily on advanced technologies, such cases are rare. This is especially true in a war between two strong opponents. Even when attritional warfare has been successful, as in the allied victory on the Western Front in World War I, too heavy a price was paid in losses of personnel and materiel, and sorely needed time. All wars of attrition are inherently protracted. Hostilities are more likely to end in a stalemate rather than in a decisive victory.

The gap between tactics on the one hand and strategy and policy on the other cannot be overcome by physical combat alone. The tactical framework is too narrow and the strategic perspective too broad to ensure the most decisive employment of employment of combat forces. Therefore, the need for an intermediate field of theory and practice between strategy and tactics seems obvious. This third component – called "operational art" today – serves as an interface and a bridge between strategy and tactics. The results of tactical actions are useful only when linked together as part of a larger design framed by strategy and orchestrated by operational art. Sound application of the tenets of operational art would ensure the most effective employment of combat forces and achieve decisive results. It would shorten the hostilities and result in the least losses for friendly forces. The smaller the one's forces, the more critical is to master operational art.

The scope of the study and practice of operational art is large. Thus, it is difficult to define in both precise and all-encompassing terms. Generically, operational art can be defined as *a component of the art of war concerned with the theory and practice of planning and executing major operations and campaigns aimed at accomplishing operational or strategic objectives in a given theater*. Operational art can be applied across the entire spectrum of warfare, from operations short of war to high-intensity conventional war. The full richness of operational art can only be appreciated in a high-intensity conventional war.

For all the well-documented and proven importance of operational art in winning wars decisively and with the least loss of resources and time, too many military theoreticians and practitioners have neglected or, even worse, ignored operational art – the components of the art of war between strategy and tactics. Emphasis is all too often given today to advanced technologies and tactics, while operational art is

DOI: 10.4324/9781003333685-1

given a short shrift. Yet by themselves, technological advances, numerical superiority and brilliant tactical performance are inadequate to achieve ultimate success in war. A sound, coherent strategy combined with operational excellence was the key to winning wars in the past, today, and will remain so for the future.

Problems of Terms

Many different terms were used in the past in referring directly or indirectly to operational art. There is a lack of consensus in both defining what constitutes operational art and its true purpose. A well-known an influential French General Count Jacques Antoine Hippolyte de Guibert was credited to be the first to use the term "grand tactics." He wrote that grand tactics is properly the science of the general in chief, because it is the compilation of all military knowledge. Napoleon I never used the term "strategy," preferring "grand tactics" or what he called "higher part of war."[1]

The Swiss-born French General Antoine-Henri de Jomini believed that the art of war consisted of five distinctive parts, including grand tactics (today, operational art), and minor tactics (today, tactics).[2] He defined grand tactics as "the art of making good combinations preliminary to battles, as well as during their progress."[3] The British General J. F. C. Fuller) also used the term "grand tactics" in referring to the intermediate field of study and practice between strategy and tactics.[4]

Until the beginning of the nineteenth century, the military in German states differentiated only between strategy and tactics.[5] The term "strategic action" (*strategische Handlung*) was used as the technical military expression for movements in a campaign that led to a major battle (*Schlacht*). It pertained to the skillful bringing of one's forces to obtain a favorable initial position and for strategic exploitation of a victory.[6] After 1866, Field Marshal Helmuth von Moltke, Sr, the chief of the Prussian (and later German) general staff, used the term "operation" (*Operation*) almost exclusively to mean the movement of large forces prior to a major battle.[7] In his view, a major battle was a matter of strategy, and not an integral part of an operation. He introduced into German military theory the term *operativ* (loosely translated as "operational"), pertaining to an operation.[8] Yet he did not elevate the "operation" to a third component of military art; rather, he considered it a part of the lower spectrum of strategy. Moltke, Sr, like Clausewitz, still accepted the division of military art into two components – strategy and tactics.[9] By the late nineteenth century, "operations" emerged as an intermediate area of study and practice in the German military.[10] An operation pertained to a movement of an army from its the base of operation to a scene of a major battle.

By the turn of the twentieth century, the Russian theoreticians used the terms grand tactics and applied strategy in referring to the intermediate level between strategy and tactics.[11] In 1907, based on the experiences of the Russo-Japanese War of 1904–1905, the Russians introduced a new term, "operatika" (pertaining to "operations"), to explain the new phenomena of armed conflict.[12] This term was attributed to A. V. Gerua and E. E. Messner. In their view, strategy dealt with waging war as a whole while "operatika" pertained to the conduct of the battle at the army level. Tactics was a domain of an army corps and lower command echelons. This early division of military art into strategy, *operatika*, and tactics was a significant development, because it helped to eventually separate operations from strategy and tactics.[13] The Polish General and theoretician Stefan Mossor used the term "operatika" in referring to the entire theory and practice of conducting operations (major operations in U.S. terms).[14] He also published book *Operatika* in 1945.

The Soviets/Russians and all formerly communist-dominated militaries used the term "operational art." In contrast, the Chinese military use the term "science of campaign." The term operational art was coined by the former tsarist general, and one of the most important Soviet military theoreticians, Aleksander A. Svechin in his book *Strategy* (*Strategiya*).[15] He explained that "tactical activity is governed by operational art (*operativnoye iskusstvo*).

> Combat operations are not self-contained. They are only the basic material from which an operation is formed. Only in very infrequent cases can one rely on achieving the ultimate objective of combat operations in a single battle. Normally, the path to the ultimate objective is broken down into a series of operations separated by more or less lengthy pauses, which take place in different areas in a theater and differ significantly from one another due to the differences between the immediate objectives one's forces temporarily strive for.[16]

For Svechin, operational art was focused on study and practice of "operations" (major operations in U.S. terms). He defined an operation as "an act of war if the efforts of troops are directed toward the achievement of certain intermediate objective in a certain theater of military operations (*teatr' voyennykh deystvii)* without any interruption."[17]

Between 1920s and the end of the former Soviet Union in 1991, the Soviets, except for some changes in wording retained Svechin's understanding of operational art. In the late 1980s, the Soviets described operational art as being

> concerned with the study of the rules, nature, and character of contemporary operations; working out the means for preparing and conducting combat operations; determining the function of large forces (fronts, armies) and formations (corps, divisions) of the armed forces; establishing means and methods for organizing and supporting continuing cooperation, security, and command and control of forces in combat; delineating the organizational and equipment requirements of large units of the armed forces; working out the nature and methods of operational training for officers and command and control organs; developing recommendations for the operational preparation of a theater of military operations; and investigating the enemy view on the conduct of operational-level military operations.[18]

The term operational art is somewhat misleading because it implies that it deals only with practice. However, like strategy and tactics, operational art is both an art and a theory.[19] Each component of military art demands reflection, study, and practical application. The art aspect means that operational art, like the other two components, is ambiguous and imprecise, requiring a creative and highly innovative approach to its application. Therefore, the term operations would be more appropriate, if not for the fact that within the U.S. military this term is so extensively and loosely used that it has lost its true meaning. The term operational warfare is perhaps the better term to use when referring to both the study and practice between strategy and tactics.

The U.S. Army's field manual (FM) 100-5 *Operations* published in 1986, defined operational art as "the employment of military forces to attain strategic goals in a theater of war or theater of operations through the design, organization, and conduct of campaigns and major operations."[20] This definition underwent some changes in 2001. Operational art was then defined as "a component of military art concerned with planning, coordinating, and sequencing the outcomes of individual tactical events into a chain of linked actions, within the context of major operations and campaigns, so as to achieve the objectives established by a theater strategy."[21] Since then, the U.S. Army revised several times definition of operational art. In 2010, operational art was described as "the application of creative imagination by commanders and staffs supported by their skill, knowledge, and experiences to design strategies, campaigns, and major operations and organize and employ military forces." Operational art "integrates ends, ways, and means across the levels of war."[22] Both the latest version of FM 3–0 *Operations* (2017) and Joint Publication (JP) 3–0: *Joint Operations* (2017) defined operational art as the "cognitive approach by commanders and staffs-supported by their skill, knowledge, experience, creativity, and judgment – to develop strategies, campaign, and operations to organize and employ military forces by integrating ends, ways, and means."[23] All these definitions contain several inaccuracies. For example, "designing strategies" is not a domain of operational art but strategy and policy. The terms "ends, ways, and means" are usually related to strategy not operational art. Operational art is not applied "across the levels of war" but only at the operational and military-/theater-strategic levels.

Emergence of Operational Warfare

The principal factors in the emergence of modern operational warfare on land were changes in society and their impact on the conduct of war, advances in science and technology, industrialization, the influence of the Prussian and German general staff and the influence of great military leaders and theoreticians. In the last decade of the eighteenth century the political and social revolution in France transformed the character of war. The advent of general conscription – the *levee en masse* – expanded the size of armies. This required extensive logistical support to sustain armies in the field. The new armies of literate citizens had a different relationship to society than the armies of the seventeenth and eighteenth centuries. The structure of the officer corps began to change, because many officers were increasingly drawn from non-aristocratic families. The French Revolution also started a trend in which nationalism and ideology played an increasing role in the preparation and conduct of war. Industrialization brought tremendous resources to the state. New armies could be created on the foundations of those that had been partially destroyed. A war required the conduct of several campaigns to come to an end.[24]

After the end of the Napoleonic Wars in 1815 the increased role of nationalism, ideology and public opinion brought a steady expansion of war's objectives. The latter included not only the defeat of the enemy's forces in the field but also the overthrow of the enemy's social system. The standing armies in Europe became larger. Instead of one army, there were several smaller armies and corps, dispersed over a large part of the theater and seldom brought physically together except on the field of battle.

Several revolutionary and numerous evolutionary advances in technology had a significant impact on both the practices and theory of land warfare in the mid- and late nineteenth century. The most important technological advances were rifled small arms, breech-loading rifled guns and machine guns; the invention of the railroad, steam propulsion and the internal combustion engine; and new means of communications (electric telegraph and telephone). The cumulative effect of these advances was the emergence of operational warfare.

By the mid and late nineteenth century, several technical developments increased the effectiveness of the infantry. In 1832, the expendable cylindro-conoidal bullet was invented, making musket loading much faster. The invention of the percussion cap significantly reduced the number of misfires. It also made the rifled musket five times more accurate than its smoothbore predecessor. Rifled small arms soon outranged smoothbore artillery. This led to the rifling of field guns. The first breech-loading gun was produced in England in 1854. Afterward, the ever-larger caliber and a more effective breech-loading mechanism were introduced into service. The major advance in the design of large-caliber guns came in the 1890s with wire winding. Another major development that subsequently had a major impact on land warfare was the machine gun, invented in 1885.

No technological advances had a larger impact on the planning and conduct of war than the railroad, steam engine and electric telegraph. The railroad revolutionized the mobilization and deployment of armies. The steam locomotive marked the greatest revolution in land transportation since the invention of the wheel. Railways were rapidly built across Europe in the 1830s and 1840s. The railroad could move troops and their supplies fifteen times faster than on foot. The railhead gradually replaced the Fortress Magazine as the collection point for military supplies. However, local forage remained important to armies beyond a certain distance from the railhead. Railroads in enemy territory had to be repaired and protected. As invading armies advanced into the depth of the enemy's territory, lines of supply were steadily extended. Their links with the nearest railhead had to be maintained and protected. The railway department became the most important section of every general staff in Europe, governing the planning of war.

By the end of the nineteenth century, large armies could not have been mobilized, maneuvered or supplied without railways. However, railways were in their essence an inflexible instrument. Their use could not be improvised, and their effectiveness was conditional on adherence to strict timetables kept under rigid centralized control. The effective use of railways during mobilization required detailed plans encompassing precise calculation of how many locomotives and railroad cars were required for each troop movement and their timeline.

The invention of the electric telegraph (in 1837) and Morse code (in 1844) allowed near-instantaneous transmission of messages over hundreds of miles between headquarters. However, on the battlefield itself signaling was still conducted with drums, bugles and flags. The telegraph fundamentally changed strategic command and control. It provided statesmen with better knowledge of the strategic situation. The speed of mobilization was increased because of the introduction of the railroad and the telegraph. The telegraph allowed the advancing army to keep in contact with distant headquarters. Instructions could be sent instantaneously to any headquarters within reach of a telegraph wire.

In the late nineteenth century, general conscription led to the establishment of large standing armies. This development, combined with the rapid advances in weapons and the advent of railroads and telegraph, led to dramatic changes in force–space–time relationships.[25] War fighting became influenced as never before by the political and economic aspects of the situation in a theater.[26] There were enormous problems in effectively controlling large armies in the field. For Moltke, Sr, answer was to further decentralize decision-making by introducing what was later known as the mission command.

Wars on land between strong opponents became progressively more protracted. Large battles and many minor tactical actions had to be fought continuously to defeat the enemy's fielded armies. The ultimate strategic objective could be accomplished only through a series of intermediate – "operational" – objectives. Increased firepower gave infantry greater scope for maneuver. With the development of open order tactics, a battle would start at a greater distance from the enemy lines and extend over a greater area. Warfare became an art to be pursued on the map, offering an infinitely greater number of possibilities to a commander than ever before. The higher commander became less a participant in a battle than a director of his force in battle.[27]

By the end of the nineteenth century the rudiments of modern operational warfare at sea also emerged because of the cumulative effects of a series of naval technological advances and concurrent changes in society and in the character of war on land. In contrast to war on land, the influence of society on war at sea was less profound than the impact of changes in military organization and advances in naval technology. The principal naval technological developments were steam propulsion, breech-loading guns, smokeless powder, new underwater weapons (mines and torpedoes) and new means of communications (electric telegraph, undersea cable, telephone and wireless telegraph). Steam propulsion allowed warships to attain higher speeds than by using sail, but initially this resulted in smaller ships with shorter range and endurance. The fleets became heavily dependent on shore bases for fuel and supplies. They required many auxiliary ships to carry coal.

By the turn of the twentieth century, advances in propulsion plants made it possible to fit powerful engines even on small ships. The introduction of torpedoes and mines led to the design of new, small platforms capable of attacking larger ships. This led to the proliferation of small warships of all types. Major navies of the day consisted of not only battleships and cruisers but also a large number of destroyers, torpedo craft, gunboats and auxiliaries. In the last decade before the outbreak of World War I, the major navies built both large and small naval vessels in great numbers. By 1914, the Royal Navy had in service 542 warships, including 68 battleships, 110 cruisers and 218 destroyers. The French Navy consisted of 357 ships (21 battleships, 30 cruisers, 83 destroyers, 153 torpedo boats and 70 submarines), the Imperial German Navy, 301 ships (37 battleships, 48 cruisers, 142 destroyers, 47 torpedo boats and 27 U-boats), the Imperial Russian Navy, 177 and the Austro-Hungarian Navy 105 warships.[28]

On the eve of World War I, the size of the potential naval battlefield had been increased due to the introduction of new platforms and improvements in radio communications. The introduction of submarines and aircraft added second and third dimensions to the battlefield. The ability to control the movements and actions of the fleet's elements over large ocean areas was enhanced by the invention of the new signaling devices and the wireless radio. Shore commands communicated with warships many hundreds of miles out at sea. Radio communications became essential for controlling large fleet elements during a battle. The cumulative effect of all these changes was that major navies of the day conducted major and diverse actions far from their basing areas. Most of the major navies planned to conduct a decisive battle (also called "general fleet action") intended to destroy the major part of the enemy fleet and thereby obtain command of the sea. Yet, radio was unable to ensure secure and reliable command and control of naval forces spread over a large ocean/sea area. In addition, the sheer size of major navies of the day with their widely dispersed home bases/installations made it increasingly difficult if not impossible to achieve decisive results by fighting a decisive naval battle.

Application of operational art was heavily influenced by the ideas of the classical military/naval thinkers. Jomini made a significant contribution to the understanding of the emerging operational level of war. He was considerably influenced by the British General and theoretician Henry H. E. Lloyd and the Prussian General and theoretician Friedrich von Tempelhoff. Napoleon I read Lloyd but had nothing good to say about his theory of war. Yet Jomini used precisely Lloyd's ideas to explain the military legacy of Napoleon I.[29]

Jomini explained that grand tactics were concerned, among other things, selecting positions to fight a battle; orders of battle or "grand" maneuvers prior to the attack; the clash of two armies on the march, or unexpected battles; and the arrangements for leading troops into a battle.[30] The main components of grand tactics were a secure base of operations, the main objective, bases of operations, strategic fronts, defense lines, zones of operations, lines of operations, lines of retreat, time-limited strategic fronts and strategic lines, objective points (or "decisive points" today) and natural and artificial obstacles. Jomini overemphasized the significance of geographic lines and terrain obstacles in the conduct of warfare.[31] He introduced into the military vocabulary the term "theater of operations." Jomini wrote that strategy conducts its operations on the strategic chessboard, the theater of war, while grand tactics uses the theater of operations as its chessboard. In contrast to Clausewitz, he almost absolutized the importance of occupying a central position and operating from interior lines.[32]

Based on his study of Napoleon I's way of warfare, Carl von Clausewitz wrote that war consists of two main phases: preparations for war and the war itself.[33] The first is influenced by strategic considerations, while the second is an integral part of combat.[34] Strategy designs the plan for a war (*Kriegsplan*) and plans of individual campaigns. Although Clausewitz did not spell out the existence of an intermediate domain between strategy and tactics, he clearly implied its existence. Among other things, Clausewitz introduced the concept of point of culmination (*Kulminationspunkt*) and "weight of main effort" (*Schwerpunkt*) or what later evolved by default into "center of gravity."

Jomini and Clausewitz differed in their understanding of an intermediate level of study and practice between strategy and tactics that emerged during the French Revolutionary and Napoleonic wars. While Clausewitz differentiated between the actions prior to and those during a major battle, he recognized only strategy and tactics. Jomini, in contrast, by introducing the term grand tactics, referred more clearly than Clausewitz did to the existence of the third domain of the study and practice of military art.

In the late nineteenth century, concurrently with the changes in society and the political environment, the introduction of new and more advanced weapons and equipment, and the new modes of transport and communications, dramatic changes occurred in the conduct of war. The most significant development (which was not readily apparent to contemporaries) was that by the late nineteenth century, military strategy, at its upper end, had become closely tied to national strategy and policy, while its lower end – concerned with the planning, preparation and movement of large forces prior to physical combat – emerged as "operations" (or operational art today).

Moltke, Sr, as chief of the general staff, was solely responsible for the conduct of operations – that is, execution of the campaign plan.[35] He made a clear distinction between the plan of an operation and war itself. Moltke, Sr, defined the objective of an operation as the enemy army as far as it protects the objective of war – that is, the territory, capital, resources and political power of the enemy.[36]

For Moltke, Sr, operations were a link between strategic planning and physical combat. In his view, there existed a logical hierarchy of thinking and levels of actions. The most important factor was the determination of strategic objectives, which provided military leadership with a practical purpose for action. These objectives could not be in any way contrary to the strategic objectives, for then they would endanger the very strategy. Moltke, Sr, wrote that "when in doubt, the higher level of thinking and actions always had priority. The same relationship existed between operational and tactical levels."[37]

Moltke, Sr, insisted that mass armies could not be led by the genius of a single commander and intuition alone. His solution was to further decentralize command and control through renewed emphasis on issuing mission-oriented orders. Moltke, Sr, firmly believed that a commander must have the ability to act as situation warrants it instead of waiting on orders. In his view, the greater the relationship among space, time and force, the greater the dependence of successful operational leadership on exercising a wide freedom of action and initiative. Only when all commanders at all levels have an ability for, and familiarity with, independent action is it possible to move the mass of one's forces with ease.[38]

By the late nineteenth century, operations emerged as an intermediate area of study and practice in the German military.[39] Field Marshal Alfred von Schlieffen, chief of the general staff, had to deal with the problem of command and control of the "million-men" armies. In his view, only by having a large mass and by exerting great effort could great things be achieved and great ideas realized. In contrast to Moltke, Sr, Schlieffen wanted to impose strict control over the actions of subordinate commanders. He was a man of the new era. The German telegraph troops were his invention. He was a strong supporter of heavy artillery. During Schlieffen's tenure, the capacity of the German railways for military purposes was dramatically increased.[40]

Schlieffen tried to solve Germany's dilemma of avoiding a two-front war by using strategic interior lines to achieve a decisive victory over France before it could be effectively helped by the Russians. He insisted that all calculations on a future war must be based on the decisive importance of the factor of time.[41] Schlieffen's name is inseparable from the terms "attack," "envelopment" and "annihilation." Yet his rigid views on the value of the offensive led to the neglect of the defensive at both the tactical and operational levels. Another negative consequence of his teaching was the underestimation of the work of statesmanship, both prior to and during a war.[42]

Schlieffen opposed a frontal attack at the operational level because he was concerned that it would lead to a war of attrition. He focused exclusively on annihilation through the envelopment maneuver. For him this was the only possibility for a numerically inferior force to achieve victory.[43] Yet this did not mean Schlieffen rejected the penetration maneuver under favorable conditions, if one could detect gaps in the enemy's defenses and then, while using deception, concentrate one's forces against such weak spots.[44]

A major problem with Schlieffen's teachings was his dogmatic insistence on inculcating common operational thinking through the solution of several tactical-strategic tasks. During numerous general staff rides, he tried to reinforce his teachings on using envelopment to achieve a decisive strategic victory. Schlieffen's use of the Battle of Cannae (in 216 BC) was a classic example of the highly suspect method of applying historical examples to modern conditions.

General Friedrich von Bernhardi, one of the leading German military theoreticians of his era, was also one of Schlieffen's fiercest critics. He believed that imposing common operational outlook had the latent danger of imposing conformism instead of fostering reliance on one's own intuitive response to the situation at hand.[45] Bernhardi was highly critical of Schlieffen and his disciples for emphasizing "material factors" and "numerical superiority" while ignoring the unquantifiable and most decisive factors – specifically, troop morale and discipline, training, and bold and brilliant leadership. Bernhardi warned the general staff against seeking firm rules that could limit freedom to act in the conduct of war.[46]

By the turn of the twentieth century, the Russian theoreticians used the terms "grand tactics" and "applied strategy" in referring to the intermediate level between strategy and tactics. They were very much influenced by the writings of Moltke, Sr, and his interpreters.[47] Several Russian theoreticians made significant contributions to the then-evolving theory of operational warfare. In fact, they laid the foundations for the Soviet theory of operational art that evolved in the 1920s and 1930s. The works of Generals Genrykh A. Leer and Colonel (later Major General) Aleksander A. Neznamov are unknown in the west. Both were teachers at the prestigious Nicholas Military Academy in Sankt Petersburg. For political and ideological reasons, the Soviets did not appreciate their contribution to the works of their theoreticians of operational art.

Leer published the first edition of his *Strategy (Strategiya)* in 1869. He divided strategy into two parts: ideal and practical. Ideal strategy dealt in the broadest sense with the philosophy of military affairs. Practical strategy in the narrow sense was more concrete, and Leer called it "tactics of the theater of military actions," which he equated with what Napoleon I called "higher tactics" (or operational art, in modern terms).[48]

Neznamov's lectures on operations dealt with the rudiments of operational art.[49] His ideas were suspect to his peers and students, because it was widely believed that he was expounding foreign, German views.[50] In his work *Modern War: The Activities of a Field Army (Sovremennaya voina: Deystviia polevoy army)*, published in 1912, Neznamov reaffirmed the primacy of strategy over operations. He insisted that strategy determines the overall military objective and attempts to achieve it through the grouping of various operations. Strategy also determines the relative importance of the various theaters of war and the forces assigned to each theater. In practical terms, this involves drawing up a plan for strategic deployment and determining strategic objectives at the beginning of a war. This in effect constitutes the first operational plan. During the war, strategy continues to regulate operations by orienting the fronts' (army group, in Western terms) axes of advance and allocating reinforcements and by otherwise reacting to changes in the overall situation.[51]

Prior to 1914, the most influential blue-water naval thinkers were Admiral Alfred T. Mahan and Julian S. Corbett. Both explained in some detail many aspects of the struggle for sea control. Mahan only made distinction between naval strategy and tactics. Like Jomini, he considered logistics as one of the components of the art of war. Mahan explained that the line between strategy and tactics was the point of contact between opposing forces. He insisted that whether engaged in strategic deployment or tactical maneuver, the correct course of action is to distribute one's force as to be superior to the enemy in one quarter, while holding the enemy in check in other quarters and for long enough to allow one's main attack to reach its full result. Operating from a central position, it is possible to mount a naval offensive along interior lines outward from the center, enabling the attacker to keep his enemy separated and therefore inferior by concentration against one unit while holding the other in check.[52] Borrowing many concepts that were applied in land warfare, Mahan adopted from Jomini three key ideas: the inherent value of a strategic central or interior position, the principle of concentration and the close relationship between logistics and combat.[53]

Corbett was the first blue-water naval theoretician who tried to delineate a strategy for a maritime power engaged in war with a superior continental power. He was also the first naval strategist who thought seriously about the direct contribution that maritime strategy could make to a war on land. In his study of maritime strategy, Corbett made a distinction between what he called *major strategy* (or grand strategy) and *minor strategy*. The first deals with the purpose of war, including international relations and economic functions. In contrast, minor strategy is concerned with particulars about waging war, including planning army, navy or combined operations.[54] He also differentiated between maritime and naval strategy. Corbett asserted that "naval strategy is but that part of it which determines the movements of a fleet when maritime strategy has determined what part the fleet must play in relation to the actions of land forces." Obviously, he confused naval strategy with what is in today's terms understood as operational warfare at sea. In contrast to other naval theoreticians of the day, Corbett correctly observed that "it is almost impossible that a war can be decided by naval action alone."[55]

Between 1871 and 1914, military thinking in Europe was influenced by tactics. The main areas of discussion dealt with weapons, the size of one's forces in relation to the size of theater, tactical mobility and methods of combat employment. This focus led to the neglect of operational thinking. In other words, what was missing was the link between national strategy and tactics that would ensure that the aims of policy and strategy were accomplished.[56]

By the eve of World War I, all the major European powers had large standing armies that could be vastly expanded through mass mobilization. Mobility was increased, and with it the speed of deployment and concentration. The introduction of longer-range, more precise and more lethal weapons, combined with the increased distance at which one's forces could be effectively controlled, caused a steady expansion of the potential battlefield. Yet narrow tactical thinking predominated in the conduct of war.[57]

In Germany, the most important and most decisive factor was the lack of recognition of the linkage between war and policy, the economy and the morale. Harmony between the spiritual side and tactical-technical conditions was a prerequisite for tactical success and hence for strategic victory. That the war resulted in positional warfare, although all the armies of the warring states were trained for and determined to conduct mobile warfare, is the best proof of the failure of the then-prevalent theory of war.[58]

After the failure of the initial German offensive in August–September 1914, the war in the west became a war of attrition. The 440-mile continuous front line extended from the Franco-Swiss border all the way to the English Channel. The enormous increase in defensive fires did not allow the conduct of operational-scale envelopments. Penetrations of the front could not be expanded to operational success due to enemy fires and the lack of one's own reserves.[59] Offensive operations became a struggle against fortifications.[60] Only in central and east European theaters were large-scale operations conducted.[61]

In World War I, armored cars were used on all fronts. Tanks were used in large numbers for the first time in the Battle of Cambrai in November 1917. In addition to infantry, field artillery and aircraft, tanks were also used in the allied offensive in the West in July 1918. Both the German and Entente offensives in 1918 represented the first rudiments of combined arms warfare. During World War I, several large-scale fleet-versus-fleet actions took place, in the North Sea, the Mediterranean, the Adriatic, the Baltic and the Black Sea. The battle of Jutland (Battle of Skagerrak for the Germans), fought on 31 May/01 June 1916, came closest to what can be considered a major naval operation. Both sides aimed to destroy major part of the opposing fleet, thereby accomplishing an operational objective in the North Sea. The battle consisted of several gunnery duels between the main forces and a series of minor actions aimed to provide combat support to the respective battle fleets.[62] The German High Seas Fleet (*Hochseeflotte*) tactically won the battle of Jutland. Operationally it was clearly a British victory. Yet the Grand Fleet was unable to detach its forces to operate beyond the confines of the North Sea.[63]

The Anglo-French amphibious landing at Gallipoli in April 1915 was the first example of a major amphibious operation. The initial and operational objective for the allies was to capture the Gallipoli peninsula. The political strategic objective was to open direct link with the embattled Russia, taking Turkey out of war and impress on Greece to openly side with the Entente Powers.[64] Neither of these objectives was accomplished. A major reason for failing to accomplish the operational objective was that the plan did not integrate naval attack with the landing of ground forces.[65] The initial attack by the Anglo-French heavy ships against the fortifications in the Dardanelles on 19 February and 18 March 1915 failed to sufficiently weaken Turkish defenses. The landing by ground troops was not conducted until 25 April 1915. This gave the Turks and their German advisors sufficient time to strengthen the beach defenses. Another, but more successful operation was the German amphibious landing in Riga Bay in October 1917 (Operation Albion). The operational objective of the landing was to open the Gulf of Riga and thereby threaten the rear of the Russian Twelfth Army, defending the Baltic coast. The Germans achieved complete surprise, and the entire operation was a resounding success. The Russian 12th Army was destroyed, and the threat to the flank of the German 8th Army was eliminated.[66]

World War I witnessed the emergence of the first maritime campaign. In 1914–1918, the Germans conducted, what can be seen in retrospect, an offensive campaign aimed to cutoff the Entente's maritime trade, while the Entente navies fought a defensive campaign to defend and protect their maritime trade. The German campaign was conducted primarily by U-boats, but surface commerce raiders took part as well.

The lessons of World War I, combined with new technological advances, led to new operational concepts in the interwar years. In land warfare, the principal debate in the interwar years concerned the future of motorization, particularly the role of tanks. In the West, the most important early proponents of new technologies and mobile warfare on land were the British theoreticians General John F. C. Fuller and Captain B. H. Liddell Hart. Fuller was the author of the revolutionary proposal

"Plan 1919," which envisioned a force of about 5,000 tanks with close air support, aimed to penetrate some twenty miles deep into German defenses, thereby paralyzing their command and control. Liddell Hart suggested a four-stage plan for the transformation of the British army into the "New Model" army, capable of operating independently off roads and railways and advancing up to some one hundred miles per day. He envisioned the need for infantry as an integral part of his proposed mechanized force. In contrast, Fuller assigned infantry a secondary role: protection of fixed bases and lines of supplies.[67]

Both Fuller and Liddell Hart were strong proponents of airpower. They insisted that aircraft should be used in support of ground troops by bombing enemy positions, thereby allowing friendly forces to penetrate enemy lines and carry out lightning-like blows. Subsequently, one's forces would conduct area-wide operations aimed to attack the enemy's critical vulnerabilities and encircling and destroy or annihilate the enemy forces. Fuller wrote that once the grand strategist has correlated and adjusted one's forces to the political object, the next step is to endow them with structure so that they can operate, and this is the responsibility of the grand tactician. The latter takes over the forces as they are distributed and arranges them according to the enemy's resistance, they are likely to meet. This arrangement constitutes the plan of the war or campaign, and if the spirit of the plan is the political object, then the heart of the plan is the military object.[68]

Fuller regarded the grand tactical object as the destruction of the enemy's plan, to reduce his will to win and thereby force his surrender or make him sue for peace. He argued that it would be a grave mistake for a grand tactician to think solely in terms of physical destruction, which is usually the purpose of minor tactics.[69]

Fuller and Liddell Hart were more influential in Germany and Soviet Russia than in Britain. The British senior military officials did not share their enthusiasm for armored warfare. Nevertheless, some experiments with armored forces were conducted in the late 1920s. The first exercise of British armored forces took place on Salisbury Plain (county of Wiltshire) in August 1927. Two years later, the first manual on armored warfare was published. Its author, Brigadier Charles Broad, envisaged the use of tanks independently and for shock action. In another British exercise in 1931, the entire force was composed of tracked vehicles.[70] Despite these encouraging beginnings, the British army did not make much progress in developing sound doctrine for armored warfare prior to the outbreak of World War II in September 1939.

In the 1920s, the Germans embarked on a systematic search for the real reasons they had suffered catastrophic defeat in World War I. During the 1920s, three different views of the future war emerged. One group of theoreticians believed that a modern war of material (*Materialkrieg*) would immediately be transformed into a war of attrition. Therefore, the outcome of a war would be decided by the will of the people and the resources "behind the front." The second group, led by popular military theoretician (and retired general) Max Schwarte, contended that the Imperial German Army had neglected to take full account of technological advances and their effect on the conduct of war. This group of theoreticians strongly insisted on increased motorization and extensive use of aircraft, chemical weapons, and modern means of communication. They believed that concentrating the technical capabilities of the nation would revive what they called "operational warfare" (*Operative Kriegsführung*) and the strategy of annihilation. The third group believed that, in addition to technology, new operational concepts must be developed. Specifically, this included concentration at the weight of main effort (*Schwerpunkt*), surprise, and other factors for conducting area-wide operations.[71] By the 1930s, this group of theoreticians had won the argument on the vision of the character of the future war.

General Hans von Seeckt, the chief of the Reichswehr, had the principal influence on the development of operational thinking in the German military in the interwar years. In 1921, the Reichswehr's first field manual, *Leadership and Combat of Combined Arms* (*Führung und Gefecht der verbundenen Waffen*, F.u.G. 1921), was issued. Among other things, this manual envisaged the employment of the army corps, armies and army groups at the operational level. F.u.G. 1921 was based on Seeckt's vision that the future war would be a war of movement, not a positional war. Seeckt did not develop any overarching concept for mobile warfare, yet he laid down the foundations for the operational employment of cavalry, combat vehicles and motorized units.[72]

Seeckt considered positional warfare (*Stellungskrieg*) and the resultant war of attrition as a perversion of military art. Hence, he emphasized the war of movement (*Bewegungskrieg*). Seeckt believed that the main objective of strategy is to employ mobile and highly capable operational forces to reach a decision.[73] He insisted that training in the Reichswehr should focus on the operational level of war. At the same time, Seeckt did not neglect the study of the effects of the new technological advances in warfare. Like Moltke, Sr, Seeckt directed that instructions (*Weisungen*), not orders (*Befehle*), be used in the employment of large formations. For Seeckt the art of operational leadership consisted of having a clear idea of the relationship among the factors of force, time and space; having the ability to think ahead, simplicity and the ability to issue short, quick orders; creating the opportunity to effect envelopment of the enemy; always trying to achieve surprise and the utmost concentration of forces at one place; acting swiftly; and having the courage to weaken forces along the long front and to make decisions in uncertainty.[74]

In 1933, the German army issued the new service regulations *Troop Leadership* (*Truppenführung*) (H.Dv. 300/1). Like F.u.G. 1921, this document emphasized the need for combined arms warfare in defense and offense, as well as the importance of leadership and simplicity of orders.[75] The war of movement was based on the effects of the new technological developments. The authors emphasized that a future war would be all-European, multifrontal and protracted in character.[76] The Germans believed that panzer attacks must be supported by infantry and motorized infantry or armored artillery, antitank guns and tactical aircraft.[77] They were fully committed to the combined arms warfare concept, in which infantry and artillery would enhance the panzers' potential. Unlike the British and French, the Germans believed that panzers would not be deprived of their mobility if they were tied to infantry.[78]

One of the early and most influential proponents of panzers in the German army was Colonel Heinz Guderian. He borrowed many ideas from Fuller and Liddell Hart. Guderian emphasized that panzer forces must include motorized infantry, self-propelled artillery, engineers and signal troops.[79] The first three panzer divisions were authorized after the Wehrmacht maneuvers in September 1935, in which experimental panzer units sufficiently impressed the most senior German officers. Afterward, the German panzer forces made remarkably rapid progress. It was not until the campaign in Poland in September 1939 that most German army officers realized the potential of these forces in conducting operational warfare.[80]

The French made some progress in their study of motorization in the 1920s. Colonel Joseph E. A. Doumenc suggested creating large, motorized units. In the 1930s, the French built a relatively large force of tanks but did little in terms of developing sound doctrine for their employment because the most senior officers argued for delay until further studies were done.[81] Colonel Charles de Gaulle was one of the French army's most forceful and articulate proponents of armored warfare. In his book *The Army of the Future* (*Vers l'armée de métier*), published in 1934, he proposed the establishment of an army corps composed entirely of motorized units. This force of about 100,000 men would include six mechanized infantry divisions, one light reconnaissance division, reserves and an air observation corps. De Gaulle's ideas were very far from the combined arms concept developed by the Germans. He was mistaken in his belief that mechanization would lead to smaller, professional armies.[82]

In the interwar years, the most influential naval thinkers in the West were German vice admiral Wolfgang Wegener and French vice admiral Raoul Castex. However, both Wegener and Castex were more concerned with naval strategy than with what can be considered operational art. It was the Soviets who laid the theoretical foundations for operational warfare at sea, or what they called "naval operational art." The *Combat Manual of 1930* (BU-30) was the first doctrinal document to spell out the rudiments of joint operational employment of naval forces and ground troops.

The first course on naval operational art was established at the Soviet Naval Academy in 1932.[83] Five years later, the Soviet navy adopted the *Provisional Combat Manual of Naval Forces* [*Vremeniy Boyevoy Ustav Morskykh Syl RKKA* (BU-MS-37)].[84] For the first time, it elaborated "naval operations" (*morskaya operatsiya*) and "day-to-day activities" (or "systematic combat actions," in Russian terms today) as the principal methods by which the combat employment of naval forces and aviation would accomplish operational objectives. The new manual envisaged a dozen types of naval operations [e.g., operations aimed at destroying the enemy forces at sea, operations against coastal "objects" (installations), operations on sea communications and blockade, operations in support of one's own sea communications and counterblockade, landing operations, antilanding operations and operations in support of the army flank]. The day-to-day activities of the fleet forces were aimed at accomplishing operational objectives by obtaining what the Soviets called a "favorable operational regime" in one's coastal waters and deployment area.[85] The Soviet views on the nature of modern warfare at sea and operational art were incorporated into the *Provisional Manual on Conduct of Operations* (NMO-40), issued in 1940.

Major Western navies paid less attention to theory than the Soviets did, focusing instead on the practical application of operational warfare through planning, war gaming, and exercises at sea. In 1927, the U.S. Naval War College adopted for the first time the study of "operational" problems in addition to "strategical" and "tactical" problems. In the war games held at the Naval War College in the 1930s, the U.S. Navy repeatedly tested its plan *Orange* for the operational employment of fleet forces in a hypothetical war with Japan. The U.S. Marine Corps embarked on an intensive effort to create its first operational doctrine for conducting large-scale amphibious landings. The result was the *Tentative Manual for Landing Operations*, issued in 1934.[86] This manual borrowed heavily from the lessons learned in the unsuccessful Allied Gallipoli landing in 1915 and the successful German landing at the Bay of Riga in 1917.[87] After 1935 this document was used as a guide for future amphibious exercises and research and development.[88]

The German Navy used extensively planning games, war games and naval exercises in preparing for a future war at sea. By the early 1930s, the German Navy introduced "operational," in addition to strategic and tactical, war games. One of the lessons of these games for the German Naval High Command was the need to occupy parts of the southern Norwegian coast and the French Atlantic coast in case of a war with France and Great Britain. The tactical concept of using U-boats at night and on the surface was first described in 1922 in a study written by two German naval officers. Their ideas were refined during war games in the early 1930s and became part of naval doctrine in 1935.[89] The concept was tested for the first time during the large-scale Wehrmacht exercises held in the fall of 1937. Admiral Karl Dönitz, Leader of the U-boats (Führer der Unterseeboote), used shortwave radio from his flagship in Kiel and directed the employment of submarine groups in the Baltic. Afterward, the concept was tested during another, larger exercise held in the North Sea. In May 1939 the U-boats operated in groups in the Atlantic off Cape Finisterre and in the Bay of Biscay for the first time.[90]

In the aftermath of World War I, major navies of the day tried to learn lessons from both the successes and failures in the war at sea. However, these efforts had mixed results. By the late 1920s and the early 1930s, the introduction of more capable surface ships, submarines and aircraft and their associated sensors (sonar, radar, etc.) enhanced the capabilities of major fleets to fight far from their own shores and for extended periods. The potential operating areas were increased with the advent of carrier-based aviation and longer-range submarines. Aircraft carriers, land-based aircraft and submarines emerged as the first platforms capable of theaterwide or operational employment. The U.S. Navy and the Imperial Japanese Navy (IJN) used fleet exercises in the late 1920s and the 1930s to further refine the employment of their carrier forces.

Advances in naval communications, especially the invention of shortwave (SW) radio in 1923, allowed secure and reliable control of the movement of fleet units and individual ships at distances up to several thousand nautical miles from the shore. High-frequency/direction-finding (HF/DF) techniques were further refined in the 1920s and 1930s. The experiences gained in World War I led all the major navies to study the problems of theaterwide command and control, as well as the deployment and concentration of fleet forces in a given theater. By the early 1930s, major navies began to move their theater fleet headquarters from the sea to the shore. Initially, these changes in command and control were made during large-scale exercises, but by the late 1930s, the theater fleet headquarters were established ashore on a permanent basis.

The roots of the airpower theories in the interwar years can be found in the limited experience conducting independent operations with air forces in World War I. Many theoreticians in Italy, the United Kingdom, and the United States drew lessons about the potential future use of airpower that were either suspect or outright dangerous. These lessons were often based on a superficial reading of the historical evidence. Sometimes the facts were overlooked or ignored. World War I offered very little practical experience in the strategic employment of aircraft.[91] Yet it was precisely that aspect of airpower that caught the imagination and enthusiasm of the early proponents of airpower.

The Italian General Giulio Douhet and the leading British and American proponents of airpower were exclusively concerned with the strategic employment of the then-nascent air forces. The Germans, in contrast, focused from the beginning on the employment of air forces in support of the army, that is, the operational level of war in the air. Several German airpower theoreticians in the late 1920s and the 1930s developed the concept of "operational air warfare" (*Operative Luftkrieg*). Another German term that gained in popularity at the time, "operational air force" (*operative Luftwaffe*), referred to the air force as an independent service of the armed forces for conducting independent offensive air warfare. Operational Luftwaffe consisted of bombers, dive bombers and heavy fighters, but not reconnaissance and observation aircraft, army-controlled AA units, and air defense units for air defense of the national territory.[92]

The Germans contemplated the use of airpower in a systematic attack on an enemy nation and against its production centers and its supply of raw materials.[93] General Helmuth Wilberg issued "Instructions for the Conduct of Operational Air Warfare" in 1926.[94] The first field manual for the employment of the new Luftwaffe (officially disclosed in March 1935), *Conduct of the Air War* (L. Dv. 16: *Luftkriegführung*), was published in 1936. This document was written by a committee headed by General Walther Wever. It remained in force, with some amendments, until 1945.[95] This short, clearly written document reflected operational thinking on the employment of airpower. The L.Dv. 16 envisaged the use of the *Luftwaffe* in support of the army and navy within the framework of national strategy.[96] Although the *Luftwaffe* had an interest in strategic bombing, the focus clearly remained on the need to cooperate with other services to bring about the collapse and defeat of the enemy.[97]

By the eve of World War II, the Germans had developed a remarkably broad and comprehensive theory for the use of Luftwaffe in support of other two services, and army in particular. The Germans contemplated the employment of Luftwaffe in defense of homeland against enemy strategic attack from the air. They created airborne troops capable of capturing and holding ground deep in the enemy's rear. They built large numbers of transport aircraft to supply ground forces in the field, and AA artillery to provide support to both troops in the field and homeland defenses.[98]

In France, the air force was considered a supporting service whose employment was subordinate to major formations of the army.[99] The French Air Force, established in 1933, failed to obtain separate service status. Its focus was on conducting air battles.[100] The mission of the air force was to take part in air operations and joint operations with the army and navy, and in ground-based air defense.[101] The Soviets developed their airpower theories in the interwar years in close cooperation with other services. Hence, the Soviet Air Force was organized and trained to fight as an integral part of the combined arms concept.[102]

In World War II, all the major belligerents conducted numerous major operations on land, at sea, and in the air, and many land and maritime campaigns. Tactical actions became more diverse and complex and were fought with the combat arms of several services. The Allied victories in the Pacific were due to a combination of realistic strategy and sound operational thinking on the part of the theater commanders. Operational brilliance was the principal factor in the early German victories. The German superb performance at the operational and tactical levels was the main reason the war lasted as long as it did, considering their country's fatally flawed strategy and policy. Despite the extensive practice of operational warfare in all theaters in World War II, after 1945 no systematic attention was given to operational warfare in the United States and other Western militaries. The only exception was West Germany, where former Wehrmacht officers continued to think and write about operations. The Korean War of 1950–1953 was the last war prior to the late 1980s in which U.S. armed forces applied the tenets of operational art. The decline in operational thinking in the United States and the West was the result of the firm belief held by most military thinkers and practitioners that the advent of nuclear weapons and ballistic missiles meant the end of large-scale conventional operations.

Western views were not shared by the Soviets, who did not discount major conventional operations even when nuclear weapons would be used.[103] The Soviets believed that future war would not be limited to a certain area, in which "fronts" (army groups, in Western terms) would seek a decision but would instead be global in scope. A decision would be obtained through the destruction of the enemy's economic and social structure by using nuclear weapons. The tactical employment of forces would take place in close coordination with nuclear fires. Instead of conducting conventional (major) operations, opponents would use nuclear weapons as a means of accomplishing strategic objectives. Hence, in global nuclear war, operational art would be of secondary importance.[104] The Soviets considered "space" as a strategic factor. This mixing of operational factors with strategic objectives meant that operations became a concrete form of strategy. The Soviets believed that even in conditions of nuclear war, the thrust into the depths of the enemy's defense and the concentration of forces would remain the principal element of warfare.[105]

In the early 1980s, the Soviets argued that it would be possible to conduct combat actions by conventional weapons in qualitatively new and more lethal ways. The operational maneuver group (OMG) concept was introduced. The emphasis was not on a single operation but on strategic operations (campaigns, in Western terms).

During such an operation, there would be several front operations related to each other and possibly conducted without a pause.[106] The Soviets concurrently intensified the study of successive operations as the key factor in the success of strategic operations.[107] After 1987 and until the end of the USSR, the Soviets still adhered to the general concept of the strategic operation. The offensive was considered the best guarantee of victory in future war. Central to the concept were the traditional aspects of deep battle and deep operations conducted by OMGs.[108]

In the 1950s, the U.S. military tended to belittle the importance of operational art. The U.S. services considered the Korean War (1950–1953) as an anomaly in the nuclear world. In the late 1950s and 1960s, the emphasis was on counterinsurgency warfare. The services did not believe in the value of joint training. Instead, they favored tactical training in the army, fleet exercises in the navy and strategic studies in the air force.[109]

The U.S. services rediscovered the importance of being ready to fight a high-intensity conventional war in the aftermath of the Vietnam War. This led to steadily increased interest in operational art. The U.S. Army's doctrinal manual FM 100-5, *Operations*, published in 1976, was the first in a series of capstone documents that would guide the evolution of operational art through the next two decades. The next edition of FM 100-5, published in 1982, adopted the AirLand Battle operational concept while highlighting the principles of war and maneuver as the dynamic elements of combat. It reflected a change from attrition-centered warfare to a war of maneuver.[110]

The U.S. Army was the first service to incorporate tenets of operational art in its doctrine in 1982. The FM 100-5 (1982) clarified the fundamentals upon which major operations rested. It explained that the operational level of war uses available military resources to accomplish strategic objectives within the theater of war. It involved the planning and conduct of campaigns. Campaigns are sustained operations designed to defeat an enemy force in a specified space and time with simultaneous and sequential battles. The disposition of forces and the selection of objectives and actions to weaken or outmaneuver the enemy all set the terms of the next battle and exploit tactical gains. They are all part of the operational level of war.[111]

A new edition of FM 100-5, published in 1986, reinforced the idea of integrating all war-fighting resources within a theater and waging simultaneous battles forward, deep and in the enemy's rear.[112] The revival of operational thinking in the U.S. military was coupled, however, with some confusion regarding an appropriate term to adequately explain the true meaning of "operations" or "operational warfare." Because of the wide abuse of the term "operations," the U.S. Army adopted the Soviet term "operational art" in the early 1980s. Afterward, the British army, the Royal Navy, the German Bundeswehr and the Swiss army incorporated operational art into their basic doctrinal documents.

Importance

Operational art potentially offers great advantages for the side that properly practices it. Among other things, it provides a larger and broader framework within which tactical combat takes place. Without operational art, war would be a series of randomly fought tactical actions, with relative attrition the only measure of success or failure. Ideally, a war should consist of only decisive actions resulting in quick defeat of the opposing forces and large losses in personnel and materiel. This is not possible unless the enemy commits grave errors in massing his forces in large numbers and thereby making them vulnerable to be decisively beaten. The single biggest challenge for the operational leaders is to prevent attrition from dominating at the strategic and operational levels of war. The knowledge and skillful application of operational art enhances the prospects for a highly capable, well-trained and skillfully led force, guided by a sound and coherent strategy, to defeat a much stronger opponent. The very essence of operational art is to win decisively in the shortest time possible and with the least loss of human lives and materiel. This is especially important in an era of smaller forces, limited resources and low tolerance of casualties by the political leadership and the public.

For all its value, the importance of numerical superiority should not be overstated. General Bernhardi, criticized the belief in the importance of numerical superiority that was so prevalent in the European armies in his time. The European military believed that armies armed and equipped with equal numbers of weapons and equipment would have similar capabilities.[113] He emphasized it should never be forgotten that moral and spiritual factors are very different in each situation. They are also often more important than numerical factors. Sometimes the spiritual strength of an army can balance other deficiencies. The influence of a single personality can increase the capabilities of the entire army and even the entire state.[114]

Inferior forces have often defeated a much larger force because of the better quality of their leaders and the better training, morale and discipline of their troops. During the Seven Years' War (1756–1763), Frederick the Great, with 41,000 men, defeated 64,000 Austro-French troops (20,000 were not engaged) in the Battle of Rossbach in November; and at Leuthen in December 1757, with an army of 35,000 men, he defeated 65,000 Austrians.[115] In the American Civil War (1861–1865), General Robert E. Lee, with fewer than 50,000 men, successfully repulsed an attack by 90,000 men of the Union Army in the Antietam (Sharpsburg) campaign (in fact a major operation) in September 1862.[116]

In the German invasion of the Benelux countries and France in May 1940, better leadership and training – not better weapons – were the principal reason for their phenomenal successes despite numerical inferiority. The Allies then had 3,740,000 soldiers, versus 2,760,000 Germans; the French had 3,254 tanks but only three armored divisions, while the Germans had organized their 2,574 tanks in ten panzer divisions. In contrast, the balance in the air did not favor the Allies. The French had only 1,090 modern aircraft (including 610 fighters, 130 bombers and 350 reconnaissance aircraft); the British provided, in addition, a total of 160 fighters and 272 bombers. The Germans had some 3,500 aircraft available.[117]

In the Six-Day War (5–10 October 1967), the Israeli Defense Forces (IDF) defeated quickly and decisively much larger Arab armies. The Israelis mobilized 250,000 men.[118] Their forces were smaller but much better led and trained than their Arab enemies. For the Israelis, the Sinai was the principal part of the theater.

This was the reason that the Israelis concentrated some 70,000 men and 750-800 tanks on their southern front.[119] Initially, the Israelis were strategically on defensive on the Jordanian and Syrian fronts.[120] In contrast, the Egyptian Army's strength was fragmented. Some 50,000 men were deployed in Yemen (because of the Egyptian intervention in a civil war) and 70,000 troops was scattered west of the Suez Canal (to deal with a potential landing of the Anglo-French forces). This left only about 100,000 troops defending the Sinai. Most of the Egyptian troops were ill-trained.[121]

An excellent and recent example of successful planning and execution of major/joint operations by a small country was the Croatian liberation of the Krajina in August 1995 (Operation Oluja-Storm). The battle line extended for some 390 miles. The area under control of the Serbian rebels was about 4,247 square miles. In the operation took part all three services (army, air force and navy) plus special police of the Ministry of Interior (MUP).[122] In addition to their 80,000-man active army, the Croats employed an additional 120,000 reservists against 40,000 Serbian rebels organized in four army corps.[123] This was one of the major reasons that the entire operation was successfully completed within only four days. The Serbs were poorly organized and had low morale. They offered weak or no resistance to the advancing Croatian troops.

Operational art highlights the need for the commanders and their staffs to fully comprehend nonmilitary (diplomatic, political, economic, financial, social, religious, etc.) aspects of the situation in a given theater. In modern times, national (or theater) strategic objectives cannot be accomplished without the properly sequenced and synchronized employment of multiservice and often multinational forces. This requires the highest degree of cooperation or jointness among the services. The employment of two or more services can neutralize the disadvantages of one service by using the advantages of other services.

A knowledge and understanding of operational art are essential for subordinate tactical commanders' success as well. To act in accordance with the operational commander's intent, major tactical commanders must understand a broader – that is, operational – picture of the situation. By understanding operational art, they can make decisions that will contribute to the accomplishment of the overall operational or strategic objective.

The study of past campaigns and major operations improves the quality of officers' professional education. The study of the history of operational warfare underscores the fact that all fundamental warfare areas are critical to success. It shows that it is false to artificially divide warfare into purely "offensive" and "defensive" categories. Operational art provides a framework for writing operational doctrine.

The application of operational art poses several difficulties which must be overcome by the operational commanders and their staff. Normally, the strategic or operational objective cannot be accomplished without the employment of two or more services of the armed forces. Therefore, the command structure and command and control are more complicated than at the tactical level of command. With the vast amount of information that needs to be collected, processed, evaluated and transmitted to users, information overload is potentially an enormous problem. Operational planning is more complex and time-consuming than planning for tactical actions. Likewise, the logistical support and sustainment of multiservice and especially multinational forces are very complicated. Because of the differences in doctrine and weapons and equipment, there are problems with interoperability. Successful operational planning and execution of campaigns and major operations require close interagency planning and coordination, which are extremely difficult to achieve even under the best of circumstances.

Theory and Practice

Like strategy and tactics, operational art is both an art and a science. The art aspect means that operational art, like the other two components, is ambiguous and imprecise, requiring a creative and highly innovative approach to its application. While theory of operational art is universal because it is based on the experiences of the militaries of all civilizations and nations. Application of that theory depends on the national or service way of war or even personalities and leadership of the individual operational commanders.

Theory and practice of operational art are inextricably linked. The practice of operational warfare normally has preceded the theory. Afterward, theory paved the way for the practice of employing combat forces. A sound theory should ensure that it is not disconnected from the operational realities; otherwise, it cannot provide the basis for a successful application of the tenets of operational art.

The value of the theoretical aspects of operational art cannot be overrated. Theory is critical to refining and improving the existing methods of applying operational art. Theory should deal with each war and each era of warfare on its own terms and should always accommodate itself to change. It is the task of theory, taking full account of new technological advances and corresponding changes in the character of warfare, to analyze, refine and further develop constituent parts of operational art.

The field of theoretical study and practical application of operational art is too large and too diverse to be neatly arranged into a "system" of thinking. Nor is there any particular order or sequence in which these components should be studied or applied. The various elements that comprise any given component of operational art are grouped together because they are related to each other and it makes it easier to discuss them and develop theory.[124]

Theory of operational art provides both a framework and a direction for the development of service and joint doctrine. Among other things, theoretical knowledge and understanding of a given aspect of warfare make improvisation easier in combat, when both time and resources are in short supply. Theory also greatly improves precision in using various key operational terms, facilitating communication between theoreticians and practitioners.

Types

There are both commonalities and some significant differences in the ways services employ their forces. These distinctions are due to the features of the physical

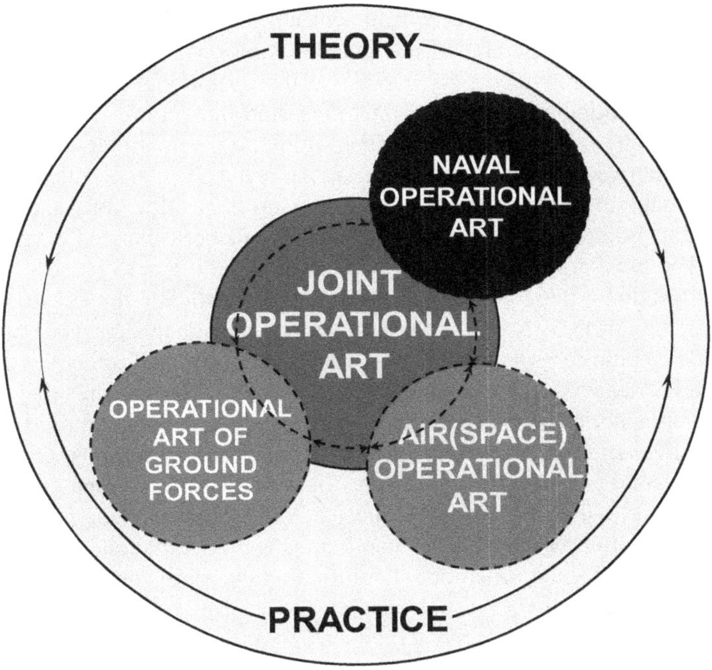

Figure 1.1 Types of Operational Art

environment (land, air/space and sea/ocean) in which each service operates. Ground troops operate in much more diverse and in many ways more difficult physical environment than air force. Their employment is heavily affected also by the situation in the air. Land warfare is conducted in two physical mediums (land and air), while war at sea is conducted in three physical mediums (surface, air and subsurface). Clearly, the objectives and methods of employment of combat forces in land, air and naval warfare are considerably different.

Because of these distinctions, each service consists of different platforms and weapons/sensors. There is also considerable difference in the ways each service is employed individually or part of multi-service (joint) forces. Hence, a distinction should be made between operational art specific for each service, and operational art for multi-service/multi-national forces. Each service should be chiefly concerned with developing theory of planning, preparing and conducting of the respective major/joint operations, while joint forces should focus on theory and practice of land or maritime campaigns.[125]

Prerequisites

Success in the practical application of operational warfare is difficult to achieve unless several prerequisites are met. Among other things, the conduct of major operations and campaigns requires sufficient physical space for combat forces to maneuver freely. The operational commander's *mastery of tactics* in his chosen specialty is a key prerequisite for success at the operational and strategic levels of war. The operational commanders and their staff must think *operationally* vice tactically. Among other things, operational thinking means to have the ability to differentiate between essential and non-essential or even trivial events. It impresses on the operational commander the need to ensure that all actions and measures must be planned and executed within a larger framework dictated by policy and strategy. The most critical element of operational thinking is to possess an operational perspective – the ability to see clearly and realistically the essentials of the military and nonmilitary aspects of the situation in a given theater. The operational commander should foresee how his decisions and actions will affect the decisions of the political leadership in achieving ultimate victory. This requires that the operational commander possess at least an approximately accurate picture of the situation several weeks or even months into the future.[126] The operational commanders should have solid knowledge and understanding of nonmilitary aspects of the situation and trends in a given part of the theater. Another prerequisite for the success of a major operation or campaign is continuous and effective operational support (operational intelligence, operational information warfare, operational logistics and operational protection).

The operational commanders and their staff should use commonly accepted and understood operational terms; otherwise, communications within a service and among services would be difficult, if not impossible. It does little good to recognize a problem and to formulate an approach to it if the language with which it is discussed is confused or uncertain. As the late Henry E. Eccles (professor at the U.S. Naval War College) so eloquently wrote,

> strict accuracy should regulate our use of language. Thoughts should be expressed with perspicuity and correctness. False logic, disguised by specious phraseology, too often gains the assent of the unthinking multitude, disseminating far and wide the seeds of prejudice and error. A misapplied or misapprehended term is sufficient to give rise to fierce and interminable disputes.[127]

Characteristics

The principal characteristics of operational warfare, as compared with tactical actions, are larger dimensions of forces, time, and space. The main reason for these differences is the much larger scale of an operational or strategic objective. Major operations and campaigns are conducted in a large part of the theater. A tactical action can last hours or even minutes, while a major operation is conducted over several days or even weeks. The duration of a typical campaign varies, from weeks to several months or even much longer. The accomplishment of operational or strategic objectives requires the employment of larger and more diverse forces than does the accomplishment of tactical objectives. In the past, large and diverse forces (numbered/theater armies, naval fleets, and air fleets) were normally required for conducting a campaign or major operation. This is not the case today, where a small force can be employed to accomplish an operational or, in some special cases, even a limited strategic objective. What counts most is not the numerical strength of one's forces but their combat potential. Another factor is the perceived strength or weakness of the enemy's forces. Therefore, even in the past there were several cases in which small forces were able to accomplish operational or even strategic objectives. The U.S. General Winfield Scott's landing at Vera Cruz, Mexico, in early March 1847 was conducted by about 12,000 troops and had a strategic effect on the outcome of the Mexican–American War.[128]

In several instances in World War II, forces no larger than an army corps conducted major operations. Just before their invasion of France in May 1940, the Germans formed their first armored operational-tactical force, Panzergruppe von Kleist (an army-size force in all but name), which, in cooperation with the Luftwaffe, could conduct major land operations. This force was the key factor in the successful German penetration of the French defenses at Sedan and in the subsequent advance to the French Channel coast that eventually cut off a major part of the French and British troops in northwestern France. The Japanese 25th Army, composed of three divisions with only 35,000 men but strongly supported by air and naval forces, conquered Malaya in just 68 days. Of the initial British force of some 70,000 troops (plus 15,000 administrative and unarmed personnel), 62,000 British, Indian and Australian troops surrendered unconditionally to the Japanese after the fall of Singapore on 15 February 1942.[129] For the invasion of the Philippines, the Japanese assigned the Southern Army, composed of only two infantry divisions plus supporting troops consisting of seven regiments (four of which were combat troops) and six battalions.[130] The main forces employed in May 1941 in the ill-fated German attempt to destroy British convoys in the northern Atlantic (Rheinűbung – Rhine Exercise)) were organized into a combat group (*Kampfgruppe*) composed of a single battleship (*Bismarck*) and a heavy cruiser (*Prinz Eugen*).[131]

Because of the larger scale of operational/strategic objectives and consequently the larger force involved, planning for campaigns/major operations is more complex and requires more time than planning for a tactical action. They are planned by using "regressive" method, in which the ultimate (strategic or operational) objective is determined first and from that, a series of intermediate military objectives (operational or major tactical) is determined. The deployment of combat forces taking part in a campaign/major operation encompasses a much larger physical space and requires far more time to complete than deployment for a tactical action. More important, errors in deploying combat forces into a theater can fatally impact the initial phase and even the outcome of a campaign/major operation. Combat forces' deployment is the very heart of a plan for a major operation or campaign. It directly affects not only the accomplishment of the next operational objective but also the subsequent decisions made in combat.

Because of the long duration of campaigns/major operations and the diversity of participating forces, logistical support and sustainment are more complex to organize and execute than they are for tactical actions. The operational commander is normally responsible for organizing, and controlling logistical support and sustainment, while a tactical commander uses logistics but does not control it.

Strategy–Operational Art–Tactics Nexus

All three components of the art of war are closely related. Their mutual relationships are highly dynamic. There is no clear-cut line separating strategy, operational art and tactics. Actions and events at the tactical level often affect strategy and policy in the most profound way. Likewise, the application of operational art affects both strategy and policy on the one hand and tactics on the other.

Strategy and policy provide a broad framework for the application of operational art and tactics. They should ensure political strategic objectives are based on operational realities. Strategy defines and imposes limitations on the employment of combat forces in accomplishing military strategic and operational objectives.

Sound understanding of operational art is necessary to convert a political into a military strategic objective. A major operation or campaign must be conducted within a framework of what is operationally and strategically possible.[132] Tactical and operational successes are irrelevant unless guided by sound and coherent strategy.

Operational art is a critical link between strategy and tactics. If that link is weak or broken, strategic success will be wanting. Operational competence is necessary to avoid more time-consuming and costly attrition at the operational and strategic levels of war. Operational victories are useless unless converted into strategic successes. Whenever the ends and means at the strategic level are seriously disconnected or mismatched, brilliance at the operational and tactical level – as the Germans consistently displayed in World War II – can only delay, but cannot prevent, ultimate defeat. The situation becomes untenable if the political leadership is unwilling to commit all available sources of power, as the example of the United States in the Vietnam War (1965–1975) shows.

Strategy should invariably dominate operational art; otherwise, the results could be fatal.[133] A major problem is when operational art began to interfere with or even worse, dominate policy and strategy either by intent or by default. The Axis campaign in North Africa in 1941–1942 was driven and dominated by operational, and sometimes even tactical, vice strategic, considerations. Field Marshal Erwin Rommel's successes against the Allies led to the steady expansion of the Axis war aims, which required more forces to be drawn into a secondary theater of operations. Rommel tried to advance to the Suez Canal and capture Middle Eastern oil fields far beyond a reasonable distance from his bases of support, resulting in a serious mismatch between ends and means.[134]

Poor application of operational art can lead to tactical defeats, which may have not only operational but also strategic consequences. The Japanese Combined Fleet suffered a decisive defeat in the Battle of Midway in June 1942 because of a flawed operational plan, despite the overwhelming Japanese superiority in forces. This demonstrates how the superiority of one's forces can easily be squandered when operational thinking is inadequate or entirely lacking.

Tactics is both the art and science of planning, preparing, and employing individual platforms, weapons/sensors, and single/combat arms/branches to accomplish tactical objectives. Moltke, Sr, observed that tactics are the basis of operations. Tactics should create the prerequisites for operational or strategic success, which the operational commander must then exploit.[135] Tactics should ensure that its results are in harmony with operational art and strategy. A tactical action should not be fought unless it directly or indirectly contributes to the accomplishment of an operational or strategic objective. Tactical victories are meaningless if they are fought outside of a given operational framework. On 25 October 1944, during the Battle for Leyte, Admiral William F. Halsey, commander, U.S. Third Fleet, won a tactical victory in the Battle off Cape Engaño (Luzon) with his Task Force 38 (TF-38) (fast carrier force) over a much smaller and weaker Japanese carrier force. However, his decision to uncover San Bernardino Strait almost led to the failure of the entire supporting major naval operation because it was fought outside the operational context. The Third Fleet's mission in the Leyte operation was to provide effective distant (operational) cover and support to the Allied forces that landed on Leyte. Only Vice Admiral Takeo Kurita's decision to turn north and leave the scene of action when his powerful heavy surface force was on the verge of defeating much weaker U.S. Task Group (TG) 77.4 (escort carriers) in the Battle off Samar on 25 October saved the Allies from suffering an ignominious defeat.

The accomplishment of operational and strategic objectives depends on the results obtained by tactics. Strategy must ensure that tactical combat is conducted under conditions favorable for accomplishing strategic objectives. Bad tactics can invalidate a good strategy and operational art. Therefore, a sufficient level of tactical competence is always required to accomplish strategic or operational objectives. The U.S. Navy did not match tactical skills with the Japanese during the protracted struggle for Guadalcanal (August 1942–February 1943). The Allies won because they matched means and ends at the operational and strategic levels better than the Japanese did. Defeats in a major operation, such as those of the Germans at Stalingrad (November 1942–January 1943) and the Japanese at the Battle of Midway (June 1942) in the Midway–Aleutians Operation not only can doom the entire major operation but also have an immediate and profound effect on the strategic situation in a given theater.

Perhaps the worst thing to do is to confuse tactics with strategy and strategy with the conduct of war, as the Imperial Japanese Navy (IJN) did in the prewar years. The Japanese were fixated on the single decisive battle, after the manner of the Battle of Jutland in June 1916. That preoccupation guided their tactical doctrine and ship designs. It resulted in having a powerful surface force that was both "one-dimensional and brittle."[136]

Tactics should never be allowed to significantly influence, much less dominate, strategy, either by design or by default. If this occurs, strategy will be defined by tactical considerations, or even applied as an afterthought.[137] The principal reasons for the "tacticization" of strategy have been the uncontrolled ambition of military field commanders and the tactical thinking of political leaders. The ever-increasing reliance on advanced weaponry and sensors, coupled with the reluctance to accept unnecessary casualties in the hostilities involving no vital national interests, has created a situation in which targeting has become a de facto substitute for sound strategy.

Tactical brilliance rarely can overcome poor operational performance. Yet no number of tactical victories can save one's forces from ultimate defeat unless they contribute to the accomplishment of a given operational objective. A lack of operational thinking invariably results in wasting the fruits of tactical victories. The German offensive on the Western Front in April 1918 resulted in heavy losses for the allies and a considerable gain of space for the Germans.[138] Despite a series of tactical victories, the Germans failed to expand their penetrations into operational breakthroughs at any part of the Western Front. The key prerequisite for reaching a strategic decision was not achieved. Moreover, the Germans' local gains in terrain resulted in a longer front line to be defended. The German army was morally and psychologically exhausted. In the summer, the Germans reverted to defense.[139] The Allies gained the initiative that eventually led to an armistice on 11 November 1918 and ended in Germany's defeat.

Tactics can sometimes be heavily influenced by politics, especially domestic policy considerations. This excessive "politicization" of one component of military art often has fatal consequences for military effectiveness. The French tactical doctrine of "offensive at any price" before 1914 affected the operational level. During World War I, as casualties mounted and the public outcry became deafening, this doctrine became increasingly disconnected from the strategic reality.[140]

Conclusion

The roots of operations or operational art are in the wars conducted by the great captains in the ancient and medieval eras. Yet the origins of modern operational warfare are in the French Revolution and the wars of Napoleon I. Afterward, changes in society and international relations, combined with the rise of nationalism and the increased role of public opinion, led progressively to the steady expansion of war's aims. The introduction of mass conscription led to the establishment of large standing armies in peacetime and million-man armies in wartime. The invention of railroads and the internal combustion engine, the introduction of ever-longer-range and more precise weapons, the invention of the telegraph and wireless radio, and the advent of aircraft and tanks led to a steady increase in the size of the battlefield. Several armies were deployed in a given theater, and their defeat could not be accomplished by fighting a single "decisive" battle, or even several such battles. World War I showed that strategic objectives in a theater could be accomplished only through a series of major operations, all sequenced and synchronized in terms of time, place and purpose. Operational or theaterwide warfare had come of age. Since then, operations or operational art on land, at sea and in the air have steadily evolved, both in practical application and in the development of theory.

Operational art is the only means of orchestrating and tying together tactical actions within a larger design that directly contributes to the objectives set by strategy. Tactical successes cannot lead to victory if they are achieved outside a broader operational framework. The sound application of operational art is the key to winning decisively in the shortest time and with the least loss of men and materiel. All past wars were won or lost at the strategic and operational levels, not at the tactical level. The history of past wars has shown that neither superior technology nor superb tactics can, by itself, ensure victory in a war. An exclusive focus on technology or tactics is likely to result in time-consuming and costly attritional warfare against a strong and resilient enemy. Moreover, relying solely on better technology would rob warfare of its "art." It would make one's forces vulnerable to an opponent who, while technologically inferior, thinks better and faster and uses his smaller forces more creatively, asymmetrically.

Notes

1 Raoul Castex, *Strategic Theories*, selections trans. and ed. with an introduction by Eugenia C. Kiesling (Annapolis, MD: Naval Institute Press, 1993).
2 Other parts were statesmanship, strategy, logistics, and engineering; Antoine-Henri de Jomini, *The Art of War* (London: Greenhill Books & Novato, CA: Presidio Press, 1992), 13.
3 de Jomini, *The Art of War*, 178.
4 J. F. C. Fuller, *The Foundations of the Science of War* (London: Hutchinson, 1925), 107.
5 Gottfried Greiner, "Militärische Operationen heute (Ein Beitrag zur Klärung)," *Wehrkunde*, no. 8 (August 1965), 393.
6 Greiner, "Militärische Operationen heute (Ein Beitrag zur Klärung)," 394.
7 Volker Wieker, *Operation, Operative Führung zwischen Taktik und Strategie – Bedeutung und Entwicklung des Begriffes seit Moltke d. Ae. – Bearbeitung aus der Sicht der Landstreitkräfteführung* (Hamburg: Führungsakademie der Bundeswehr, 1987), 8; John English, "The Operational Art: Development in the Theories of War," in *The Operational Art: Developments in the Theories of War*, ed. B. J. C. McKercher and Michael A. Hennessy (Westport, CT: Praeger, 1996), 8; Georg von Sodenstern, "Operationen," *Wehrwissenschaftliche Rundschau* 1 (January 1953), 53; Jörg Besterhorn, *"Freie Operationen"-zur Bedeutung eines neuen Begriffes im Rahmen der operativen Schulung des Heeres* (Hamburg: Führungsakademie der Bundeswehr, January 29, 1999), 7.
8 Carl-Gero von Ilsemann, "Das operative Denken des Älteren Moltke," in *Operatives Denken und Handeln in deutschen Streitkräften im 19. und 20. Jahrhundert*, ed. Militärgeschichtlichen Forschungsamt (Herford/Bonn: Verlag E. S. Mittler & Sohn, 1988), 22.
9 Hans Hitz, "Taktik und Strategie. Zur Entwicklung kriegswissenschaftlicher Begriffe," *Wehrwissenschaftliche Rundschau* 11 (November 1956), 616; von Sodenstern, "Operationen," 53.
10 Norbert Hanisch, *Untersuchen Sie die operativen Ideen Mansteins hinsichtlich Schwerpunktbildung, Überraschung, Initiative und Handlungsfreiheit an den Beispielen Westfeldzug 1940 (Sichelschnitt-Plan) und Operation Zitadelle* (Hamburg: Führungsakademie der Bundeswehr, January 15, 1988), 4; Roland G. Förster, "Das operative Denken Moltkes des Älteren und die Folgen," in *Ausgewählte Operationen und ihre militärhistorischen Grundlagen*, ed. Hans-Martin Ottmer and Heiger Ostertag (Herford/Bonn: Verlag E. S. Mittler & Sohn, 1993), 254–55.
11 Jacob W. Kipp, *Mass Mobility, and the Red Army's Road to Operational Art, 1918–1936* (Fort Leavenworth, KS: Soviet Army Studies Office, U.S. Army Combined Arms Center, 1987), 5.

12 A. A. Kersnovskiy, *Filosofija voyni* (Belgrade: Izd. Tsarskogo Vestnika, 1939), 31; Other sources claim that the term "operatika" was coined in 1912; I. Mariyevskiy, "Formation and Development of the Theory of Operational Art (1918–38)," in *The Evolution of Soviet Operational Art 1927–1991: The Documentary Basis. Vol. I: Operational Art, 1927–1964*, trans. Harold S. Orenstein (London: Frank Cass Publishers, 1995), 245.
13 Richard W. Harrison, *The Russian Way of War. Operational Art, 1904–1940* (Lawrence, KS: University of Kansas Press, 2001), 29.
14 Jovo Ninković, *Teorijske osnove operatike* (Belgrade: Vojno Izdavački Zavod, 1985), 171.
15 The Soviet military theoretician Nikolai Varfolomeev (1890–1939) credited Svechin with the first use of that term as early as 1922; Nikolai Varfolomeev, "Strategiya v akademicheskoy postanovke," *Voyna i revolutsiya* (November 1928), 84.
16 Aleksander A. Svechin, *Strategy* (Minneapolis, MN: East View Press, 1992), 68–69.
17 Svechin, *Strategy*, 69.
18 David M. Glantz, *Soviet Military Operational Art: In Pursuit of Deep Battle* (London: Frank Cass, 1991), 10–11.
19 "Science" is the possession of knowledge through study, while "art" is the performance of actions as acquired by experience, study, or observations; *Webster's Third New International Dictionary* (Springfield, MA: Merriam-Webster, 1981), 2032; Carl von Clausewitz wrote that the object of science is knowledge, and the object of art is creative ability; Carl von Clausewitz, *On War*, ed. and trans. Michael Howard and Peter Paret (Princeton, NJ: Princeton University Press, 1976), 122; J. F. C. Fuller wrote in 1926 that to deny a science of war and then to theorize on war as an art is pure military alchemy, a process of reasoning that for thousands of years has blinded the soldier to the realities of war and will continue to blind him until he creates a science of war upon which to have his art; Fuller, *The Foundations of the Science of War*, 21.
20 Headquarters, Department of the Army, *FM 100-5: Operations* (Washington, DC: U.S. Government Printing Office, May 1986), 10.
21 Stephen E. Runnels, "A Different Approach," *Military Review* 10 (October 1987), 47.
22 Headquarters, Department of the Army, *FM 100-5: Operations* (Washington, DC: U.S. Government Printing Office, March 2010), GL-7.
23 Field Manual (*FM*), *3-0: Operations* (Washington, DC: Headquarters, Department of the Army, October 2017), Glossary-13; Joint Publication 3-0, *Joint Operations* (Washington, DC: Joint Staff, the Pentagon, January 17, 2017 Incorporating Change 1, 22 October 2018), GL-13.
24 Wallace P. Franz, *The Character of Modern War: Theory-Doctrine-Practice at the Operational Level* (Carlisle Barracks, PA: Strategic Studies Institute, U.S. Army War College, December 15, 1981), 3.
25 Wieker, *Operation, Operative Führung zwischen Taktik und Strategie*, 3.
26 Jürgen Fehling, *Das Problem der Ūberträgbarkeit taktisch/operativer Erfahrungen aus der Kriegsgeschichte* (Hamburg: Führungsakademie der Bundeswehr, 1986), 8–9.
27 Franz, *The Character of Modern War: Theory-Doctrine-Practice at the Operational Level*, 3.
28 Ronald B. St. John, "European Naval Expansion and Mahan, 1889–1906," *Naval War College Review* 23 (March 1971), 74.
29 John Shy, "Jomini," in *Makers of Modern Strategy: From Machiavelli to the Nuclear Age*, ed. Peter Paret (Princeton, NJ: Princeton University Press, 1986), 148–49.
30 Gérard Chaliand, *The Art of War in World History: From Antiquity to the Nuclear Age* (Berkeley/Los Angeles/London: University of California Press, 1994), 738–39.
31 Thomas Will, *Operative Führung: Versuch einer begrifflichen Bestimmung im Rahmen von Clausewitz' Theorie 'Vom Kriege'* (Hamburg: Verlag Dr. Kovac, 1997), 8.
32 Will, *Operative Führung*, 8.

33 Carl von Clausewitz, *Vom Kriege*, ed. with comments by Dr. Werner Hahlweg, 16th ed. (Bonn: Dümmlers Verlag, 1952), 174.
34 Hitz, "Taktik und Strategie. Zur Entwicklung kriegswissenschaftlicher Begriffe," 612.
35 Greiner, "Militärische Operationen heute (Ein Beitrag zur Klärung)," 394.
36 Cited in Militärgeschichtliches Forschungsamt, ed., *Operational Thinking of Clausewitz, Moltke, Schlieffen, and Manstein*, 2nd ed. (Freiburg im Breisgau: Militärgeschichtliches Forschungsamt, 1989), 25.
37 Martin Kutz, *Operative Führung als Denkfigur und Handlungskonzept der Heeresführung der Bundeswehr. Politische und militärfachliche Implikationen und Gefahren der aktuellen Diskussion* (Hamburg: Führungsakademie der Bundeswehr, February 1989), 11.
38 Wieker, *Operation, Operative Führung zwischen Taktik und Strategie*, 7.
39 Hanisch, *Untersuchen Sie die operativen Ideen Mansteins hinsichtlich Schwerpunktbildung*, 4; Roland G. Főrster, "Das operative Denken Moltkes des Älteren und die Folgen," in *Ausgewählte Operationen und ihre militärhistorischen Grundlagen*, ed. Hans-Martin Ottmer and Heiger Ostertag (Herford/Bonn: Verlag E. S. Mittler & Sohn, 1993), 254–55.
40 Friedrich von Cochenhausen, "Schlieffen. Zu seinem 100-jährigen Geburtstage 28. February 1933," *Militärwissenschaftliche Mitteilungen (Vienna)* 2 (February 1933), 19.
41 Wieker, *Operation, Operative Führung zwischen Taktik und Strategie*, 11.
42 Hermann Foertsch, "Die gedanklichen Grundlagen und die Organisation der deutschen Kriegführung bis zum 2. Weltkriege," in *Die Gedanklichen Grundlagen des OKH und deren Auswirkungen auf seine Organisation. Ein Schlusswort zur Gesamtarbeit "OKH," P-041kk, ZA/1 1935*, ed. Oldwig von Natzmer (Freiburg im Breisgau: Studien der Historical Division Headquarters, United States Army Europe, Foreign Military Branch, Bundesarchiv/Militärarchiv (BA-MA), December 1949), October 1948, 15.
43 Hermann Foertsch, "Die gedanklichen Grundlagen und die Organisation der deutschen Kriegführung bis zum 2. Weltkriege," 14.
44 Wieker, *Operation, Operative Führung zwischen Taktik und Strategie*, 12.
45 Wieker, *Operation, Operative Führung zwischen Taktik und Strategie*, 14.
46 Robert G. Baker, "Friedrich von Bernhardi: Clausewitz Disciple and Prussian Militarist," *Marine Corps Gazette* 5 (May 1991), 97–98.
47 Jacob W. Kipp, *Mass Mobility, and the Red Army's Road to Operational Art, 1918–1936* (Fort Leavenworth, KS: Soviet Army Studies Office, U.S. Army Combined Arms Center, 1987), 5.
48 Harrison, *The Russian Way of War: Operational Art, 1904–1940*, 24.
49 Harrison, *The Russian Way of War: Operational Art, 1904–1940*, 28.
50 Cited in Kipp, *Mass Mobility, and the Red Army's Road to Operational Art, 1918–1936*, 5.
51 Harrison, *The Russian Way of War: Operational Art, 1904–1940*, 29–30.
52 Cited in Philip A. Crowl, "Alfred Thayer Mahan: The Naval Historian," in *Makers of Modern Strategy. From Machiavelli to the Nuclear Age*, ed. Peter Paret (Princeton, NJ: Princeton University Press, 1986), 457–58.
53 Cited in Crowl, "Alfred Thayer Mahan: The Naval Historian," 456.
54 Barry M. Gough, "Maritime Strategy: The Legacies of Mahan and Corbett as Philosophers of Sea Power," *RUSI Journal* (Winter 1988), 58.
55 Julian S. Corbett, *Some Principles of Maritime Strategy* (London: Longmans, Green and Co., 1918), 11–12.
56 Fehling, *Das Problem der Übertrağbarkeit taktisch/operativer Erfahrungen aus der Kriegsgeschichte*, 10.
57 Fehling, *Das Problem der Übertrağbarkeit taktisch/operativer Erfahrungen aus der Kriegsgeschichte*, 9.

58 Günther Blumentritt, "Die Gedanklichen Grundlagen des Alten O.K.H.," in *Die Gedanklichen Grundlagen des OKH und deren Auswirkungen auf seine Organisation. Ein Schlusswort zur Gesamtarbeit "OKH," P-041kk, ZA/1 1935*, ed. Oldwig von Natzmer (Freiburg im Breisgau: Studien der Historical Division Headquarters, United States Army Europe, Foreign Military Branch, Bundesarchiv/Militärarchiv (BA-MA), December 1949, 3–4.
59 Wieker, *Operation, Operative Führung zwischen Taktik und Strategie*, 17.
60 Fehling, *Das Problem der Ůberträgbarkeit taktisch/operativer Erfahrungen aus der Kriegsgeschichte*, 9.
61 Erich Vad, "Operative Fűhrung: Grundlagen, Merkmale und Perspektiven," *Ősterreichische Militärische Zeitschrift* 2 (March–April 1998), 130.
62 Adolf Feil, ed., *Grundlagen der Theorie der operativen Kunst der SSK*, trans. from the Russian *Operativnoye iskusstvo voyenno-morskogo flota* (Dresden: Military Academy Friedrich Engels, Study Material, March 13, 1989), 11.
63 Staff Presentation, *Operations for Securing Command of Sea Areas* (Newport, RI: Naval War College, July 8–9, 1938), 17.
64 James B. Agnew, "From Where Did Our Amphibious Doctrine Come?" *Marine Corps Gazette* (August 1979), 53.
65 Tim Travers, *Gallipoli 1915* (Stroud, Gloucestershire/Charleston, S.C.: Tempus, 2001), 21.
66 Agnew, "From Where Did Our Amphibious Doctrine Come?" 56; Carl O. Schuster, "Baltic Assault. Operation Albion," *Command Magazine*, July–August 1993, 56–57; Randolf Kugler, *Das Landungswesen in Deutschland seit 1900* (Berlin: Oberbaum Verlag, 1989), 35.
67 Brian Bond and Martin Alexander, "Liddell Hart and De Gaulle: The Doctrine of Limited Liability and Mobile Defense," in *Makers of Modern Strategy: From Machiavelli to the Nuclear Age*, ed. Peter Paret with the collaboration of Gordon A. Craig and Felix Gilbert (Princeton, NJ: Princeton University Press, 1986), 601–2.
68 Fuller, *The Foundations of the Science of War*, 107.
69 English, "The Operational Art: Development in the Theories of War," 13.
70 Bond and Alexander, "Liddell Hart and De Gaulle," 606.
71 Wieker, *Operation, Operative Führung zwischen Taktik und Strategie*, 18.
72 Karl-Volker Neugebauer, "Operatives Denken zwischen dem Ersten und Zweiten Weltkrieg," in *Operatives Denken und Handeln in deutschen Streitkräften im 19. und 20. Jahrhundert*, ed. Militärgeschichtlichen Forschungsamt (Herford/Bonn: Verlag E. S. Mittler & Sohn, 1988), 102.
73 Vad, "Operative Fűhrung: Grundlagen, Merkmale und Perspektiven," 130.
74 Besterhorn, *"Freie Operationen"-zur Bedeutung eines neuen Begriffes im Rahmen der operativen Schulung des Heeres*, 7.
75 Robert M. Citino, *The Path to Blitzkrieg: Doctrine and Training in the German Army, 1920–1939* (Boulder, CO/London: Lynne Rienner, 1999), 223.
76 Michael Geyer, "German Strategy in the Age of Machine Warfare, 1914–1945," in *Makers of Modern Strategy: From Machiavelli to the Nuclear Age*, ed. Peter Paret, with the collaboration of Gordon A. Craig and Felix Gilbert (Princeton, NJ: Princeton University Press, 1986), 567.
77 Citino, *The Path to Blitzkrieg*, 226.
78 Williamson Murray, "Armored Warfare: The British, French, and German Experiences," in *Military Innovation in the Interwar Period*, ed. Williamson Murray and Allan R. Millet (Cambridge, UK: Cambridge University Press, 1998), 40–41.
79 Murray, "Armored Warfare," 42.
80 Murray, "Armored Warfare," 42–43.
81 Murray, "Armored Warfare," 33.
82 Bond and Alexander, "Liddell Hart and De Gaulle," 613, 620.
83 A. Gakkel, "The Soviet Art of Naval Warfare in the Period between the Wars (1921–1941)," *Morskoy Sbornik* 10 (October 1976), 14–15.

84 The manual was written by a group of instructors of the Naval Academy led by Captain 1st Rank S. Stavitskiy. The final version of the manual was written by a commission headed by Captain 1st Rank I. S. Isakov; Gakkel, "The Soviet Art of Naval Warfare in the Period between the Wars (1921–1941)," 15.
85 A. V. Basov, *Flot v Velikoy Otechestvennoy Voyne 1941–1945. Opyti operativno-strategicheskogo primeneniya* (Moscow: Nauka Press, 1980), 50–52.
86 Allan R. Millet, "Assault from the Sea: The Development of Amphibious Warfare between the Wars: The American, British, and Japanese Experiences," in *Military Innovation in the Interwar Period*, ed. *Williamson* Murray and Allan R. Millet (Cambridge, UK: Cambridge University Press, 1998), 75.
87 Agnew, "From Where Did Our Amphibious Doctrine Come?" 58–59.
88 Millet, "Assault from the Sea," 75.
89 Holger Herwig, "Innovation Ignored: The Submarine Problem; Germany, Britain, and the United States, 1919–1939," in *Military Innovation in the Interwar Period*, ed. *Williamson* Murray and Allan R. Millet (Cambridge, UK: Cambridge University Press, 1998), 234.
90 Karl Doenitz, *Memoirs: Ten Years and Twenty Days*, trans. R. H. Stevens in collaboration with David Woodward (Annapolis, MD: Naval Institute Press, 1990), 18, 19, 21.
91 Scot Robertson, *The Development of RAF Strategic Bombing Doctrine, 1919–1939* (Westport, CT: Praeger, 1995), 7, 15, 19.
92 Michael Forgret, "Die Zusammenarbeit zwischen Luftwaffe und Heer bei den französischen und deutschen Luftstreitkräften im zweiten Weltkrieg," in *Luftkriegführung im Zweiten Weltkrieg. Ein internationaler Vergleich*, ed. Horst Boog (Herford/Bonn: Verlag E. S. Mittler & Sohn, 1993), 488–89.
93 Williamson Murray, "Strategic Bombing: The British, American and German Experiences," in *Military Innovation in the Interwar Period*, ed. *Williamson* Murray and Allan R. Millet (Cambridge, UK: Cambridge University Press, 1998), 131.
94 Forgret, "Die Zusammenarbeit zwischen Luftwaffe und Heer bei den französischen und deutschen Luftstreitkräften im zweiten Weltkrieg," 488.
95 Forgret, "Die Zusammenarbeit zwischen Luftwaffe und Heer bei den französischen und deutschen Luftstreitkräften im zweiten Weltkrieg," 489.
96 Murray, "Strategic Bombing," 130.
97 Murray, "Strategic Bombing," 131.
98 David E. Johnson, *Fast Tanks and Heavy Bombers: Innovation in the U.S. Army, 1917–1945* (Ithaca, NY/London: Cornell University Press, 1998), 11.
99 Forgret, "Die Zusammenarbeit zwischen Luftwaffe und Heer bei den französischen und deutschen Luftstreitkräften im zweiten Weltkrieg," 481.
100 Forgret, "Die Zusammenarbeit zwischen Luftwaffe und Heer bei den französischen und deutschen Luftstreitkräften im zweiten Weltkrieg," 485.
101 Forgret, "Die Zusammenarbeit zwischen Luftwaffe und Heer bei den französischen und deutschen Luftstreitkräften im zweiten Weltkrieg," 485.
102 Von Hardesty, "Die sowjetischen Luftstreitkräfte: Einsatzgrundsätze, Gliederung, technische Ausstattung," in *Luftkriegührung im Zweiten Weltkrieg. Ein internationaler Vergleich*, ed. Horst Boog (Herford/Bonn: Verlag E. S. Mittler & Sohn, 1993), 245.
103 Christian Hettfleisch, *Operatives Denken in den sowjetischen Streitkräften – die Entwicklung von der Oktoberrevolution bis heute* (Hamburg: Fűhrungsakademie der Bundeswehr, 1987), 14–15.
104 Hettfleisch, *Operatives Denken in den sowjetischen Streitkräften*, 14.
105 Hettfleisch, *Operatives Denken in den sowjetischen Streitkräften*, 16.
106 Hettfleisch, *Operatives Denken in den sowjetischen Streitkräften*, 17.
107 David M. Glantz, "The Intellectual Dimension of Soviet (Russian) Operational Art," in *The Operational Art: Developments in the Theories of War*, ed. M. A. Hennessy and B. J. C. McKercher (Madison: University of Wisconsin, 1996), 139.
108 Glantz, "The Intellectual Dimension of Soviet (Russian) Operational Art," 137.
109 L. D. Holder, "Educating and Training for Theater Warfare," *Military Review* (September 1990), 86.

110 John Romjue, *From Active Defense to AirLand Battle: The Development of Army Doctrine 1973–1982* (Fort Monroe, VA: TRADOC Historical Monograph, 1996), 44.
111 Shimon Naveh, *In Pursuit of Military Excellence: The Evolution of Operational Theory* (London: Frank Cass, 1997), 300–301.
112 English, "The Operational Art: Development in the Theories of War," 16.
113 Friedrich von Bernhardi, *Vom Heutigen Kriege, vol. 1: Grundlagen und Elemente des modernen Krieges* (Berlin: E. S. Mittler & Sohn, 1912), 84–85.
114 von Bernhardi, *Vom Heutigen Kriege*, 94.
115 R. Ernest Dupuy and Trevor N. Dupuy, *The Encyclopedia of Military History from 3500 B.C. to the Present*, 2nd and revised ed. (New York, NY: Harper & Row, Publishers, 1986), 669–71.
116 Dupuy and Dupuy, *The Encyclopedia of Military History from 3500 B.C. to the Present*, 879.
117 Eliot A. Cohen and John Gooch, *Military Misfortunes: The Anatomy of Failure in War*, 1st ed. (New York: Vintage Books, 1991), 201, 206.
118 Trevor N. Dupuy, *Elusive Victories. The Arab-Israeli Wars: 1947–1973* (Garden City, NY: Military Book Club, 1978), 231.
119 Dupuy, *Elusive Victories. The Arab-Israeli Wars: 1947–1973*, 244.
120 Dupuy, *Elusive Victories. The Arab-Israeli Wars: 1947–1973*, 233.
121 Dupuy, *Elusive Victories. The Arab-Israeli Wars: 1947–1973*, 236.
122 Rajko Rakić and Branko Dubravica, *Kratak Pregled Vojnih Djelovanja U Domovinskom Ratu 1991–1995* (Zagreb: Hrvatsko Vojno Učilište Petar Zrinski, 2009), 259; Davor Domazet-Lošo, *Hrvatska I Veliko Ratište* (Zagreb: Udruga Sv. Jurja, 2002), 119.
123 Rakić and Dubravica, *Kratak Pregled Vojnih Djelovanja U Domovinskom Ratu, 1991–1995*, 261; Croatian Helsinki Committee For Human Rights, *Military Operation Storm And Its Aftermath Report* (Zagreb: HHO, 2001), 11–12.
124 Jay Luvaas, "Thinking at the Operational Level," *Parameters* 1 (Spring 1986), 34.
125 The Soviets, in contrast to Western theoreticians, believed there is only a single military strategy, while each service develop and practice its own operational art; hence, they differentiated operational art of the strategic missile troops, ground forces, air forces, troops of air defense, and naval forces, respectively; *Einführung in die operative Kunst der Landstreitkräfte* (1974), 10.
126 Erich von Manstein, *Lost Victories* (Novato, CA: Presidio Press, 1982), 409.
127 Henry E. Eccles, *Military Concepts and Philosophy* (New Brunswick, NJ: Rutgers University Press, 1965), 5.
128 Dwight L. Adams and Clayton R. Newell, "Operational Art in the Joint and Combined Arenas," *Parameters* (June 1988), 35.
129 I. C. B. Dear et al., eds., *The Oxford Companion to World War II* (Oxford/New York: Oxford University Press, 1995), 1009–10.
130 Louis Morton, *United States Army in World War II: The War in the Pacific. The Fall of the Philippines* (Washington, DC: Office of the Chief of Military History, Department of the Army, 1953), 59.
131 *Fuehrer Conferences on Naval Affairs, 1939–1945*, foreword by Jak P. Mallmann Showell (Annapolis, MD: Naval Institute Press, 1990), 201.
132 David Jablonsky, "Strategy and the Operational Level of War," in *The Operational Art of Warfare Across the Spectrum of Conflict* (Carlisle Barracks, PA: U.S. Army War College, 1987), 14.
133 David Jablonsky, "Strategy and the Operational Level of War: Part I," *Parameters* (Spring 1987), 73.
134 Meinhard Glanz, *Rommels Entschluss zum Angriff auf Alexandria* (Hamburg: Führungsakademie der Bundeswehr, 1963), 5; Jablonsky, "Strategy and the Operational Level of War: Part I," 74.
135 Führungsakademie der Bundeswehr, *Arbeitspapier, Operative Führung* (Hamburg: Führungsakademie der Bundeswehr, August 1992), 17–18.

136 David C. Evans and Mark R. Peattie, *Kaigun: Strategy, Tactics, and Technology in the Imperial Japanese Navy, 1887–1941* (Annapolis, MD.: Naval Institute Press, 1997), 515.
137 Michael I. Handel, *Masters of War: Classical Strategic Thought*, 3rd and expanded ed. (London: Frank Cass, 2001), 355.
138 Jürgen Bassenge, *Zur Zielsetzung der Deutschen Angriffe in Frankreich, Frühjahr bis Sommer 1918. Überlegungen zum Problem des Kulminationspunktes* (Hamburg: Führungsakademie der Bundeswehr, 1964), 19–20.
139 Wilhelm Meier-Dornberg, "Die grosse deutsche Frühjahrsoffensive 1918 zwischen Strategie und Taktik," Military Historical Office, *Operative Denken und Handeln in deutschen Streitkräften im 19. und 20. Jahrhudert* (Herford/Bonn: Verlag E. S. Mittler & Sohn, 1988), 73.
140 Ludwig Beck, *Studien,* ed., Hans Speidel (Stuttgart: K. F. Köhlers Verlag, 1955), 79; Führungsakademie der Bundeswehr, Arbeitspapier, *Operative Führung* (Hamburg: Führungsakademie der Bundeswehr, August 1992), 15; Jablonsky, "Strategy and the Operational Level of War," 73.

2 Objectives

Any war is fought to achieve certain political objectives. The *political objective* can be described as the one whose accomplishment would secure national or alliance/coalition important interests in a certain part of a theater. When aimed to achieve vital national interests, a political objective is strategic in scale. Its accomplishment would have a radical effect on the course and outcome of a war. In his seminal work *On War*, Carl von Clausewitz wrote that

> no one starts a war – or rather, no one in his senses do so – without first being clear in his mind what he intends to achieve by that war and how he intends to conduct it. The former is the political purpose; the latter its operational objective.[1]

He observed that "the political object – the original motive for war – will thus determine both the military objective to be reached and the amount of effort it requires."[2] Political objectives can be purely political. However, they are often combined with ideological, geopolitical, economic, financial, social, ethnic, religious and other non-military objectives.

Political objectives are usually achieved by using military power. The *military objectives* comprise the very heart of both the theory and the practice of war. Without a clearly stated and attainable objective, the entire military effort becomes essentially pointless.[3] This is particularly true at the operational and strategic levels of war because the stakes are so much higher than at the tactical level. A military objective should be expressed clearly and concisely. A lack of clarity could lead to waste of one's time and sorely needed resources. Failing to define clear objectives or frequently changing objectives during combat would lead to confusion, lack of focus and, often, lack of success.

Generically, an objective can be understood as the "purpose" of the actions one carries out within a specific space and time. The determination of a military objective is the first and most critical step in any military undertaking. Once the objective is determined, the entire problem becomes greatly simplified (but not necessarily easy to resolve). The accomplishment of a military objective would require the *use of lethal force* and would lead to a *radical change* of a situation in the area larger than the one where opposing forces are in combat.

DOI: 10.4324/9781003333685-2

Almost all aspects of warfare are related, either directly or indirectly, to a given military objective. The objective determines the level of war/level of command, type of planning, method of combat forces' employment, the size of the operating area, scale of the elements of combat force employment (e.g., deployment/redeployment, concentration/dispersal, maneuver, center of gravity, point of culmination point, deception and reserve) and operational/combat support (e.g., intelligence, information warfare, fires, logistics and protection).

A military objective can be offensive, defensive or a combination of these two. An *offensive objective* is aimed to capture a territory/position or destroy/annihilate/neutralize the enemy forces. A *defensive objective*, in contrast, is aimed at defending, holding or retaining control of a certain territory/position by weakening the enemy's forces over time. A military objective can be predominantly physical (or tangible) or "abstract" (or intangible). Any military objective consists of a combination of tangible and intangible elements.

Optimally, the commander should determine a single *main objective* and one or more *alternate objectives*. By having an alternative objective, the commander on the scene can act quickly instead of waiting on orders from the higher commander. It is also possible to deceive the enemy about the true main objective. B. H. Liddell Hart observed that by having an alternative objective "you will achieve objective that he (enemy) guards the least." An alternative objective would provide "the opportunity of gaining an objective, whereas a single objective, unless the enemy is helplessly inferior, means the certainty that you will not gain it – once the enemy is no longer uncertain as to your aim."[4] There is probably no greater mistake than to formulate and simultaneously try to accomplish multiple objectives. Such a course of action could only be taken if one's forces possess overwhelming strength against any conceivable combination of the enemy's forces.

Actions intended to accomplish an offensive objective differ significantly from those aimed at achieving a defensive objective. An offensive objective is usually accomplished by destruction, annihilation or neutralization of the enemy forces. The enemy is *destroyed* when the very core of his forces suffers such losses that he cannot continue the fight.[5] It also means that a target is so damaged that it can neither function as intended nor be restored to a usable condition.[6] The enemy is *annihilated* when he is left with no sources of power to offer any serious resistance. *Neutralization* means that the enemy is rendered ineffective and cannot prevent one's forces from accomplishing their assigned objective.[7] Defensive objectives are expressed in terms of prevent, contain, defend and retain or deny control, in regard to a certain territory/position or sea/ocean area.

A military objective is normally divided into several constituent parts – called *tasks* – that, when achieved, will collectively lead to the objective's accomplishment. Tasks are those measurable entities that allow a commander to properly determine the rate of progress toward the accomplishment of a given military objective. Note that a task issued by a higher commander is converted into a respective objective by a subordinate commander. The linkage between the objective and its constituent tasks cannot be arbitrarily broken up without serious consequences for one's ability to accomplish a given objective. The larger the objective, the more tasks it encompasses. Thus, the larger and more diverse the forces and more time is required to accomplish a given objective. A large military objective is usually divided into several related *main* (or principal) tasks; each of these is composed of a number of *component* (or partial) *tasks*. Generally, determining too few tasks would lead to partial or even complete failure to accomplish a given military objective. Adding new tasks without changing or modifying the objective or having larger or more capable forces leads to so-called *mission creep.* If that happens, then a serious disconnect between the objective and forces/resources available would occur, resulting in the ultimate failure of a given mission.

The accomplishment of any military objective would result in a complete or partial *combat success*. Depending on the scale of military objectives, tactical, operational and strategic successes are differentiated. A given combat success must be *consolidated* by a series of actions and measures. When on the offensive, consolidation is ensured by destroying or neutralizing the remainder of the enemy forces. In a defensive, consolidation is aimed to preserve one's gains and fend off enemy attacks by further weakening of the enemy forces. Without successful consolidation of a combat success, all the gains achieved in the initial combat phase could be easily lost and thereby resulting in the worsening of the situation for one's forces. In the third and last phase, combat success must be *exploited.*

Scale

Tactical, operational and strategic objectives are differentiated according to their scale. Clearly, the larger the objective, the greater the adverse consequences of not accomplishing it.[8]

A distinction is made between the *national* (or alliance/coalition) *political strategic objective* and the military strategic objective.

The political objective should dominate the strategic objective, and the strategic objective should drive the determination of the operational and tactical objectives.[4] A *military strategic objective* can be defined as the one whose accomplishment would lead to the end of the enemy's organized resistance and thereby set the stage for the post-hostilities (or stabilization) phase of a war. In case of the continental size countries, such as the United States or the Russian Federation, or the oceanic theaters (e.g., Atlantic and Pacific), the possibility exists of having a war in two or more theaters. Then for each of them, a single military strategic objective must be determined. In World War II, the U.S. had two national military strategic objectives; unconditional surrender of the Axis powers in Europe (Nazi Germany and Italy) and in the Pacific (Imperial Japan).[9]

Normally, an *offensive* political objective would require the accomplishment of a single or more offensive military strategic objectives. In the American Civil War (1861–1865), the northern states had an offensive strategic objective – to defeat the southern states and thereby force them to remain in the Union. This objective could be accomplished only by going on a strategic offensive.[10] In all three wars of the German unification: against Denmark in 1864, Austria in 1866 and France in 1870–1871, Prussia had offensive military strategic objectives. For the "First

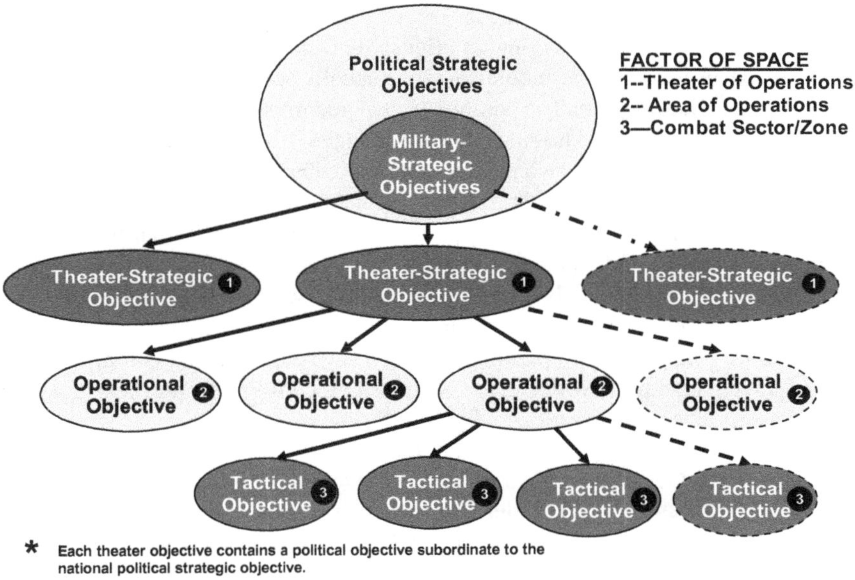

Figure 2.1 Hierarchy of Military Objectives

Operational Stage of the War" in 1941–1942, the Japanese selected offensive military strategic objectives – to gain mastery of the Far East area by (1) destroying the U.S. power in the western Pacific and the British in the Far Eastern waters, (2) cutting their respective sea communication with these areas and (3) land communications from India to China (the Burma Road).[11] In November 1941, the Japanese Central Army–Navy Agreement specified that the war objectives were "reduction of foundation of U.S., British, and Dutch power in Eastern Asia, and occupation of Southern Areas."[12] In the 1st phase of the war, the Japanese would destroy the U.S. Pacific Fleet in Pearl Harbor and conduct a simultaneous invasion of Malaya and air strikes on the Philippines, followed by the invasion of the Philippines, the Netherlands East Indies (NEI) and Southeast Asia.[13]

Sometimes political objectives were offensive, but they were not supported by offensive military strategic objectives. Russia's political strategic objectives in its war against Japan in 1904–1905, were clear – maintain control over Manchuria and decisively repel Japanese advances. However, the Russian military strategic objective was defensive – retaining control of the positions they already held: Port Arthur, the Trans-Siberian railroad, Vladivostok, and other concessions on the Yalu River.[14] Yet Russia's proper military strategic objectives were the destruction of the Japanese forces in Manchuria and obtaining/maintaining control of the Yellow Sea and the Sea of Japan.

Sometimes, the weaker side had a defensive political objective but the only way of accomplishing it was by going strategically on the offensive. In the American Civil War (1861–1865), the Confederate States had a defensive political objective – forcing the North to recognize their independence. However, this could be accomplished only by selecting an offensive military strategic objective.[15]

Defensive military strategic objectives are usually selected by the side on strategic defensive. They are usually a combination of geographic and military objectives. On 3 January 1942, the American–British–Dutch–Australia (ABDA) Command was directed to accomplish two military strategic objectives: (1) "hold Malay barrier (the Malay Peninsula, Sumatra, Java, North Australia) as basic defensive positions of ABDA area; and (2) hold Burma and Australia as essential to support of China and to defense of India."[16] The defense of the Malay Barrier was the most important theater-strategic objective while the defense of Burma and Australia was considered as a secondary objective. Directive to ABDA's Command also included (what can be considered an operational objective), reestablishing communications through Dutch East India with Luzon, supporting the Philippine garrison and maintaining essential communications within ABDA's area.[17]

An attacker can select a combination of *offensive and defensive military strategic objectives* as Germany did in its invasion of Denmark and Norway in April 1940 (Unternehmen Weserübung). Hitler in his instructions of 1 March 1940 stated that objectives were "to occupy Norway and Denmark, prevent English assault on Scandinavia and in the Baltic, secure our iron ore basis in Sweden and extend base of operation for the Luftwaffe and the Kriegsmarine in war against England."[18] The Japanese strategic objectives in the "First Operational Stage of the War" would be accomplished in two offensive phases separated by a defensive phase. Specifically, in phase I, the Japanese would destroy the U.S. Fleet in Hawaii & U.S. Army Air Forces in the Philippines and capture strategic areas and key positions for establishing a perimeter for the defense of the Southern Resources Area and the Home Islands. In phase II, the Japanese planned to consolidate and strengthen their newly established defensive perimeter (line the Kurile Islands-Marshall Islands & Wake-the Bismarck Archipelago-Timor-Java-Sumatra-Malaya-Burma & Nicobars/Andamans). In phase III they aimed to destroy (the remaining) enemy forces threatening the defense perimeter and "destroy" the U.S. will to fight.[19]

In a very large theater of war would exist two or more *theater-strategic objectives* – the one whose accomplishment would result in the destruction/neutralization of a major part of the enemy forces or successful defense/denial of a large part of one's territory or sea/ocean area in a given *theater of operations*.

In a global war, in each theater of war, there might be several theater-strategic objectives, some offensive and some defensive, to be accomplished in sequence or simultaneously. Their accomplishment would have a radical change of a situation in a given *theater of war*. It would also signify a major phase in a war.

In the offensive phase of the war in the Pacific (after August 1942) the Allies had in the Pacific Ocean Area (POA) three theater-strategic objectives: (1) defense of Alaska/Aleutians and the U.S. northwestern littoral; (2) capturing the Solomons archipelago; and (3) capturing the Japanese strongpoints in the central Pacific. In the Southwest Pacific Area (SWPA), the Allies had three identifiable theater-strategic objectives: (1) defense of Australia and New Zealand; (2) capturing New Guinea; and (3) liberating the Philippines. The final theater-strategic objective for the POA

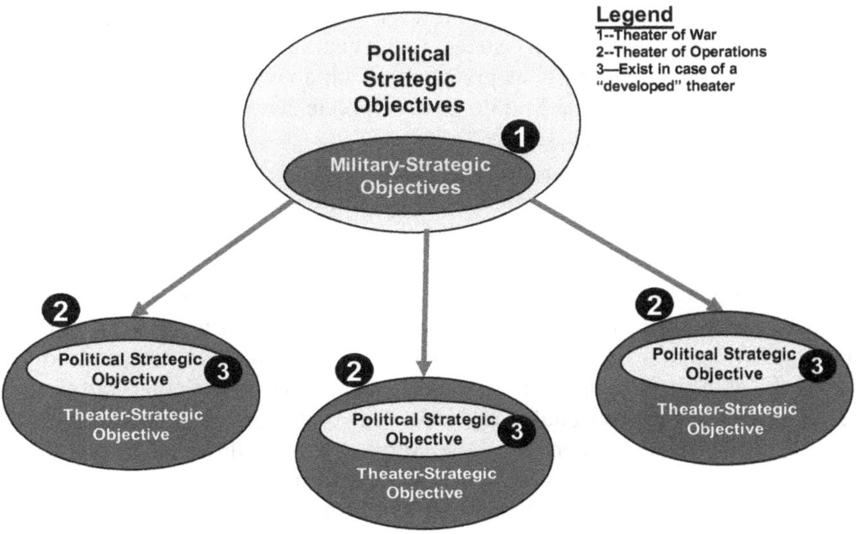

Figure 2.2 Theater-Strategic Objectives

command was capturing/neutralization of the southern approaches to the home islands (Formosa, Iwo Jima, and Ryukyus) and then jointly with SWPA's forces to assault and occupy home islands. This part of the theater-strategic objectives was made unnecessary after the dropping of the atomic bombs Hiroshima and Nagasaki in August 1945.

Directive to Admiral Chester W. Nimitz, CINCPOA/U.S. Pacific Fleet issued on 30 March 1942 by the Combined Chiefs of Staffs (CCS) stated the following objectives:

> (a) Hold the island positions between the United States and the Southwest Pacific Area necessary for the security of the line of communications between those regions; and for supporting naval, air and amphibious operations against Japanese forces; (b) Support the operations of the forces in the Southwest Pacific Area; (c) Contain Japanese forces within the Pacific Theater; (d) Support the defense of the continent of North America; (e) Protect the essential sea and air communications; and (f) Prepare for the execution of major amphibious offensives against positions held by Japan, the initial offensives to be launched from the South Pacific Area and Southwest Pacific Area.[20]

The U.S. Joint Staffs in their directive of 02 July 1942 to General Douglas MacArthur, Supreme Commander of SWPA, stated that his ultimate (theater-strategic) objective was "seizure and occupation of New Britain-New Ireland-New Guinea area."[21] The Allies determined an offensive theater-strategic objective in their invasion of North Africa in November 1942. On 14 August 1942, directive to U.S.

General Dwight Eisenhower, CINC, Allied Expeditionary Forces for landing in North Africa (operation Torch) stated that "combined military operations be directed against Africa as early as practicable with a view to gaining, in conjunction with Allied Forces in the Middle East, complete control of North Africa from the Atlantic to the Red Sea."[22]

Like a political objective, a military-/theater-strategic objective can be unlimited or limited. An *unlimited* military-/theater-strategic objective will be selected if the political objective is to overthrow the enemy's government and/or social system or capture a major part of the enemy's territory. In a case of war between two strong opponents, the accomplishment of an unlimited military strategic objective will usually result in a long war requiring maximum exertion of all spiritual and material resources of a nation or an alliance/coalition, as an example of the war on the eastern front in 1941–1945 illustrate. Selecting unlimited military- or theater-strategic objectives could very easily lead to the removal of all limits as far as political objectives are concerned. Military leaders would usually press for all political decisions and actions to be subordinate to military objectives.[23] Hitler's Directive No. 21 issued on 18 December 1940 for the invasion of Soviet Russia (Case Barbarossa) stated that during the initial phase, "the bulk of the Russian army stationed in western Russia will be destroyed in a series of daring operations spearheaded by armored thrusts; the organized withdrawal of intact units into the vastness of interior Russia must be prevented." In the next phase, the ultimate objective of the operations was to "screen Europe against Asiatic Russia along the course of the Volga River and then along a general line extending northward toward Archangelsk. The Luftwaffe will be in a position to neutralize the last industrial region remaining in Russian hands situated in the Urals."[24]

In other cases, a much stronger side might accomplish its offensive and unlimited political and military strategic objectives relatively quickly, as examples of the German invasion of Poland in September 1939, Denmark/Norway in April–June 1940 and Yugoslavia and Greece in April–May 1941 show. Germany's military strategic objectives in the invasion of Poland in September 1939, as stated by the Army's High Command (OKH – Oberkommando des Heeres) (or general staff) on 13 June 1939, were "the Polish army will be destroyed in the western part of the country and reserves would be prevented from mobilizing or concentrating to slow German advance."[25]

The Allies had unlimited military-strategic objectives in their war against Nazi Germany and Imperial Japan in World War II. Yet sometimes it is possible to achieve unlimited political strategic objectives without the complete destruction of the enemy forces, as was the case in World War I, when both Germany and Austria-Hungary signed an armistice with the Entente and Associated Powers on 11 November 1918.

In its invasion of Ukraine, Russia initially selected offensive and unlimited political and military strategic objectives. In his speech on 21 February 2022, Putin rejected the very idea of Ukraine as a nation-state, casting it instead as "an inalienable part of our history, cultural, and spiritual space sundered from Russia by the Bolsheviks and the time of revolution and then bolstered with territory seized from Hungary, Poland, and Romania under Stalin"[26]. On 28 February 2022, Russia's objectives in Ukraine were leaked by accident by the government agency RIA Novosti. Putin stated that "'Greater Russia' (including Ukraine and Belarus) has returned to its 'rightful place' as a world power." This implied that Russia's political objective was to gain control of the entire Ukraine.[27] The U.S. officials and some analysts believe that initially, the Russian objective was to "decapitate" the Ukrainian government. Putin reportedly believed that the Ukrainian military would be ineffective and that the Ukrainian political leadership could be easily replaced.[28] Putin expected to capture Ukraine's capital Kyiv quickly and install a compliant government.[29] The Russian plan was to capture Kyiv in just two days. The rapid takeover of Ukraine would present the West with *fait accompli*.[30] The Russians made another major error in assuming that there would not be any significant resistance by the Ukrainian population.[31]

In a war fought for limited political objectives, the military-/theater-strategic objectives will be also *limited*. One normally does not risk all for limited political objectives nor does one commit all his sources of power in such a war.[32] The accomplishment of a limited military-/theater-strategic objective would require low to modest use of military power, efforts, and time. A state might not need to pursue an unlimited military strategic objective by trying to destroy the enemy's forces and seek their surrender. Liddell B. H. Hart wrote that a state that is seeking not conquest but the maintenance of its security would accomplish its military strategic objective if the threat is removed and "if the enemy is led to abandon his purpose."[33] Or it

> may desire to wait until the balance of forces can be changed by the intervention of allies or by referring forces from another theater. It may desire to wait, or even to limit the military effort permanently, while naval or economic action decides the issue.[34]

A limited military- or theater-strategic objective is often selected by a weaker side to gain time in changing the correlation of forces to its favor. Then, the stronger side's strength is weakened over time. The enemy's forces should be then weakened at a faster rate than one's forces. Disproportionately higher losses can be inflicted by local attacks and raiding the enemy's supplies. Liddell Hart observed that a war with limited "political aim imposes a strategy of limited aim, and a decisive aim should be only adopted with the approval of government, which alone can decide whether it is worth the candle."[35]

The Russo-Japanese War of 1904–1905 was a war with limited political and military strategic objectives for both sides. Both Russia and Japan disputed control of the area, which did not belong to either of them. Japan was unable to completely defeat Russia but that was also unnecessary. This was also the case for Russia. Neither Japan nor Russia wanted to fight to the very end. Thus, they were unwilling to commit utmost efforts and sacrifices which might lead to a complete exhaustion.[36] The Russian Tsar and his inner circle argued for land acquisition while the Russian prime minister Sergey Witte was more interested in commercial expansion in the

Far East. Japan felt humiliated and double-crossed by the Russian acquisition of Port Arthur from 1895 onward. It was staunchly opposed to the growing Russian influence in Manchuria. By going to war, Japan claimed that its aim was to "liberate" Manchuria from the Russian grasp.[37] The Japanese military objectives were in consonance with their political objectives. Specifically, the Japanese aimed to capture Korea and then destroy the Russian army in Manchuria. Preconditions for this was to obtain control of the Yellow Sea and ensure security of land and sea communications between Korea and Manchuria.[38]

The Gulf War of 1990–1991 was a war with limited political and military strategic objectives. The U.S.-led coalition never intended to defeat the Iraqi armed forces as a whole and occupy the country. The coalition objectives were immediate, complete and unconditional withdrawal of all Iraqi forces from Kuwait, restoration of the legitimate Kuwaiti government, the security and stability of Saudi Arabia and the Persian (Arabian) Gulf, and the safety and protection of American citizens abroad.[39] The United States aimed to remove Saddam Hussein by his domestic opposition, but without endangering Iraqi territorial integrity. The coalition did not intend to defeat Iraq so completely that the ensuing power vacuum would be exploited by Iran and spark further turmoil in Iraq.[40] The United States was unwilling to pursue its objectives directly and did not intend to be involved in the nation-building and humanitarian relief that would surely follow the overthrow of the Iraqi regime. A serious disconnect still existed between the more ambitious ends and modest means to be used by the United States and its coalition partners. The termination of the Gulf War of 1990–1991 was not only confused and ambiguous but had unintended and adverse consequences for U.S. national interests.[41]

The geographical separation of centers of power of the belligerents plays an important role in a war for limited military-/theater-strategic objectives. This is especially the case when there is lack of overland link between the two main belligerents due to an ocean or neutral states or if land area is so large to make it difficult or impossible for either belligerent to exert their full strength against the other.[42] The Spanish-American War of 1898 was fought for a limited objective by both belligerents. Spain wanted to retain its control of Cuba and Puerto Rico and the Philippines. Initially, the U.S. objectives were also limited – liberate Cuba from the Spanish rule. However, these war aims were expanded to also include control of Puerto Rico, Philippines and Guam.[43] This is an example of an expansion of the initial limited objective. The nations have gone to war in defense of an object which was by no means vital to their national welfare, but their prestige or national honor would not permit them to relinquish them.

A stronger side might be forced to change its military strategic objective *from unlimited to limited* because of a series of military setbacks or defeats in the field. By early April 2022, the Russian offensive in Ukraine stalled. The Russians were unable capture Kyiv and Kharkiv. The Russian forces began to withdraw from the vicinity of Kyiv and were redeployed to self-declared Donetsk and Luhansk People's Republics. Once the Russians realized that their political objectives could not be achieved, they for the time being reduced both political and military strategic objectives. This was formally announced on 26 April 2022. Afterward, the

Russians launched an offensive to fully occupy the Donetsk sand Luhansk provinces and strengthen their control in southern Ukraine.[44] That offensive failed to achieve its stated objectives. By 30 December 2022, the Russian forces were generally on defensive except for some limited ground assault against selected positions in eastern Kharkiv, Donetsk and Luhansk districts, and southern area.[45]

In contrast, Ukraine's initial political and military strategic objectives were defensive and limited to preserving territorial integrity, protection of Kyiv and major cities, and to survive until western support arrived.[46] Because of the battlefield successes the Ukrainian military strategic objectives were changed in the spring 2022. The Ukrainian forces went on offensive and recaptured relatively large part of eastern Ukraine including city of Kherson in southern Ukraine in November 2022. By 12 November, the Ukrainians liberated 28,742.000 square miles of their sovereign territory (the Russians still control 17,165.000 square miles).[47] Their political objective remains essentially defensive, but the military strategic objectives were expanded to the recovery of the territories lost to Russia in 2014 and 2022.[48]

Sometimes a side on a strategic defensive might go on the offensive but would select a limited military-/theater-strategic objective, as the example of the UN forces in Korea in the summer of 1950 illustrates – defeating and forcing North Korean forces to withdraw north of the 38th Parallel. However, the success of a counter-offensive might lead the political and military leadership to change military strategic objective from *limited to unlimited*. After the UN amphibious landing in Inchon on 15 September 1950 (operation Chromite), the Joint Chiefs of Staff (JCS) directed General MacArthur on 27 September 1950:

> Your military objective is the destruction of the North Korean armed forces. In attaining this objective, you are authorized to conduct military operations, including amphibious and airborne landings or ground operations north of the 38th parallel in Korea provided that at the time of such operations there has been no entry into North Korea by major Soviet or Chinese Communist forces, no announcement of intended entry, nor a threat to counter our operations militarily in North Korea.[49]

Because of its scale and complexity, a military strategic or theater-strategic objective cannot be accomplished by a single act; several intermediate objectives – *operational objectives* – must be accomplished.

An operational objective could be defined as one whose accomplishment would result in destruction/neutralization or annihilation of a *major part* of the enemy forces or holding/maintaining control of a major part of one's territory/important position. It would result in a radical change in the situation in a given *theater of operations*.

The true meaning of the term operational objective is often not well understood by commanders and planners. Yet without that understanding it is not possible to prepare a sound plan for a campaign or major operation. In recent years, it has been very common in the U.S. military to confuse some purely political activities (e.g.,

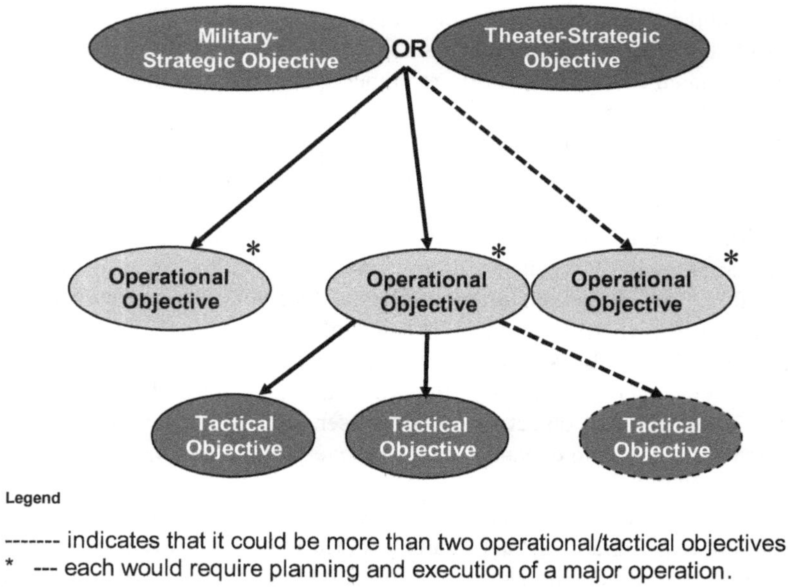

Figure 2.3 Operational Objectives

securing regional support) or elements of combat employment of one's forces (e.g., alerting, posturing or deploying) as operational objectives.[50] Another problem is general lack of understanding of the process for determining an operational objective. These are not trivial matters because it makes it more difficult to make operational decisions and prepare operation plans.

The operational objectives could be located on land, at sea, and in the air. For each of them it is necessary to select and assess from enemy's perspective operationally important elements of the factors of space, time, and force. The focus in analyzing *factor of space* should be on location and the size of the operating area, distance to basing areas, terrain, mountain passes/valleys; road/rail hubs, population density, urbanization, oceanography/hydrography and the weather/climate. In evaluating the *factor of time*, the commander and the staff should consider the time required for the enemy's forces to be deployed, concentrated and reinforced; the time required to commit enemy reserves; and the time required for regenerating combat power. The main consideration in the *factor of force* should include the size/composition of enemy forces attacking or defending a given area/position, their mobility, logistical support and sustainment, combat readiness, morale and discipline, skills of the commanders, soundness of doctrine and anticipated enemy's resistance. This list is not all-inclusive. It is critically important to evaluate not only tangible but also intangible enemy's elements of the factor of force. Obviously, which elements of these factors should be evaluated depends on whether operational objectives are located on land, at sea or in the air. Afterward, the commander and the staff should determine the size of the operating area, the time required for accomplishing the given operational objectives and the required size/composition of friendly forces.

In selecting an operational objective, perhaps the most important is its location and the distance to various militarily important points in the surrounding area. The Allied selection of the 2,437 square miles island of Leyte as the initial operational objective into what evolved to the Philippines campaign (1944–1945) was based on several elements of the factor of space. Leyte was located in the center of an extensive network of enemy airfields. The enemy aircraft based on Luzon and Mindanao and on the other Visayan Islands could easily support enemy forces on Leyte. Because of the short distances and the presence of numerous islands and islets, the Japanese could covertly bring troop reinforcements to Leyte from the Visayas, Luzon and Mindanao by using barges and smaller vessels and under the cover of darkness and in bad weather.[51] The selection of Leyte also incurred several serious disadvantages. The island was located some 635 miles from the Allied airfields on Morotai (in the Halmahera). This distance was beyond the effective range of Allied land-based fighters based on Morotai. This means that in the initial phase of the landing, the Allies had to rely primarily on carrier aircraft for close air support and to prevent the enemy from reinforcing their troops on Leyte. The success of the entire operation hinged on the ability of Allied naval forces to keep the enemy from building up its strength on Leyte and the adjoining island of Samar, and to prevent enemy naval forces from attacking friendly shipping in the beachhead area.[52]

Another major reason for selecting Leyte was the belief that the island with its large and fertile Leyte Valley offered a favorable place for the construction of airfields from which could be launched large-scale attacks against Luzon and the rest of the Philippines and China's mainland. However, the subsequent events proved that Leyte was unsuitable for the construction of many airfields and air facilities because of poor drainage and soil. The Allied planners correctly believed, however, that Leyte would provide an excellent anchorage together with sites for supply bases.[53]

In determining the location of a major amphibious landing, the planners should select such a place that would force the enemy to react operationally. General MacArthur, Commander of the UN forces in Korea, selected Inchon (Incheon today), in his view, the most vulnerable place for the North Korean forces, then occupying southern part of Korea. By capturing Inchon, it would be possible to attack the South Korean capital, Seoul. The key road and rail link for the North Korean troops besieging UN troops at Pusan (Busan today) (140-mile-long defensive line) led from Seoul.[54] Inchon was some 150 miles away from the Pusan Perimeter.[55] MacArthur's maneuver into the operational depth of the enemy's defenses was too deep to allow the North Korean Army the option of rapid response and counterattack.[56]

MacArthur made a bold decision bordering with a gamble by landing some 70,000 men of the X Corps in the operational depth of the enemy defenses.[57] The intent was to conduct a turning movement to cut off supply lines from Seoul to the south.[58] Such a maneuver would force the enemy forces to abandon their current positions and move to meet the new threat. The selection of Inchon and the idea of the landing was solely due to MacArthur. Despite opposition from the highest military officials in Washington, D.C., MacArthur was able to get approval for the landing. He received credit for the "conduct of operations, determination to carry it through, and full acceptance of hazards of the operation."[59]

In a war on land an operational objective could be a major enemy force, territorial/geographic, enemy's capital, large city/urban area, major port/naval base, airfield cluster, an important geographic position, or military-economically important area/production facility. The Soviets accomplished an operational objective with their encirclement and destruction of the German Sixth Army in November 1942–January 1943. Out of the original strength of the Sixth Army of 250,000 men, 47,200 were killed and 91,000 captured (including 24 generals and 2,500 other officers).[60] The Germans lost 20 divisions in the encirclement plus six divisions were virtually destroyed outside the encirclement.[61] The Germans evacuated by the air some 10,000 specialists and 25,000 wounded. The Luftwaffe lost 488 transport aircraft in its resupply efforts.[62] The Soviets employed up to 1.0 million men in the Stalingrad (major) operation.[63] The Soviet losses were substantial but not verified by neutral sources. The Soviet sources claimed that between 19 November 1942 and 02 February 1943, the Soviet losses were 485,777 (154,885 killed and 330,892 wounded).[64] The Axis' losses were 800,000 (including 580,000 killed, wounded and missing plus 220,000 captured).[65] Destruction of the Sixth Army was the largest German defeat up to that point in World War II. It represented a turning point in the war on the Eastern Front.

In World War II, the Allied objectives in the European theater in World War II were primarily *geographic* or *territorial*.[66] In North Africa, the American inexperienced troops engaged initially poorly equipped French forces. They looked to threaten and eventually to bottle up in Libya the Axis forces and very vaguely intended to expel the Axis forces from all of North Africa.[67] In the invasion of Sicily in July 1943, the Allies regarded the island not as an opportunity to destroy Axis defenders but only as a stepping stone to southern Europe.[68] Sicily was large enough to hold sufficient troops to threaten further action in another place – mainland Italy, Corsica, Sardinia, or southern France.[69] The Allies were apparently not interested to trap or eliminate the Axis forces in Sicily. To do that the Allies had to reach Messina before the Axis forces did.[70]

The Allies were convinced that capturing cities, especially the capitals, was the most certain way to win the war.[71] In purely military terms, *enemy's capital* is an operational objective to be captured or defended. In a highly centralized state, the capture or defense of the capital might have a large strategic effect. The efforts to capture the country's capital led to a waste of time. Yet experience shows that the most effective and fastest way to end the hostilities was by destroying enemy forces.[72]

An important political or military event could sometimes diminish the importance of capturing the enemy's capital. After landing on Italy's mainland in September 1943, the Allies were trapped by circumstances to fight in a terrain overwhelmingly favoring the defense. Rome became a magnet drawing Allied forces northward.[73] Churchill exaggerated the stakes, however, writing that "if Rome were not captured, the world would 'regard our campaign as a failure.' . . . [W]hoever holds Rome holds the title deeds of Italy." He wrote to General Mark Clark, the

Fifth Army's commander, that without Rome the campaign would "peter out ingloriously."[74] The capture of Rome in June 1944 had little importance for the progress of war on Italy's mainland. It was "newsworthy and a source of some excitement. It gave more emotional and psychological than military advantage to the Allies."[75]

In the Allied invasion of Sicily in July 1943 (Operation Husky), the British Eighth and U.S. Seventh armies landed at two separate sectors in the southeastern part of the island. Their initial operational objectives were to capture the *ports/airfields.* The Seventh Army was directed to capture the ports of Gela and Licata and surrounding areas, while the Eighth Army aimed to capture the ports of Syracuse and Augusta and the airfield cluster at Gerbini.[76] The Allies selected the wrong operational objective after they obtained control of Sicily in August 1943. Instead of landing on the mainland coast much closer to Rome, they decided to capture the port of Naples. This required landing at Salerno Bay on 15 September 1943 and then advancing to Naples (a distance of some forty-five miles).[77] After some heavy fighting against the German troops, the Allies entered Naples on 01 October 1943. Other more promising operational objectives than Naples but ignored by the planners were capturing Sardinia or Corsica.

Some elements of country's military-economic potential, such as transportation system, electricity grid, oil production/refining facilities, rare metals, ball-bearing plants and chip production plants, can be considered as operational objectives if their destruction or defense has a direct effect on the combat employment of enemy ground and/or air forces. In World War II, ball-bearing plants were considered critical for the production of military items such as airframe control, guidance systems, artillery, armored vehicles, U-boats and other weapons and equipment. The German industry consumed 2,395,000 ball bearings annually. Five ball-bearing plants at Schweinfurt alone produced 52.2 percent of the total output of ball-bearings in Germany. In contrast to other German ball-bearing plants, they produced a whole range of ball bearings.[78] Hence, they became the key target for the Allied heavy bombers.

Because of Germany's economic weaknesses in continuing the war against the economically more powerful enemy coalition, Hitler gave extraordinary importance to securing the economic resources of Soviet Russia. The most important economic resources were in the eastern part of Ukraine (Donbas; from Donets Basin) and in the Caucasus. The capture of the Caucasian oil fields would present a mortal threat to the Soviet regime. The most important oil center was Baku, some 745 miles away from the southernmost part of the German front. Another problem was that the oil fields must fall undamaged into German hands.[79] According to the British estimates in 1942, the Soviet oilfields in the Caucasus – Maikop and Grozny – each produced 18.3 million barrels of oil annually or 10 percent of Soviet total oil production.[80] The German capture of these oilfields would allow it to continue war.[81] Hitler also intended to capture Chiaturi, Transcaucasia which produced 1.5 million tons of manganese ore or over 50 percent of the Soviet total production.[82]

Operational objectives in a war at sea vary from destroying a major part of the enemy fleet at sea/bases, carrier group, amphibious task force, capturing/defending a strait or large island, securing control of one's basing/deployment area or denying the same to the enemy, and destruction or defense of the military-economic potential at sea.

The Japanese Midway–Aleutians Operation – or, as it is popularly known in the United States, the battle of Midway – represented a turning point in the Pacific War of 1941–1945. The Imperial Japanese Navy (IJN) suffered the greatest defeat in its proud history. After June 1942 Japan was forced onto strategic defensive and was never able to regain the initiative before its unconditional surrender in August 1945. Japanese losses in the Midway operation were extremely high. They had lost four frontline carriers, 253 aircraft and one heavy cruiser. In addition, one heavy cruiser had been heavily damaged, and one destroyer suffered moderate damage, while one battleship, destroyer and oiler had suffered slight damage. Some sources claim that the Japanese lost 332 aircraft, including 280 that went down with the carriers.[83] About 150 Japanese pilots were saved. The Japanese had lost some 3,500 men. In contrast the United States lost one fast carrier and ninety-two officers, and 215 men killed. However, three U.S. carrier air groups had been decimated.[84] American losses in aircraft had been heavy – 147 machines were shot down.[85]

In the Normandy landing (06 June 1944), the initial operational objective was to *secure lodgment*, a large area containing ports, airfields and maneuver room for subsequent advance to Germany. Five divisions landed on four contiguous beaches and one outlier on the Contentin Peninsula.[86] Some 57,500 troops landed in the American sector and 75,215 in the British sector.[87] These troops were transported by 4,126 landing ships/craft (including 236 LSTs).[88] By 2200 on D-day, some 155,000 Allied troops, 1,550 tanks and 12,500 motor vehicles were ashore.[89]

Capturing control of a major port by an attack across the sea is usually an operational objective. General MacArthur organized X Corps formed around the 1st Marine Division and the 7th Infantry Division plus Republic of Korea (ROK) Marine units to capture Inchon and then drive inland and capture Seoul and sever the main enemy's line of communication to the south.[90] Amphibious objective area was an arc extending some thirty miles inland. Close air support was provided by one fast carrier group (TF-77) and air support group (TF 90.5), each. The U.S. Fifth Air Force provided one-week general air support in isolating the objective area starting on D-10. By 17 September Kimpo airfield was secured. The six-day battle to capture Seoul started on 20 September.[91]

Large archipelagos can be selected as operational objectives in case they are strongly defended by enemy forces. In the Central Pacific Campaign (November 1943–September 1943) the Allies selected four successive operational objectives: (1) the Gilberts (November 1943); (2) the Marshalls (February 1944); (3) the Marianas (June–July 1944); (4) western Carolines (September 1944) while one operational objective (central Carolines) was neutralized.

In July 1942, the German Luftwaffe in cooperation with U-boats achieved an operational objective by destroying a major part of the Allied PQ17 convoy to northern Russia in the Barents Sea. Out of 33 ships, 22 (or 66 percent) were sunk.[92] These ships carried 430 tanks, 210 aircraft, 3,350 motor vehicles plus 99,316 tons of other cargo.[93] Between 2 and 10 July, the 5th Air Fleet employed 130 Ju 88, 43 He 111 (20 aborted) and 29 He 115 (6 aborted).[94] The Germans lost only five aircraft in these attacks. In many cases, the U-boats sunk heavily damaged ships hit initially by the Luftwaffe.[95] In the aftermath, all Allied convoys to Russia via the Arctic route were suspended for almost two months. The next convoy for northern Russia did not sail until 2 September 1942. During the next two years convoys ran only during the long dark months of winter. This, in turn, resulted in much smaller losses than in 1942; only four ships were lost (three in 1944 and one in March 1945).[96]

Operational objectives in the air can range from obtaining or denying control of the air over a large part of airspace, destroying or suppressing enemy air defense in a major part of the theater, destroying an airfield cluster, or destroying a large part of the land-based airspace infrastructure, and destroying major industrial complexes or sources of energy.

The Israeli Air Force accomplished the initial operational objective by obtaining control of the air on 05 June – the first day of the Six-Day War of 1967. Its order of battle on that day consisted of 250 aircraft (including 65 Dassault Mirage F1 fighter interceptors, 35 Dassault Super Mystère IV fighter-bombers, 50 Dassault Ouragan fighter, 10 Sud Aviation S.O. 4050 Vatour II light bombers, and 45 Fouga Magister CM-170 training aircraft).[97] The Israeli plan was to eliminate the enemy air forces on the first day of the war. It envisaged joining a dozen squadrons over eleven targets in the 20- and 45-min intervals. This plan was extremely complicated and dangerous. The Israeli airspace was left practically defenseless. The Israelis believed that the Egyptian air force could be destroyed within three hours.[98]

The results of the Israeli major air offensive operation (Operation Moked) were catastrophic for the Egyptian Air Force. Out of 420 combat aircraft, 286 were destroyed (including 30 Tu-16, 27 Il-28 medium bombers, 12 Su-7 fighter bombers, 90 MiG-21 fighter interceptors 20 MiG-19s, 75 MiG-17s, 32 transport planes and helicopters); 13 air bases were made inoperative plus 23 radar stations and AA sites.[99] Some 20 Egyptian 12 MiG-21s and eight MiG-1s that flew from Hurghada following the initial strikes were either destroyed in the air or forced to crash land because of damages to the runways; only eight MiG-21 engaged Israeli aircraft but all of them were shot down. Some 100 Egyptian pilots were killed – at that time there were only about 350 qualified pilots.[100] In effect, the Egyptian Air Force ceased to exist.[101]

Aircraft were used on many occasions for destroying enemy sources of economic strength. In World War II, the U.S. Eighth Air Force conducted the first massive attack on the ball-bearing plants in Schweinfurt and the Messerschmidt aircraft works in Regensburg on 17 August 1943.[102] The attacking force consisted of 376 B-17s escorted by fighters.[103] After fighters turned back, the Luftwaffe's fighters and AA guns shot down some 60 B-17s; eleven B-17s were damaged beyond repair; 605 crewmembers were lost (the majority as POWs). The Luftwaffe lost only 27 fighters.[104] The B-17s scored 80 hits on the ball bearing plants reducing their production to 1/3 or pre-raid level.[105] The Messerschmidt factory was out of operation for only several weeks (because machinery was not destroyed). This prevented 1,000 fighters from being built; damage to ball bearing plants was not critical and machinery survived – but the raid forced dispersal of production.[106]

On 14 October 1943, the second Schweinfurt raid was conducted with 291 B-17s. The losses were heavy; 77 B-17s with 642 crewmen were shot down. The Luftwaffe lost only 32 fighters.[107] Both raids were a great setback to daylight strategic bombing. As a result of these two raids, the Eighth Air Force put daylight bombing on hold for four months.[108]

An operational objective consists of two or more intermediate – major or minor *tactical* – objectives. The accomplishment of a *major tactical objective* would represent a major step toward accomplishing the respective operational objective. It should lead to a radical change in the situation in a given *area of operations*. Examples of major tactical objectives are the destruction of a surface action group (SAG), amphibious task force (ATF), or large convoy, and the capturing or holding of a large naval base, port or airfield complex. In the war on land, a major tactical objective might be destroying or neutralizing a major element of the enemy division or capturing or successfully defending a large city, a critical major road junction or mountain pass or some controlling heights. In an amphibious operation, major tactical objectives could include capturing a beachhead or some important island facing the main landing beach or capturing an important airfield.

The accomplishment of a *minor tactical objective* would directly contribute to achieving the respective major tactical objective. It would also lead to a radical change in the situation in a given *combat zone/sector*. Examples of minor tactical objectives are the capture or defense of some dominating features, such as a hill, valley, mountain pass, river crossing, port/anchorage, village/town and road junction. The destruction or neutralization of force elements, such as an army company or battalion, and one or a group of ships or aircraft, also constitutes a minor tactical objective. In practice, it is often difficult to precisely differentiate between major and minor tactical objectives. It is often a matter of the commander's judgment and experience.

The Objectives and Effects

The commander and the staff must give some thoughts to anticipating the possible effects of the accomplishment of a given military objective. They should try to realistically foresee the possible effects of the accomplishment of a given objective fully, only partially, or in case of its failure. Much depends on the commander's knowledge and understanding of the enemy and all aspects of both the military and nonmilitary situation. There are many pitfalls in the process, which makes any prediction on the possible effects tenuous at best.

Generally, the effects of accomplishing a military objective can be positive (desired or intended) or negative (undesired or unintended). Effects are not limited to the enemy and friendly sides but are also felt by neutrals. They can be military or nonmilitary in their character. They can be tangible or intangible (or both). One should bear in mind that in most cases the type of effects and their strength and duration cannot be accurately predicted, much less expressed in quantifiable terms. This is not a science, as many information warfare advocates believe, but rather an art.

In terms of the factor of space, effects can be limited to a given part of the theater where the military action took place, or they can be felt far beyond the confines of that theater. The duration of effects can vary greatly. The effect of accomplishing or failing to achieve the objective might not be immediately recognized by the enemy or friendly or neutral sides; it might be some time before the effect of one's action is fully understood. The effects of success or failure in accomplishing the objective can reinforce or neutralize each other. The larger the scale of the objective, the more likely the effects will be felt in a much larger area and over a longer time. These effects often lead to changes in the political, diplomatic, economic, informational, psychological and other aspects of the situation in a theater, or even beyond the theater where the military action took place.

The accomplishment of an operational objective can sometimes have highly adverse effects at the strategic level, as the Japanese surprise attack on the U.S. Pacific fleet at Pearl Harbor on 7 December 1941 illustrates. In the planning for the attack on Pearl Harbor, the Japanese operational objective was to destroy the striking force (fast carriers and battleships) of the U.S. Pacific fleet so that the Japanese forces could capture the Philippines without serious opposition by the U.S. Navy. At the strategic level, the Japanese hoped that by the time they had consolidated their position in Southeast Asia, the United States would accept their conquests as a fait accompli. The strategic effect of their unprovoked attack on the U.S. territory was just the opposite. The attack brought together divisive elements of the U.S. government and galvanized political and public support for declaring war on Japan. In other words, the action had strategic effects that were unintended and undesirable for Japanese interests.[109]

A decisive victory over a numerically much larger force can have profound effects at both operational and strategic levels, as the U.S. Navy's victory in the battle of Midway on 4–6 June 1942 illustrates. The U.S. Navy's force, composed of two carrier groups (three carriers, eight cruisers and fourteen destroyers), sank all four Japanese fast carriers. The Japanese suffered their first defeat at sea since the sixteenth century; 253 aircraft and one cruiser were sunk, and 3,500 men were killed, including 100 pilots.[110] Admiral Isoroku Yamamoto, CINC of the Combined Fleet, still had huge superiority in surface forces, but immediately (and properly) realized that without carriers any attempt to sustain Japanese forces on Midway would eventually end in disaster, and made a sound decision to withdraw from the scene of action.

The true effect of the operational defeat at Midway was not recognized by Japanese naval and military leaders for several months. The Japanese army took the lead in preparing several offensive plans. The army general staff decided to capture Port Moresby, Papua New Guinea but overland. Allegedly, the news of Field Marshal Rommel's advance in North Africa emboldened the Japanese to revive their plans to capture Ceylon (Sri Lanka today) and carry the offensive further into the Indian Ocean. On 29 June 1942, the army general staff issued orders to execute that operation. It was the Japanese navy that was unwilling to support the action. Instead, the Japanese planned a new raid in the Bay of Bengal for August 1942. However, that operation was canceled because of the Allied landing at Guadalcanal, Lower

Solomons, on 7 August 1942. Afterward, the Japanese faced progressively larger requirements for the defense of their outposts in the Pacific but had steadily weakened forces.[111]

Perhaps no operational failure had so many significant and enduring effects as the German defeat in the battle of Stalingrad, November 1942–January 1943. The loss of the Sixth Army, with 20 divisions, plus the destruction of the four armies of their allies was a huge defeat for the Germans.[112] For the Soviets the battle of Stalingrad was an operational victory that had significant effects at both operational and strategic levels. After Stalingrad, the strategic initiative began to shift to the Soviets. Defeat at Stalingrad marked the culmination of the process of shrinking German options for achieving victory in the east.[113] Yet the defeat at Stalingrad was not as decisive as some claimed, because the combat strength of the German army was not yet broken, as the German counteroffensive in southern Russia and in February–March 1943 and at Kursk in July 1943 showed. It was only after the battle of Kursk in July 1943 that the strategic initiative finally went over to the Soviets.

Stalingrad was much more than a military defeat. It had a huge political and psychological effect on both the victors and the losers. Soviet morale rose tremendously. At the same time, the defeat at Stalingrad caused feelings of despair among the German population. There was also a widespread impression that the defeat represented a turning point in the war. Even Hitler became the target of discontent among the Germans.[114] Despite Nazi minister of propaganda Joseph Goebbels' call for total war, the German people lost confidence in Hitler's "genius." Some young staff officers posted on the eastern front began to organize resistance to Hitler, which eventually led to the failed assassination attempt in July 1944. Hitler's basic mistake in planning the summer offensive in Russia was that he confused policy, strategy and propaganda. The consequences were bad not only for Hitler's standing among the Germans but for the German people as well.[115]

The foreign policy effects of the German defeat at Stalingrad were substantial.[116] Stalingrad had a major effect on the attitudes and policies of Germany's closest allies and of neutrals as well. Finland, Hungary, and Romania almost immediately sensed that the war against Soviet Russia had been lost and started to look for ways to ensure their own salvation. Mussolini also became concerned about the war's outcome and proposed making a separate peace with the Soviet Union. The Romanian Dictator Marshal Ion Antonescu proposed a separate peace with the Western Allies. In early 1943, the Finns, Romanians and Hungarians all started to have some contact with the Western Allies to mitigate the consequences of a Soviet victory in the war.[117]

A tactical action should be decisive in its ultimate results; otherwise, it might lead to an operational failure. Very often, losses in men and materiel are not the only or even the most important factor in determining which side won or lost in some tactical action or a major operation. In the battle of Jutland in June 1916, the numerically much smaller German High Seas Fleet (*Hochseeflotte*) inflicted heavier losses on the British Grand Fleet than they received.[118] The Germans could plausibly claim that they had won a great tactical victory despite their large inferiority in numbers and materiel. However, in operational terms, the British Grand Fleet was the clear victor, because the operational situation in the North Sea remained as it was prior to the battle. The High Seas Fleet remained unable to operate beyond the narrow confines of the North Sea. Another, in retrospect, much greater effect of the Battle of Jutland was that the German admiral staff and the army's high command persuaded Emperor Wilhelm II and his chancellor, Theobald von Bethmann Hollweg that the only way for Germany to break out of the enemy blockade was to launch unrestricted U-boat warfare.[119] This decision had extremely bad consequences for Germany because it eventually led to America's entry into the war on the side of Germany's enemies and ultimately to Germany's defeat.

The Levels of War

The concept of the levels of war is relatively new. Some U.S. theoreticians asserted that Clausewitz was the first who identified two levels of war: strategy and tactics.[120] However, Clausewitz referred to strategy and tactics not "levels of war." Like other theorists of his era, Clausewitz considered strategy and tactics as components of the art of war. Claims were made that three levels of war were invented by the Prussians in the aftermath of the Franco-Prussian War of 1870–1871. Yet field marshal Helmuth von Moltke, Sr used the term "operation" in describing the movement of large forces from their basing area to the scene of a major battle.[121] Some sources say that the Imperial Russian General and Soviet theorist Aleksander A. Svechin proposed the concept of the operational level of war.[122] However, Svechin and other Soviet theorists did not use the term "levels of war." Instead, their focus was theory of strategy, operational art and tactics as components of the military art (*voyenno iskusstvo*) In the U.S. military, the term "level of war" appeared in the Army's field manual (FM 100-5 *Operations* issued in 1982. This field manual made a distinction among three levels of war: strategic, operational, and tactical. They were defined as "broad divisions of activity in preparing for and conduct in war."[123] However, the FM 100-5 of 1982 failed to fully describe the operational level of war "as the linchpin between strategy and tactics nor it clearly differentiated tactical and operational warfighting."[124] The strategic level of war was falsely equated with military strategy. It was described as a level that "employs the armed forces of a nation to secure the objectives of national policy by applying force or the threat of force. Military strategy sets the fundamental conditions for operations." The operational level of war was defined as a level "using available military resources to attain strategic goals within a theater of war. Most simply it is the theory of larger unit operations. It also involves planning and conducting campaigns." The tactical level of war was described as "the specific techniques smaller units use to win battles and engagements which supports operational objectives."[125] The U.S. joint doctrine still essentially equates levels of war with the components of the military art.[126]

After 1982, U.S. theorists provided several interpretations of the levels of war. In one definition, the strategic level of war was defined as a level "concerned with the art and science of employing national power."[127] Strategic level of war is a level at which a nation or group of nations determine national or alliance security objectives and develop use of national resources to accomplish these objectives; determines strategic objectives, sequence the objectives, define limits and assess risks for the use of military and other instruments of power; develop strategic plans to achieve objectives and provide armed forces and other capabilities in accordance with the strategic plans.[128]

The operational level of war was concerned with the planning and conduct of campaigns. It is at this level that military strategy is implemented by assigning missions, tasks and resources to tactical operations.[129] In another definition, the operational level of war was described as a level at which

> campaigns and major operations are planned, conducted, and sustained to accomplish strategic objective within theater or areas of operations. Activities at this level link tactics and strategy by establishing operational objectives needed to the strategic objectives and initiating actions and applying resources to bring about and sustain those events. These activities imply a broader dimension of time and space than do tactics. They ensure the logistic and administrative support of tactical forces and provide means by which tactical successes are exploited to achieve strategic objectives.[130]

Tactical level of war is concerned with the "planning and conduct of a battle and is characterized with the application of concentrated force an offensive action to gain objectives."[131] It is

> the level at which battles and engagements are planned and executed to accomplish military objectives assigned to tactical units. Activities at this level focus on the ordered arrangement and maneuver of combat elements in relation to each other and to the enemy to achieve combat objectives established by the operational level commander.[132]

The strategic, operational and tactical levels of war "link tactical actions to the achievement of national objectives. There are no finite limits or boundaries to these levels, but they help commanders design and synchronize operations, allocate resources and assign tasks to the appropriate commands."[133] Common to all these definitions were confusing levels of war and components of the military art or art of war. This ignores the fact that each of these components consists of theory and practice.

In the proper understanding of the terms, each level of war is directly linked to the corresponding *military objective* to be accomplished not to a corresponding component of the military art. The larger the military objective, the higher the level of war. The *strategic level of war* can be divided into two sublevels: national-strategic and theater-strategic. The *national-strategic level of war* is conducted in all three mediums for a country, or for groups of countries of both sides in a war. In World War II, U.S. President Franklin D. Roosevelt, his closest advisers, and the Joint Chiefs of Staff (JCS) were chiefly responsible for determining strategy, providing strategic direction and sufficient personnel and materiel, and coordinating strategic actions in multiple theaters of war. Their counterparts in

Great Britain were Prime Minister Winston S. Churchill, the War Council, and the British Chiefs of Staff. The U.S. and British Combined Chiefs of Staff (CCS) operated at the coalition's strategic level. Adolf Hitler and his Supreme Command of Wehrmacht (OKW) and Josef Stalin and the High Command (Stavka) operated at the national-strategic level. In the Gulf War of 1990–1991, the U.S. president, his national security advisers and the JCS operated at the national-strategic level.

The *theater-strategic level of war* is conducted in a specific theater (of war). In the case of a global war such as World War II, in each theater of war, there were several theater-strategic objectives to be accomplished. Normally, the highest politico-military leadership would assign to each theater-strategic commander a single national or alliance/coalition strategic objective. In World War II, the China-Burma-India (CBI) Theater, the Pacific Ocean Area (POA) and the Southwest Pacific Area (SWPA) represented theater-strategic levels of war, in today's understanding of the term.

The *operational level of war* is linked to a single theater-strategic objective. It is conducted in a declared or undeclared theater of operations. In World War II, the operational level of war was conducted in the northern, central and southern POAs, respectively. Likewise, the operational level of war was conducted in Papua New Guinea and adjacent areas in 1942–1944 and in the Philippines in 1944–1945. The operational level of war took place in the southern Atlantic during the Falklands/Malvinas Conflict of 1982; in Afghanistan and surrounding areas in 2001–2002; and in the Iraqi Theater of Operations (ITO) during the U.S.-led invasion of Iraq in March 2003.

The *tactical level of war* is conducted in a small part of the theater, ranging from a combat zone/sector to an area of operations. This is the level of war at which friendly and enemy forces are engaged to accomplish their respective major or minor tactical objectives. These actions take place in all three physical mediums.

The levels of war and levels of command are related but not identical. A level of war exists only in the case of employing one's forces in combat. In contrast, a level of command exists across the entire spectrum of conflict. It is a prerequisite for conducting war at a given level. Levels of command are clearly differentiated from each other. In contrast, the boundaries between individual levels of war often overlap, sometimes considerably. In a physical sense, there are, of course, no "levels," but only different sizes of physical space in which friendly and enemy forces operate. The levels of war refer to the actual employment of military and nonmilitary sources of power to accomplish given military objectives in a specific part of a theater. One's combat forces can be applied offensively at one level and defensively at another level.

A thorough knowledge and sound understanding of the levels of war are critical for selecting the proper methods of combat employment of one's forces. The levels of war determine the scale of the commander's decisions. Without the understanding and knowledge of what constitutes a given level of war, it is hard to understand the true differences in the concepts of critical factors and centers of gravity, culmination point and maneuver, to name just a few elements of operational warfare. The higher the level of war, the more important the role each of these elements plays in the planning and execution of a given military action. The higher the level of war, the more complex the situation the military commanders and their staffs must understand, evaluate, and synthesize. Both military and nonmilitary aspects of the situation are critical for success at the operational and strategic levels across the spectrum of conflict.

Levels of war are not identical to the respective components of military art. Each component of military art encompasses the fields of study and practice, while the levels of war encompass only the field of practice. To complicate the matter, strategy and operational art span several levels of war. The operational art is largely applied from the operational-tactical to the theater-strategic levels of war. Likewise, strategy is applied at the operational, theater and national-strategic levels.

Conclusion

The military objective is the single most important part of operational warfare. Almost all its aspects are related in one way or another to, or dependent on, the military objective to be accomplished. Among other things, the objective determines the levels of war to be conducted, methods of employment of combat forces and establishment of the corresponding levels of command. The commander and the staff should have full knowledge and understanding of distinctions among military-/theater-strategic, operational and tactical objectives, respectively. Otherwise, it is not possible to select the proper method for combat employment of forces. One of the main responsibilities of military strategic leadership is to convert a political to a military-strategic objective. The operational commanders bear the responsibility to convert a military-strategic to a theater-strategic objective. In both cases, this is largely an art, not a science. The military strategic leaders must inform the highest political–military leadership about the purpose and scope of the military- or theater-strategic objective; otherwise, there is a great danger that these objectives are not aligned with the aims of policy and strategy. Ideally, an operational commander should be solely responsible for determining operational objectives. Except in very rare cases, political leaders should not interfere in selecting operational objectives. The military considerations, not economic, or other aspects of the situation should have a dominant role in determining operational objectives.

The commanders and planners should always try to think ahead and to anticipate what effects the accomplishment of a given objective will have on both military and nonmilitary aspects of the situation in a theater. Many of these effects are hard or impossible to quantify. Hence, the commander should avoid the temptation to embark on the time-consuming and labor-intensive but ultimately futile effort of trying to accurately predict what these effects might be.

Mastery of operational art is difficult to achieve without comprehensive knowledge and understanding of each level of war and the mutual relationships among the levels. Because of the difference in the scope of the military objectives to be accomplished, each level of command and level of war requires commanders and their staff to have a different perspective. The higher the level of command, the broader that perspective. The narrowest perspective, in terms of both physical space and the elements of the situation, is at the tactical level. At each level above the tactical level, the perspective becomes progressively larger and more complex.

Notes

1 Cited in Joel E. Hamby, "Striking the Balance," *Armed Forces & Society* 30, no. 3 (Spring 2004), 334.
2 Carl von Clausewitz, *On War*, ed. and trans. Michael Howard and Peter Paret (Princeton, NJ: Princeton University Press, 1984), 81.
3 C. R. Brown, "The Principles of War," *Proceedings* 6 (June 1949), 624.
4 B. H. Liddell Hart, *Strategy*, 2nd revised ed. (New York, NY: Frederick A. Praeger Publishers, 1967), 348–49.
5 Hanisch, *Untersuchen Sie die operativen Ideen Manstein hinsichtlich Schwerpunktbildung, Überraschung, Initiative und Handlungsfreiheit an den Beispielen Westfeldzug 1940 (Sichelschnitt-Plan) und Operation Zitadelle* (Hamburg: Führungsakademie der Bundeswehr, 15 January 1988), 33; Translated from the Russian by F. Beer et al., *Part 1: Die Operative Kunst der Luftstreitkräfte* (Dresden: Military Academy "Friedrich Engels," 1976), 119.
6 Joint Publication 1-02, *Department of Defense Dictionary of Military and Associated Terms* (April 12, 2001; As Amended through April 14, 2006), 370.
7 Department of Operations, *Sound Military Decision (Including the Estimate of the Situation and the Formulation of Directives)* (Newport, RI: U.S. Naval War College, 1938), 158.
8 Ash Irwin, *The Levels of War, Operational Art and Campaign Planning, Occasional Paper No. 5* (Amberley/Surrey, England: Strategic and Combat Studies Institute, 1993), 6.
9 Joint Chiefs of Staff Memorandum: Strategic Plan For the Defeat of Japan. Approved By The Combined Chiefs of Staff, 19 May 1943 (JXS 287/1 and CCS 220), in Louis Morton, *United States Army In World War II. The War in the Pacific. Strategy and Command: The First Two Years* (Washington, DC: Center of Military History, United States Army, 1989), 644.
10 Günther Korten, *Erfahrungen über Wehrmachtsführung aus dem amerikanischen Sezessionkrieg, Lecture at the Kriegsakademie, RW 13/v. 34* (Freiburg im Breisgau: Bundesarchiv-Militärarchiv (BA-MA), March 1936, 13.
11 Ministry of Defence (NAVY), *War with Japan, vol. I: Background to the War* (London: Her Majesty's Stationery Office, 1995), 84.
12 Appendix E, "Japanese Operational Plan for Commencing Hostilities," in *Ministry of Defence (NAVY), War with Japan, vol. I: Background to the War* (London: Her Majesty Stationery Office, 1995), 134–35.
13 Appendix E, "Japanese Operational Plan for Commencing Hostilities," 135.
14 Hamby, "Striking the Balance," 336.
15 Korten, "Erfahrungen über Wehrmachtsführung aus dem amerikanischen Sezessionkrieg," 13.
16 Louis Morton, *Strategy and Command: The First Two Years* (Washington, DC: Office of the Chief of Military History, Department of the Army, 1962), 607.
17 Appendix 20: "Directive to the Supreme Commander ABDA Area dated 3rd January 1942," in *The War Against Japan, vol. I: The Loss of Singapore*, ed. S. Woodburn-Kirby, et al. (London: Her Majesty's Stationery Office, 1957), 529–30.
18 "Weisung für 'Fall Weserübung'," March 1, 1940, in Walther Hubatsch, *Hitlers Weisungen für die Kriegführung 1939–1945. Dokumente des Oberkommandos der Wehrmacht*, 2nd ed. (Koblenz: Bernard & Graefe Verlag, 1983), 41.
19 Ministry of Defence (NAVY), *War With Japan, vol. I: Background to the War*, 84.
20 Morton, *Strategy and Command: The First Two Years*, 617–18.

21 Appendix E, "Joint Directive for Offensive Operations in the Southwest Pacific Area Agreed Upon by the United States Chiefs of Staff, 2 July 1942," in *United States Army In World War II. The War in the Pacific. Strategy and Command: The First Two Years*, ed. Louis Morton (Washington, DC: Center of Military History, United States Army, 1989), 619.
22 George F. Howe, *United States Army in World War II. The Mediterranean Theater of Operations. Northwest Africa: Seizing the Initiative in the West* (Washington, DC: Center Of Military History, United States Army, 1991), 16; Directive to Commander-in-Chief, North Africa, C-0-S.(42) 90th Mtg(O) Annex 1. 14th August 1942.
23 Lothar Gall, *Bismarck. The White Revolutionary, vol. 1: 1851–1871* (London: Allen & Unwin, 1898), 300.
24 Walther Hubatsch, ed., "Weisung Nr. 21 Fall Barbarossa," 85; Center for Military History, *The German Campaign in Russia. Planning and Operations (1940–1942)* (Washington, DC: Department of the Army Pamphlet No. 20-261a, March 1955), 22.
25 Robert M. Kennedy, *The German Campaign in Poland (1939)* (Washington, DC: Department of the Army, April 1956), 60.
26 "A Bad Beginning," *The Economist*, February 26, 2022, 20.
27 Rob Johnson, "The First Phase of the Russian Invasion of Ukraine 2022," 2, accessed at https://static1.squarespace.com/static/55faab67e4b0914105347194/t/62320286e1a1bd2dcf8915e0/1647444630378/The+First+Phase+of+the+Russian+Invasion+of+Ukraine+CCW+2022.pdf
28 Cited in Andrew S. Bowen, *Russia's War in Ukraine: Military and Intelligence Approach* (Washington, DC: Congressional Research Service, April 13, 2022), 11.
29 Cited in Bowen, *Russia's War in Ukraine: Military and Intelligence Approach*, 11.
30 Rob Johnson, "Dysfunctional Warfare: The Russian Invasion of Ukraine 2022," *Parameters* (Summer 2022), 5, 8.
31 "Putin's Fehler," *Der Spiegel, Nr. 10*, March 5, 2022, 22.
32 Bruce B. G. Clarke, *Conflict Termination: A Rational Model* (Carlisle Barracks, PA: Strategic Studies Institute, U.S. Army War College, May 1, 1992), 11.
33 Liddell Hart, *Strategy*, 338.
34 Liddell Hart, *Strategy*, 334.
35 Liddell Hart, *Strategy*, 335.
36 Otto Groos, *Seekriegslehren im Lichte des Weltkrieges. Ein Buch für den Seemann, Soldaten und Staatsmann* (Berlin: E.S. Mittler & Sohn, 1929), 23; Hein-Peter Weyher, *Der Begriff 'Seestrategie' Und Seine Deutung in Den Westlichen Kriegstheorien des 20 Jahrhunderts* (Hamburg: Führungsakademie der Bundeswehr, July 1967), 11.
37 Ian Nish, "The Russo-Japanese War: Planning, Performance and Peace-Making," in *Der Russisch-Japanische Krieg (1904/05)*, ed. Josef Kreiner (Bonn: Bonn University Press, V&R Unipress, 2005), 12.
38 Hamby, "Striking the Balance," 335.
39 Cited in Jerome V. Martin, *Victory from Above: Air Power Theory and the Conduct of Operations Desert Shield and Desert Storm* (Maxwell AFB, AL: Air University Press, Research Report No. AU-ARI-92-8, June 1994), 24.
40 Mark Garrard, "War Termination in the Persian Gulf," 5, accessed at www.airpower.maxwell.af/mil/airchronicles/apj01/fal01/garrard.html
41 Stanley T. Kresge, *Gulf War Termination Revisited* (Maxwell AFB, AL: Air War College, Air University, April 1999), 22–23.
42 G. H. Bowdey, *The Nature of Naval Warfare* (Newport, RI: Staff Presentation, Naval War College, 1938), 12.
43 Bowdey, *The Nature of Naval Warfare*, 11.
44 Philip Wasiliewski, *The Evolving Political-Military Aims in the War in Ukraine After 100 Days* (Philadelphia, PA: Foreign Policy Research Institute, Eurasia Program, June 2022), 2.
45 *Russian Offensive Campaign Assessment* (Washington, DC: Institute for the Study of War, December 30, 2022), 3–6, accessed at www.understandingwar.org/backgrounder/russian-offensive-campaign-assessment-december-30

46 Wasilewski, *The Evolving Political-Military Aims in the War in Ukraine After 100 Days*, 5.
47 Martin Armstrong, *Liberated Ukrainian Territory* (Washington, DC: Institute for the Study of War, Statista Research, November 18, 2022), 1, accessed at www.statista.com/chart/28748/ukraine-territory-control-status-distribution/
48 Wasiliewski, *The Evolving Political-Military Aims in the War in Ukraine After 100 Days*, 5.
49 Malcolm W. Cagle and Frank A. Manson, *The Sea War in Korea* (Annapolis, MD: United States Naval Institute, 1957), 116.
50 In preparations for the U.S. invasion of Iraq in March 2003 (Operation Iraqi Freedom), U.S. Central Command (CENTCOM)'s planners listed among operational objectives for Phase I-Preparations, secure regional and international support, alert, deploy, and posture forces for shaping operations, set conditions to neutralize Iraqi forces; for Phase II—Shape the Battlespace-- maintain international and regional support for ops, continue to deploy and posture forces for decisive operations, deter WMD employment, suppress Iraqi TBM/WMD delivery/storage systems, suppress TBM/WMD production processes and deny access to WMD facilities; CentCom Power-Point Slides Briefed to White House and Rumsfeld in 2002, Tab K, 4-5; obtained by National Security Archive through Freedom of Information Act, National Security Archive Electronic Briefing Book No. 214, ed. by Joyce Battle and Thomas Blanton; in another example, as the initial operational objective was (falsely) considered "to allow follow-on forces from the United States, allies, and partners to flow toward Taiwan . . . . to regain sea control in the approaches to China."Scott Tait and Anthony Lavopa, "It All Comes Down To Sea Control," *Proceedings* (December 2023), 33.
51 Walter Krueger, *From Down Under to Nippon. The Story of Sixth Army in World War II* (Washington, DC: Combat Forces Press, 1953), 145; Douglas MacArthur, *Reminiscences. General of the Army Douglas MacArthur* (New York, NY: McGraw-Hill Book Company, 1964), 212.
52 Krueger, *From Down Under to Nippon*, 145; MacArthur, *Reminiscences*, 212.
53 M. Hamlin Cannon, *United States Army in World War II, the War in the Pacific, Leyte: The Return to the Philippines* (Washington, DC: Center of Military History, United States Army, [1954] 1987), 3, 1, 15.
54 Curtis A. Utz, *The U.S. Navy in the Modern World Series, No. 2: Assault From the Sea. The Amphibious Landing at Inchon* (Washington, DC: Naval History and Heritage Command, Department of the Navy, 2020), 18.
55 Lynn Montross, "The Inchon Landing-Victory over Time and Tide," *Marine Corps Gazette* 7 (July 1951), 27; Malcom W. Cagle, "Inchon – the Analysis of a Gamble," *Proceedings* 1 (January 1954), 49.
56 Robert O. Brunson, *The Inchon Landing: An Example of Brilliant Leadership* (Carlisle Barracks, PA: U.S. Army War College, April 2003), 7.
57 Wilson A. Heefner, "The Inchon Landing," *Military Review* 2 (March–April 1995), 69–70.
58 Malcolm W. Cagle and Frank A. Manson, *The Sea War in Korea* (Annapolis, MD: Naval Institute Press, 1957), 77–78.
59 Cagle and Manson, *The Sea War in Korea*, 81.
60 Earl F. Ziemke, *Stalingrad to Berlin. The German Defeat in the East* (Washington, DC: Center for Military History, United States Army, 1987), 79; Albert Seaton, *The Russo-German War 1941–45* (Novato, CA: Presidio, 1971), 335.
61 Seaton, *The Russo-German War 1941–45*, 336.
62 Alfred Philippi and Ferdinand Heim, *Der Feldzug gegen Sowjetrussland 1941–1945* (Stuttgart: W. Kohlhammer Verlag, 1962), 192–93; Heinz Magenheimer, *Kriegswenden in Europa 1939–1945: Führungsentschlüsse, Hintergründe, Alternativen* (Munich/Landsberg am Lech: Günter Olzog Verlag, 1995), 124.
63 Seaton, *The Russo-German War 1941–45*, 337.
64 David M. Glantz and Jonathan M. House, *Stalingrad* (Lawrence, KS: University of Press of Kansas, 2017), 497.

65 Glantz and House, *Stalingrad*, 497.
66 Martin Blumenson, "A Deaf Ear to Clausewitz: Allied Operational Objectives in World War II," *Parameters* (Summer 1993), 16.
67 Cited in Blumenson, "A Deaf Ear to Clausewitz: Allied Operational Objectives in World War II," 17.
68 Blumenson, "A Deaf Ear to Clausewitz: Allied Operational Objectives in World War II," 19.
69 Blumenson, "A Deaf Ear to Clausewitz: Allied Operational Objectives in World War II," 19.
70 Blumenson, "A Deaf Ear to Clausewitz: Allied Operational Objectives in World War II," 19.
71 Blumenson, "A Deaf Ear to Clausewitz: Allied Operational Objectives in World War II," 16–17.
72 Blumenson, "A Deaf Ear to Clausewitz: Allied Operational Objectives in World War II," 17.
73 Blumenson, "A Deaf Ear to Clausewitz: Allied Operational Objectives in World War II," 21–22.
74 Rick Atkinson, *The Day of Battle: The War in Sicily and Italy 1943–1944* (New York: Henry Holt, 2007), 323.
75 Blumenson, "A Deaf Ear to Clausewitz: Allied Operational Objectives in World War II," 23.
76 C. J. C. Molony, F. C. Flynn, H. L. Davies, and T. P. Gleaves, *History of Second World War, vol. 5: The Mediterranean and Middle East* (London: Her Majesty's Stationery Office, 1973), 28–29.
77 Blumenson, "A Deaf Ear to Clausewitz: Allied Operational Objectives in World War II," 21.
78 Greg A. Grabow, *Schweinfurt Raids and the Pause in Daylight Strategic Bombing* (Fort Leavenworth, KS: Command and General Staff College, 2008), 38–39.
79 Heinz Magenheimer, *Kriegswenden in Europa 1939–1945: Fűhrungsentschlűsse, Hintergrűnde, Alternativen* (Munich/Landsberg am Lech: Günter Olzog Verlag, GmbH, 1995), 100.
80 Shawn P. Keller, *Turning Point: A History of German Petroleum in World War II and Its Lessons for the Role of Oil in Modern Warfare* (Maxwell AFB, AL: Air Command and Staff College, Air University, 2011), 14.
81 Keller, *Turning Point*, 14; cited in Joel Hayward, "Too Little, Too Late: An Analysis of Hitler's Failure in August 1942 to Damage Soviet Oil Production," *The Journal of Military History* 64, no. 3 (July 2000), 772.
82 Hayward, "Too Little, Too Late," 772.
83 Mitsuo Fuchida and Masatake Okumiya, *Midway: The Battle That Doomed Japan – the Japanese Navy's Story* (Annapolis, MD: Naval Institute Press, 1955), 250.
84 Ministry of Defence (NAVY), *War with Japan,* vol. 2: *Defensive Phase* (London: Her Majesty Stationery Office, 1995), 164.
85 Fuchida and Okumiya, *Midway: The Battle That Doomed Japan*, 249.
86 Bryon Greenwald, "Why Normandy Still Matters," *Joint Force Quarterly* (4th Quarter 2019), 60.
87 Specifically, on UTAH beach landed 23,250, OMAHA-34,350; GOLD-24,970; JUNO-21,400 and SWORD-beaches 28,845 troops, respectively; Jûrgen Rohwer and Gerhard Hûmmelchen, *Chronology of the War at Sea 1939–1945*, 2nd revised and expanded ed. (Annapolis, MD: Naval Institute Press, 1992), 281.
88 Christopher D. Yung, *Gators of Neptune. Naval Amphibious Planning for the Normandy Invasion* (Annapolis, MD: Naval Institute Press, 2006), 76.
89 Greenwald, "Why Normandy Still Matters," 60.
90 Chapter II: Operation Chromite, *Joint Military Operations Historical Collection* (Washington, DC: The Joint Chiefs of Staff, July 15, 1997), II–6.

91 Chapter II: Operation Chromite, *Joint Military Operations Historical Collection*, II–11.
92 Naval Staff, *The Royal Navy and the Arctic Convoys. A Naval Staff History*, preface by Malcolm Llewellyn-Jones (London: Rutledge, Taylor & Francis Group, 2007), 53.
93 Stephen W. Roskill, *History of the Second World War, The War at Sea, vol. II: The Period of Balance* (London: Her Majesty's Stationery Office, 1956), 143.
94 Bodo Bullwinkel, *Kooperation Luftwaffe-Kriegsmarine dargestellt am Luftkrieg űber See im Nordmeer 1942* (Hamburg: Fűhrungsakademie der Bundeswehr, Abteilung Marine, January 5, 1971), 44.
95 Bullwinkel, *Kooperation Luftwaffe-Kriegsmarine dargestellt am Luftkrieg über See im Nordmeer 1942*, 44.
96 "Convoys to Russia, 1942," *Supplement to the London Gazette of Friday*, October 13, 1950, 5139.
97 Michael B. Oren, *Six Days War. June 1967 and the Making of the Modern Middle East* (Oxford, England: Oxford University Press, 2002), 171.
98 Oren, *Six Days War. June 1967 and the Making of the Modern Middle East*, 172.
99 Oren, *Six Days War. June 1967 and the Making of the Modern Middle East*, 176.
100 Trevor N. Dupuy, *Elusive Victory. The Arab-Israeli Wars: 1947–1974* (Garden City, NY: Military Book Club, 1978).
101 Oren, *Six Days War. June 1967 and the Making of the Modern Middle East*, 176.
102 Grabow, *Schweinfurt Raids and the Pause in Daylight Strategic Bombing*, 2.
103 Cited in Grabow, *Schweinfurt Raids and the Pause in Daylight Strategic Bombing*, 3.
104 Cited in Grabow, *Schweinfurt Raids and the Pause in Daylight Strategic Bombing*, 3.
105 Alex Coetzer, *Aiming for Strategic Effect: A Re-Examination of the Allied Assault on Schweinfurt*, "Davidson History Journal, Vol. VIII, 2022," 5; accessed at https://www.dhj.davidsonlocal.org/volume-viii-2/
106 *The Schweinfurt-Regensburg Mission: The American Raids on 17 August 1943*, by Martin Middlebrook, reviewed by David F. Crosby, *Air Power History* 60, no. 4 (Winter 2013), 58.
107 Grabow, *Schweinfurt Raids and the Pause in Daylight Strategic Bombing*, 3–4.
108 Grabow, *Schweinfurt Raids and the Pause in Daylight Strategic Bombing*, 4.
109 Donna Lucchese, *The Relationship of Center of Gravity Analysis, Targeting for Effect, and Measuring Success* (Carlisle Barracks, PA: Army War College, 1998), 7.
110 Jürgen Rohwer, "Die See-Luftschlacht bei Midway 1942," in *Entscheidigungss-chlachten des zweiten Weltkrieges, ed.* Hans-Adolf Jacobsen and Jürgen Rohwer (Frankfurt am Main: Verlag für Wehrwesen Bernard & Graefe, 1960), 226.
111 Jürgen Rohwer, "Die See-Luftschlacht bei Midway 1942," 226–27.
112 Heinz Magenheimer, *Kriegswenden in Europa 1939–1945: Führungsentschlüsse, Hintergründe, Alternativen*, 125.
113 Bernd Wegner, "The Road to Defeat: The German Campaigns in Russia 1941–43," in *Decisive Campaigns of the Second World War*, ed. John Gooch (London: Frank Cass, 1990), 123.
114 Wegner, "The Road to Defeat," 121.
115 Walter Gőrlitz, "Die Schlacht um Stalingrad 1942–1943," in *Entscheidigungss-chlachten des zweiten Weltkrieges,* ed. Hans-Adolf Jacobsen and Jürgen Rohwer (Frankfurt am Main: Verlag für Wehrwesen Bernard & Graefe, 1960), 311.
116 Gőrlitz, "Die Schlacht um Stalingrad 1942–1943," 311.
117 Cited in Earl F. Ziemke and Magna E. Bauer, *Moscow to Stalingrad: Decision in the East* (Washington, DC: Center of Military History, United States Army, 1987), 505.
118 The Germans sank 14 British ships with 115,025 tons and lost 11 of their ships with 61,180 tons. About 6,100 British officers and sailors were killed, 510 wounded, and 177 taken prisoner by the Germans. The Germans had 2,550 men killed and about 510 men wounded; V. E. Tarrant, *Jutland: The German Perspective (A New View of the Great Battle, 31 May 1916)* (Annapolis, MD: Naval Institute Press, 1995), 247–48;

Keith Yates, *Flawed Victory: Jutland, 1916* (Annapolis, MD: Naval Institute Press, 2000), 216.

119 Tarrant, *Jutland*, 251.

120 Cited in Andrew S. Harvey, "The Levels of War as Levels of Analysis," *Military Review* (November–December 2021), 76.

121 Wieker, *Operation, Operative Führung zwischen Taktik und Strategie*, 8; English, "The Operational Art: Development in the Theories of War," 8; von Sodenstern, "Operationen," 53; Besterhorn, *"Freie Operationen"-zur Bedeutung eines neuen Begriffes im Rahmen der operativen Schulung des Heeres*, 7.

122 Jacob Kipp, "Soviet Military Doctrines and the Origins of Operational Art, 1917–1936," in *Soviet Doctrine from Lenin to Gorbachev, 1915–1991*, ed. William C. Frank and Philip S. Gillette (Westport, CT: Greenwood, 1992), 56.

123 Headquarters, Department of the Army, *FM 100-5: Operations* (Washington, DC: Headquarters, Department of the Army, August 20, 1982), 2–3.

124 William R. Richardson, "FM 100-5: The Airland Battle in 1980," *Military Review* (January–February 1997), 174.

125 Headquarters, Department of the Army, *FM 100-5: Operations* (1982), 2–3.

126 Joint Publication 3-0, *Joint Operations*; 17 January 2017 Incorporating Change 1, 22 October 2018, I-12 thru I-14.

127 Cited in Martin Dunn, "Levels of War: Just a Set of Labels?" *Research and Analysis*, no. 10 (October 1996), 2, accessed at www.defence.gov.au/lwsc/issue10.htm

128 Cited in Gordon R. Peskett, *Levels of War: A New Canadian Model to Begin the 21st Century* (Toronto: Canadian Forces College, AMSC 5/CSEM 5, 2002), 10.

129 Cited in Dunn, "Levels of War: Just A Set of Labels?" 2.

130 Cited in Peskett, *Levels of War: A New Canadian Model to Begin the 21st Century*, 10.

131 Cited in Dunn, "Levels of War: Just A Set of Labels?" 2.

132 Cited in Peskett, *Levels of War: A New Canadian Model to Begin the 21st Century*, 10.

133 Harvey, "The Levels of War as Levels of Analysis," 76.

3 Operational Factors

The main components of any military situation are factors of space, time and force (or sources of power). Each of them consists of many critically important or less important elements. The higher the level of war, the larger the factors of space, time and force in terms of their scale and complexity. Hence, tactical, operational and strategic factors of space, time, and force are differentiated. Those related to a single operational or military-/theater-strategic objective are called "operational factors." The Indian scholar Kautilya in his work *The Arthashastra*, compiled probably around 150 A.D., pointed out the importance of what he called the factors of power, time, and place.[1] He wrote that the most important factor is power.

> One does not foolishly attack a stronger enemy or the economic strength backing it. The intellectual power which enables a king to make an objective Yesanalysis and arrive at the correct judgment is the most important. The three most important constituents of power, in decreasing order of importance, are intellectual power, military might, enthusiasm, and morale. The next two factors to be taken into account are place and time; the nature of terrain where the battle will be waged; and two aspects of time: time means the season when battle is likely and the expected duration of battle itself. The first three factors are interdependent. Only when a king is sure that he is superior in power, space, and time shall he proceed to consideration of the other factors. Equally, when a king finds himself superior, he shall not waste time. One should proceed against one's enemy whenever, by so doing, the enemy can be weakened or crushed.[2]

The rudiments of what the Germans call "operational factors" go to the writings of a well-known and highly influential General Gerhard Johann David von Scharnhorst. In his discussion on operations, Scharnhorst repeatedly referred to the factors of space (*Raum*), time (*Zeit*), forces (*Kräfte*), and means (*Mitteln*).[3] In 1865, during preparations for a general staff ride, the Prussian chief of general staff General (later field marshal) Helmuth von Moltke, Sr, directed that the calculation of space (*Raum*), time (*Zeit*) and means (*Mitteln*) be the foundation of all general staff work.[4]

In determining operational or military-/theater-strategic objectives, a comprehensive knowledge and understanding of the factors of space, time and forces from

DOI: 10.4324/9781003333685-3

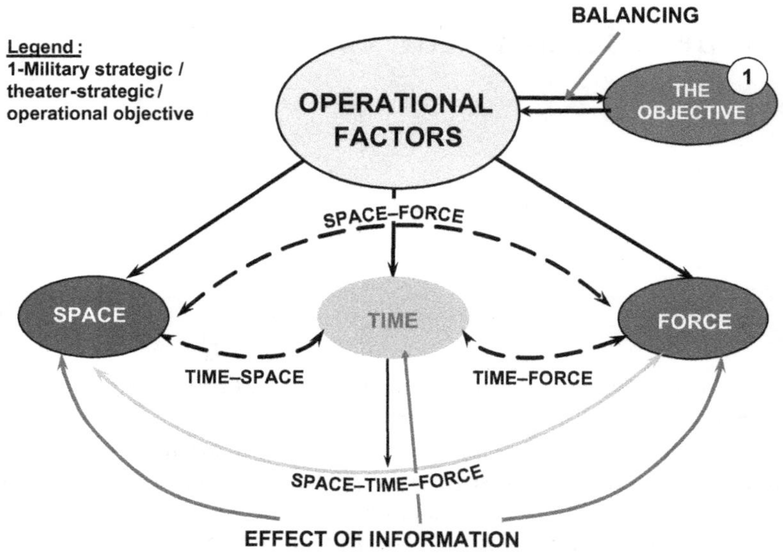

Figure 3.1 Operational Factors versus the Objective

the enemy's perspective is necessary. Afterward, the operational commander and the staff must evaluate friendly operational factors of space, time, and forces to test whether a given operational or military-/theater-strategic objective is *militarily achievable*. This requires determining selected elements of each of the operational factors to be harmonized or balanced against a given operational or military-/theater-strategic objective. General Scharnhorst alluded to this process by stating that "strategic-operational objective must be in harmony with existing forces and means."[5]

All successful commanders – Napoleon I is perhaps the best example – were known for their superb ability to properly balance the factors of space, time, and force with a given objective. The art of war is to obtain and maintain freedom of action – the ability to make critically important and diverse decisions to accomplish assigned military objectives. One's freedom of action is obtained by sound balancing the factors of space, time and forces.[6]

The German Army's current doctrine points out that factors of force, space and time should be used for the "coordination of each operation and tactical activity. The necessary or resulting effects in the information environment are a central orientation point of this coordination."[7]

The Factor of Space

The factor of space encompasses land, sea and airspace with all their features, which affects the employment of land, sea, air and special forces. Military history is replete with examples of campaigns and major operations that failed because the factor of space was either neglected or unrealistically assessed. Space in itself is both a means and an objective. It is the means because sufficient space is needed to successfully conduct military operations. It is the objective because to conduct military operations it is necessary to control a given space.[8]

Force assigned to accomplish operational or strategic objectives should have sufficient physical space to deploy, concentrate, maneuver and fight. This also implies the existence of sufficient space for the establishment and maintenance of infrastructure. The factor of space must be controlled to the extent that the assigned military objectives can be achieved.[9]

The factor of space encompasses not only the physical environment and weather/climate but also the "human-space." The human-space includes such elements as the political system and type of government, population size and density, economic activity, transportation, trade, ideologies, ethnicity, religions, social structure and traditions, culture and technology. All these and other elements of human space increasingly influence the planning, preparation and execution of a campaign or major operation. Obviously, human aspects of the factor of space play a much larger role in land warfare than in naval and air warfare. To enhance chances of success in combat, the operational commanders and their staffs must properly evaluate both the human and physical elements of the factor of space.

The success of a major operation or campaign depends on the free movement of one's forces over a large part of a given theater.[10] Without the ability to conduct large-scale movements on land, at sea and in the air, operational warfare is essentially an empty concept.[11] In the era of mass armies, it was difficult for one side to outflank or outmaneuver a strong opponent on land.[12] The reason was that the opposing forces were deployed along hundreds of miles of terrain and provided few, if any, gaps in the defense line.

In land warfare, a large space usually offers a greater range of offensive options. Among other things, it facilitates envelopment maneuvers and turning movements. In contrast, a small space compels combat forces to conduct frontal assaults.[13] In an attack on land, a large space allows wide-ranging movements of one's forces. It provides the operational commander with more options for employing friendly forces.[14] An attack into the enemy's depth, threatening his lines of supply, can often have dramatic results.[15] Generally, the farther the attacker advances into the enemy's territory, the more the defender must shift to a defensive posture – temporarily or permanently – and by smaller or larger withdrawals preserve his freedom of action.[16]

On land, large space provides the defender with sufficient room for maneuvering, both laterally and in the depth of one's rear. Large space allows the defender to trade space for time, if necessary. It facilitates the regrouping, reinforcement and redeployment of combat forces for an offensive action. The Russians used defense in depth to frustrate the French invaders in 1812. In 1941–1942, the Soviets conducted a scorched-earth policy to slow down the German advances. They also dismantled almost one-quarter of the country's manufacturing capacity and moved to the east of the Ural Mountains before the Germans overran the rest of the territory.[17]

For both sides in a war, having too much space usually makes it more difficult to seek a decision. In World War I, maneuvering space on the eastern front was so large that neither the Germans and their Austro-Hungarian allies nor the Russians were able to force a decision for most of the war.[18]

Generally, a large space can also put certain limitations on the operational commander's freedom of action. In a war on land, the need to protect ever-lengthening flanks and lines of communication requires steadily larger forces to secure the rear area and to control a hostile populace, thus reducing the attacker's striking power. In Napoleon I's invasion of Russia in 1812, the French logisticians could not keep up with the advances of the French troops because of the ever-extending supply lines. The space factor, with its effects on time and forces, resulted in Napoleon I's inability to maintain the operational tempo necessary to bring the Russians to a decision.[19]

A vast maritime space is difficult to successfully defend against a strong opponent at sea. This fact became painfully clear in the modern era after the advent of submarines and aircraft. The area in which Japan conducted active operations in 1941 stretched about 4,000 miles north to south, from the Aleutian Islands to Australia, and about 5,000 miles east to west, from the Andamans in the Bay of Bengal to the Hawaiian Islands.[20]

One of the main prerequisites for a country or group of countries is a favorable geostrategic position for the employment of military forces as a whole. In general, in relation to the adjacent sea and land area, a country may occupy central, semicentral, peninsular and insular positions. When a country or territory does not border the sea or ocean, such as Afghanistan, Switzerland or Hungary, then it is said to occupy a *central position*. If a country is located on the rim of a continental landmass and borders the sea or the ocean, it is said to occupy a *semicentral position*. Russia occupies a central geostrategic position on land but a semicentral position in regard to the surrounding seas and oceans. It stretches today for about 8,000 miles and ten time zones across two continents. Russia's seaboard encompasses four regions, each widely separated from the others. This makes it necessary for Russia to maintain a separate fleet in each of the four far-flung maritime theaters. Tsarist Russia and its successor, the Soviet Union, tried repeatedly (and unsuccessfully) to overcome the natural disadvantages of their maritime position by obtaining access to the warm waters of the Mediterranean and the Indian Ocean.

Today, the Russian home-based fleets' access in wartime to the waters of the North Atlantic from the Barents Sea is highly unfavorable, because potentially hostile powers control the stretch of water from Greenland to Norway. In contrast, Russia's access to the open ocean in the Far Eastern theater is relatively more favorable, because the Kamchatka peninsula borders the open waters of the northern Pacific. However, the peninsula, with its naval and air bases, lacks well-developed road and railroad links with the mainland and is for all practical purposes an island, with all the inherent advantages and disadvantages accruing.

A country having a longer frontier on the sea than on the land occupies a *semi-isolated* or *peninsular position*. Denmark occupies a favorable position at sea because it controls the only exit from the Baltic, the Danish Straits (Skagerrak and Kattegat). The Korean peninsula dominates all sea and air routes between

Japan and Manchuria, and sea routes from the Yellow Sea to and from the Sea of Japan. Within the Mediterranean, Italy's mainland enjoys a commanding geostrategic position in the central Mediterranean. One leg of the Apennine peninsula faces the Messina Strait, while the other leg borders the Strait of Otranto, the only exit from the Adriatic Sea.

A country situated on a large island or several large islands occupies an *insular position*. England offers perhaps the best example of the advantages that accrue from occupying an insular position. The British Isles form a 600-mile-long barrier off the coast of Western Europe. They border the open waters of the North Atlantic and three narrow seas – the North Sea, the English Channel and the Irish Sea. A country's shape can sometimes greatly affect its geostrategic position when other factors are considered.

The country's shape can be differentiated as compact, elongated (or attenuated), prorupted (or protruded) and perforated. A country is said to have a *compact shape* when all the points of its boundary lie at the same distance from the country's geometrical center. Germany, the People's Republic of China (PRC), Poland and Belgium all seem to have nearly round or rectangular boundaries. A country with a large but compact shape, such as Poland, offers the shortest possible boundary in view of the area enclosed. Also, because such a country does not have peninsulas, large islands or other protruding parts, the communications network is generally easier to establish.

A country or territory that is at least six times longer than it is wide, such as Chile or Norway, is said to have an *elongated* shape. Such a shape, in combination with the narrowness of its land area, usually makes defense against enemy invasion difficult. Chile's territory extends for about 2,500 miles along its north–south axis. It is nowhere wider than 250 miles. The high Andes mountain chain guards Chile's eastern frontiers.[21] When a country or territory, such as Croatia, is compact in shape but has a "corridor" leading away from its main body, it is said to have a *prorupt shape*. Such a country generally has a long seaboard and narrow coastal area and is highly vulnerable to an attack from the sea or land. A country completely surrounding another and smaller country, as is the case of South Africa in regard to Lesotho is said to have a *perforated* shape.

A country could be physically concentrated or fragmented. If a country stretches across a land area, either on the rim of a continent or an island, it is said to be physically *concentrated.* This position is normally favorable for defense against external attack. A country is concentrated even if it extends across two or even more islands, if each is separated from the other by a relatively narrow channel or strait, as in the case of the British Isles. Denmark enjoys a high physical concentration because its 600-odd islands are near the 210-mile-long Jutland Peninsula and each other. To the east is Bornholm, the only large island of note that lies relatively far away, some eighty miles from Sjaelland. However, if a country or territory spreads over too many widely separated islands, as is the case with Greece, Indonesia and the Philippines, it lacks physical concentration; it is *physically fragmented.* The Philippines extend about 1,150 miles from north to south, and about 700 miles from east to west. The archipelago encompasses about 7,100 islands, of which only 460 are larger than one square mile.

The distance between home bases and the combat employment area is an integral part of the space factor and is closely related to the factor of time. It is a critical aspect of the factor of space. Long distances between various points within a given theater imply large physical space and thereby allow the deployment and employment of large combat forces. Short distances, in contrast, imply a smaller operating area. Therefore, they pose great difficulties for the movements of large forces. Among other things, long distances between the home base and the operating area complicate the timely employment of one's combat forces significantly. Long distances also mean long lines of communication. These require long-haul transportation and larger forces for protection. One's forward deployment can to a certain degree alleviate these problems, but it cannot eliminate them.[22]

Space is a highly dynamic factor once hostilities start. Changes in the size of the area controlled by both sides occur often. Generally, gaining space is a prerequisite for the successful conduct of war, especially if a campaign is not completed quickly and space is too small.[23] In gaining space, the movement of supplies over ever-greater distances requires ever-larger transportation requirements. If a decision is not reached and the attack continues, the attacker will most likely overshoot his culmination point. The longer the attacker continues with his attacks, the more likely that his combat power will erode to such an extent as to allow the defender to counterattack.[24]

In a war, gaining space is in most cases an advantage. The farther the attacker advances into the defender's-controlled territory, the more he deprives the defender of the means of conducting the war.[25] Thereby, the defender has a smaller possibility of conducting a successful defense. Likewise, the loss of space is a material and moral disadvantage. Yet by giving up the space, one can sometimes obtain a better position for defense.

A very large country could lose or even abandon a large part of its territory to the enemy without lasting damage to the overall effort, as was the case with Soviet Russia in World War II. In contrast, for a small country like Israel, giving up space could be fatal to its very survival.[26] Prior to 1967, Israel encompassed about 8,000 square miles of territory (less than that of Massachusetts). Its coast stretched for only about hundred miles. North of Tel Aviv the Israeli territory was no more than twelve miles wide. At its widest area, south of Beersheba, it was just sixty-five miles from the southern tip of the Dead Sea. Israel's irregular boundaries stretched for about 600 miles. Only the seventy-mile-long border with Lebanon was somewhat secure.[27] Israel's waistline was only about eight miles wide, or a three-hour march for Jordanian troops on foot.[28]

The size of the space available for military operations can be reduced for both the potential attacker and defender by delineating spheres of influence in peacetime. Russia and England agreed in 1907 to have a sixty-two- to ninety-three-mile-wide zone to avoid direct conflict between their financial, trade and administrative interests in Afghanistan.[29] The Yalta Agreement of 1945, by delineating spheres of influence between the Soviet-dominated communist bloc and Western democracies, resulted in a large gain of space for any potential offensive action by the Soviet-led forces against Western Europe.

Neutral zones can play a significant role for both offense and defense, by either restricting or allowing the deployment of forces. Sometimes they can encompass a part of the territory of both belligerents and the entire area of a third country. An alternative to a neutral zone is to divide the entire area through military occupation, thereby enhancing political influence and control of the area by both belligerents.[30] The stability of a neutralized area or zone depends on the balance of forces between the potential enemies. Where the sides are approximately equal in strength, any advantage one side tries to obtain will be neutralized through the counteraction of the other side. The greater the symmetry, the longer the tradition of stability must be if a neutral zone is to be protected or observed by both sides. The opposite is also true.[31]

A war between two or more countries can indirectly involve a third country that is not interested in the outcome of the war but wants to remain on friendly terms with both sides.[32] Then, the space for the employment of forces of both sides would be reduced. However, in some cases such a situation could limit attacker's options, thereby providing advantages to the defender. All hostile acts in neutral territory, waters, or airspace would be prohibited.[33] The amount of time belligerent warships can stay in a neutral port or anchorage is usually limited. When one side knows that the neutralized area has limited value for the other, then it can be confident that the other side will not try to change the status quo. If a neutral zone has relatively small military, political or economic importance, it is more likely that its neutrality will be preserved.[34]

Once hostilities start, the territories of neutral countries could be violated by one or both sides. Thus, various neutral or exclusion zones may be declared by one or both belligerents or by third parties. Generally, space becomes progressively larger as important neutrals are drawn into a war. The military need for space can become so critical that limitations on the conduct of operations cease to be persuasive or politically important.[35]

Space for military actions can also be reduced by the existence of countries or territories in a state of nonbelligerency. This was the situation in Europe in World War I when, among others, the Netherlands, Switzerland, and the Scandinavian countries all declared their neutrality in the war.

In a maritime context, a neutral country can insulate itself from hostilities by declaring a *neutrality zone.* Such a zone can be effective if the diplomatic situation forces each side in a war to placate neutrals. This happened with the South American Neutrality Zone declared in Panama in October 1939. The zone encompassed all the countries of South and North America (except Canada) plus the 300-mile-wide sea/ocean area off their coasts.[36]

The Factor of Time

Time is one of the most precious commodities in the conduct of warfare. Napoleon I said that "the loss of time is irreparable in war . . . operations only fail through delays."[37] The factor of time is closely related to the factor of space. However, compared with the factor of space, time is far more dynamic and changeable.[38] For one thing, space lost can be regained; time lost can never be recovered.

Any military action requires a certain amount of time for planning, preparations, and execution. Large forces do not just suddenly appear in a theater. They sometimes need several days or even weeks to complete movement from one area to another. Strategy and operational art are concerned with greater spans of space and time than tactics are. Sometimes even a single unforeseen incident can disrupt the sequencing or synchronization of the employment of combat forces and thereby adversely affect the outcome of a major operation or even a campaign. Time lost or wasted in conducting daily routine activities can sometimes indirectly influence, to a great degree, the outcome of a combat action.

The relative value of the factor of time has gradually changed over time. In the era of two-dimensional wars, actions took place over a much smaller space and at a relatively slower pace than today, allowing the commander much more time to prepare for combat or react to unforeseen events. The rapid technological advances in the past century have considerably reduced the amount of time required to acquire and transmit information. However, the sheer amount of information to be processed and analyzed requires much more time than in the past.

Generally, time is nonmaterial, without content, and infinite.[39] Mastering the factor of time in combat means acting faster than the opponent. Then the key to success is to shorten the time for estimating the situation, making a decision, and deploying and maneuvering one's combat forces. At the same time, one should try to prolong the time the enemy needs for these processes.

A lack of time can sometimes force the operational commander to act in haste. This can lead to incomplete or poor planning and preparation for combat. The lack of time at the higher command echelon has, in most cases, a negative effect on subordinate commanders and staffs. Thus, operational commanders should pay great attention to how time is used. Above all, they should prevent losing time.

The superiority of attack over defense depends on the time required for the attacker to carry out his offensive. The attacker should try achieving his ultimate objective in as short a time as possible. If victory is too long in coming, the attacker can overshoot his point of culmination.[40] A quick advance on land by the attacker usually prevents the defender from conducting an orderly withdrawal. In addition, the enemy's morale is diminished much more if the one reaches the decision quickly.

Sufficient time is required for the attacker to consolidate his gains of space; this includes the time required to secure control of the newly acquired areas and to militarily organize the space.[41] The defender, in contrast, should delay the decision as long as possible and then try to revert to the counteroffensive to defeat the attacker.[42] He can gain time by conducting a delaying defense until reinforcements are brought in, thereby creating the conditions for eventual victory.[43]

The factor of time can have tactical, operational or strategic importance. As the terms imply, the tactical and operational aspects of the factor of time pertain to all elements of the planning, preparation, and execution of tactical actions and major operations/campaigns, respectively. The strategic aspect of time is related to, among other things, the probable duration of the future war, the time between the declaration of war and the opening of hostilities, warning and reaction time, mobilization and demobilization time, and the time required to prepare the country's armed forces for war and to establish the industrial base. These matters are normally the responsibility of the highest political-military leadership of the country or alliance/coalition. The operational commanders and their staff have little or no influence on these time-related decisions. In contrast, the factor of time in dealing with operational decisions and the planning, preparation and conducting a campaign or major operation is directly influenced by the actions of the operational commander's staff and the actions, or lack of actions, of subordinate tactical commanders.

At the national-strategic level one of the key issues to be resolved successfully is the sound appreciation of the *duration of the future war*. This is one of the main tasks of the military leadership in peacetime.[44] Yet experience shows that in many cases, before the beginning of hostilities, one or both sides falsely envisioned the duration, of the future war. This invariably had a significant negative effect on one's ability to both prepare and conduct the war successfully. Among other things, false anticipation of war's duration has often had serious consequences on force planning, the development of service/joint doctrine and the eventual outcome of the war itself.

A reasonably realistic picture of the duration of a future war would help national leaders create or expand the existing industrial base and establish the forces necessary to accomplish the political objectives. The more time available for preparing for war, the more thorough and therefore more successful these preparations will be. This is especially important for organizing mobilization and the national logistics system. Correctly estimating the duration of a war is also critical to anticipating the extent and type of vulnerabilities of one's country or of the alliance or coalition. Normally, a short war draws only upon ready stockpiles, and therefore the destruction of manufacturing facilities does not affect its outcome.[45]

A long war, in contrast, requires a significant increase in the country's manufacturing capacity, the conversion of some facilities, and the construction of new ones. All this requires a relatively long time to complete. The expansion of war-waging capacity competes with current peacetime production for manpower, materials and physical plants and machinery.[46]

Prior to 1914, all major European powers hoped to conduct a short war by using brute force at the beginning, thereby avoiding a war of attrition.[47] These views were based largely on the lessons from the German Wars of Unification (1864, 1866; and 1870–1871). Among the lessons learned from these wars was that any future war would be a war of movement and therefore decisive and short. The Germans did not pay serious attention to the lessons of the American Civil War (1861–1865) because they believed it to have little military relevance for the conditions in Europe.[48] Yet the American Civil War and the Russo-Japanese War of 1904–1905 were the real precursors of World War I – that is, positional wars of attrition, and not the Moltkean wars of movement of 1866 and 1870–1871.

In the past, too many of the highest political and military leaders committed grave errors in predicting whether or when hostilities would start. Also, the lack of a realistic appreciation of the strategic situation has often led political and military leaders to plan for a short war instead of a longer one. Hitler expected to defeat the Soviets in only eight to twelve weeks.[49] However, the war lasted almost four years and ended with a catastrophic defeat for Nazi Germany.

During the nineteenth century and the first decade of the twentieth century, one of the main challenges for the top leadership of great powers was the shortening of the *time between the formal declaration of war and the actual opening of hostilities*. This depended on the speed of mobilization and movement of troops from their mobilization posts to the area of concentration. Sixteen days passed between the French declaration of war against Prussia and the first combat in 1870.[50] By August 1914, the time to react to the enemy's mobilization in case of war between major European powers had been reduced to only a few days. Since World War II, most regional wars have started without a formal declaration of war. Hence, either the victims of aggression did not have sufficient time to prepare their defenses or the signals of war were lost in a maze of conflicting information, or the attacker achieved a strategic surprise.

Obviously, the greater the time for preparation, the higher the probability of success. Adequate *preparation time* is critical for combat employment of one's forces. The aim should be to prevent the potential enemy from having sufficient time to complete his preparations. Otherwise, the duration of the hostilities will invariably be longer than planned. In general, there is a great advantage in completing one's mobilization and deployment before the enemy does. Then one's forces can go on the offensive before the enemy is fully ready. They can also possibly disrupt the enemy's plans for mobilization, deployment and concentration. An early attack can also force the enemy to withdraw without a fight, thereby at least gaining space for the morale of one's forces. Another advantage of an early start of the operation is that the deployment and concentration area for the enemy's forces will be reduced because of the resultant loss of space.

The defender can also gain a significant advantage by completing his preparation early and then going onto a limited offensive to create a more favorable operational situation. Then one option for the defender is, by timely completing predeployment, to gain some parts of terrain that are highly favorable for defense, as the Boers did in their war with Britain in 1899–1902.

A campaign usually requires the large-scale mobilization of forces of all services and often the forces of allies or coalition partners as well. Obviously, the *time for mobilization* should be as short as possible to enhance one's chances of acting before the opponent is ready. The amount of warning time is closely related to the timely mobilization of one's forces. Among other things, the mobilization time depends on the size of the objective to be accomplished, the size and type of forces to be mobilized, and the diversity, mode and overall efficiency of the transportation system. In the late industrial era, the speed of mobilization was one of the key factors in achieving initial victories on the battlefield.

A favorable geostrategic position can considerably extend the *warning time* of an impending enemy attack. In the modern era, time for the defender can be gained through an effective and long-range early warning system.[51] An attacker always needs a certain amount of time to start deployment and build up his forces in an area of concentration, and then for the forces to move to attacking positions. Before the beginning of the hostilities in the Yom Kippur/Ramadan War in October 1973, the Israeli planners erroneously believed that they would have about two days' warning time – that is, sufficient time to mobilize Israel's reserve forces and strike a preemptive blow.[52]

Closely related to warning time is the *reaction time* to the enemy's attack or unforeseen action. Obviously, the reaction time is shorter if one's forces are on full alert or at a higher state of combat readiness. Once the hostilities begin, the reaction time, among other things, is a function of the decision-making cycle, command and control (C2) process and theaterwide communications. The side that reacts faster than its opponent would generally improve its chances of accomplishing its assigned military objectives. Reaction time can be shortened, especially in a crisis or a sudden outbreak of hostilities, by deploying one's forces in forward areas or in the areas of potential hostilities. Prepositioning weapons and equipment at various strategic locations worldwide also significantly reduces the time needed to react to any major or minor contingency requiring the employment of military forces.

The *decision-making cycle* at the operational and strategic levels should be as short as possible. This is accomplished by having a simple command structure with a fast and reliable command and control supporting system; sound intelligence; and fast and reliable communications. Human factors, especially the personality traits of the operational commander, also play a critical role in making sound decisions quickly. The decision cycle is highly dependent on the number and efficiency of the segments in a decision loop. The speed of making a decision depends on the speed of the slowest element in the cycle. Generally, the more elements the decision cycle has, the slower and less effective the entire decision-making process will be. The aim of an attacker today is to render the defender unable to respond to his actions in a timely fashion. The U.S.-led coalition forces reportedly neutralized Iraqi decision-making in the first hours after the start of the air offensive on 17 January 1991. The top Iraqi leadership could not respond in any meaningful way to these massive air attacks, although other components of Iraqi military power had not yet been destroyed or neutralized.[53]

In recent years, the greatest impact on intelligence has been from the huge increase in the volume and diversity of information acquired. Although the processing and evaluation time has been considerably reduced, the sheer amount of information far exceeds the ability to process, evaluate and distribute it. This makes it hard, if not impossible, to distinguish accurate and relevant information from inaccurate and nonessential information. The heavy reliance on computerized processing today creates potentially serious vulnerabilities that can be exploited by an agile and resourceful enemy. The expansion of the battlefield in the future will put a premium on obtaining and distributing highly accurate and timely information. However, information is worthless if it is irrelevant to the task at hand. It is too often forgotten that information is merely a means to an end, not an end in itself.

The larger the objectives to be accomplished, the larger the scale of combat actions that must be envisaged, and, consequently, the more time will be required for planning, preparation, and execution. *Planning time* is another critical element that affects the success of a major operation or campaign. The time can be greatly reduced if the planning is conducted parallel rather than sequentially. The simpler the planning process, the more speed is enhanced. The more experienced the operational commanders and planners, the faster they will be at improvising and preparing operation plans. However, today's trend of increasing reliance on computers and artificial intelligence (AI) as decision aids is fraught with potential dangers. While computer networks can considerably reduce the time needed to carry out routine planning tasks, they cannot replace the human brain and the skills and experience of operational commanders and their staffs. Moreover, computers are prone to breakdowns and are not immune to penetration by computer "hackers" or other hostile actions.

The *time between consecutive major operations* or *campaigns* should be short so that a high operational tempo is maintained. The time to mount a consecutive major operation or campaign depends on the forces available, the anticipated attrition rate of one's forces, the quality and sufficiency of logistical sustainment, and the effectiveness of operational command and control.

The *time for deployment* of one's forces is much longer at the operational or strategic level than at the tactical level of war. Obviously, the longer the distance to the employment area, the longer the time for completing the transit to the assigned areas of concentration or combat. The time needed to deploy forces to conduct a major operation or campaign will depend primarily on the distance from the home base or current operating area to the deployment area, the size and mobility of one's forces and the mode of transportation used.

The *duration* of a military action depends directly on the scale of the objective, the size and mix of one's forces, the methods of applying combat potential and the physical characteristics of the area of employment. Obviously, the larger the objective, the more time is required to accomplish it. Also, the larger an enemy force is, the more likely it is to offer a stronger defense, which will, in turn, result in a longer duration for the major operation or campaign.

An operation plan should be flexible enough to provide sufficient *reserve time* if something goes wrong or the actions take more time than anticipated. The more objectives or tasks assigned; the more time is required to accomplish them. Therefore, it is critical to focus on the most essential objectives or tasks that will collectively ensure the accomplishment of the ultimate operational or strategic objective.

The duration of a campaign or major operation depends on the number and length of "operational pauses." These depend primarily on the number of intermediate (operational or tactical) objectives. The larger the number of intermediate objectives, the more likely the need to slow the actions of the force elements to regenerate one's combat potential. The longer each operational pause, the more opportunity there is for the enemy to recover from his losses and regenerate his own combat potential/power. In May 1940, after the first great battles on the Franco-German border, the Germans were forced to halt their advance for about ten days on the Somme River.[54] This could have been sufficient time for a more agile opponent than the British or French of 1940 to mount a counterattack and achieve more than a local success.

The *time to commit operational reserves* is a critical factor in the conduct of a major operation or campaign. In general, the more mobile and smaller, but combat-ready, the operational reserve, the shorter the time needed to commit it at the point of the main attack (or defense). Also, the distance to the employment area, the transportation network and the terrain features considerably affect the time required for the commitment of an operational reserve in a major operation or campaign.

Often political leaders and sometimes even operational planners do not appreciate the need to have sufficient *time to master new weapons and equipment* acquired during hostilities. The more complex the weapons system usually the more time is needed for training in its proper use.

Timing is closely related to the factor of time. It consists of selecting the right moment to do something or issue some pronouncement. It is associated with a specific decision and not a lengthy procedure. It is less a matter of routine and more one of skilled judgment. The proper timing of a campaign or major operation requires a great deal of foresight. The timing of when a campaign is to start can sometimes have a decisive influence on the outcome of the entire military enterprise. Napoleon I postponed his invasion of Russia until summer, apparently overlooking the fact that the Russian summer is short.[55] The results of that mistake are well known.

The timing of a campaign or major operation can be drastically changed because of unforeseen political or military changes in a situation that, in some cases, requires a heavy commitment of one's forces. This can lead to a long delay in the start of the originally planned campaign or major operation. The German timing of the invasion of Soviet Russia had to be changed in the aftermath of the military coup in Yugoslavia on 27 March 1941. The OKH believed that the invasion of Soviet Russia must be postponed by at least four weeks. The deployment of the German forces was disrupted because of the need to use two higher headquarters (Generalkommando), six infantry divisions, and an additional three divisions from the army general staff's reserve (originally intended for the Barbarossa campaign), in the planned invasion against Yugoslavia. These forces could not be replaced by other forces. All forces, especially panzer and motorized divisions, required at least three weeks for rest and reequipment in their home areas and close to facilities from which spare parts could be supplied. The total time required for all these measures was estimated at five to six weeks. Eventually, because of poor weather conditions in Russia, the campaign against Russia was delayed from 15 May to 22 June. Yet despite some contrasting views, the campaign in Yugoslavia caused no more than one to two weeks of delay of the German invasion.[56]

The Factor of Force

The terms "force" and "means" are often used alternatively, although they do not mean the same thing. The term "force" in its narrowest meaning pertains to both tangible and intangible elements of military sources of power. The main source of national military power is its "armed forces" – a force the state creates and maintains for the protection of its way of life, its existence and its territory.[57] The term "means," in the military sense, encompasses all the physical and human resources required for the accomplishment of a certain military objective. This term is sometimes used in national-strategic terms to refer not only to armed forces but also to the political and economic resources of the entire state or group of states.[58]

At the national-strategic level, the greater the superiority of the factor of force relative to the opponent's, the more freedom of action the country's top politico-military leadership has in conducting strategy and policy and in determining national-strategic objectives.[59] Likewise, the greater the factor of force at the operational level, the greater the operational commander's freedom to act.

Several terms are often used loosely and interchangeably in referring to the military aspects of the factor of force. The term "combat power" is used regardless of whether forces are in combat or not. Theoretically, two distinct yet related terms are used to evaluate the factor of force: combat potential and combat power. *Combat potential* pertains to the assumed potential of a military force to accomplish the assigned mission. Theory differentiates between designed and available combat potential. *Designed combat potential* is the theoretical capability of a force to accomplish assigned missions against a specifically designed threat.[60] It is based on the standardized table of organization and equipment (TO&E) and doctrine. In contrast, the *available combat potential* is that part of the designed combat potential that is available to the commander for combat. Because of the differences between prescribed and actual TO&E, available combat potential is usually less than the designed combat potential. It includes all the elements that could enhance or reduce combat potential, such as leadership, manning, training, morale and discipline, weapons and equipment, tactics and doctrine. In general, the higher the quality of forces in these and similar categories, the higher the available combat potential. Because many elements are intangible, it is extremely difficult to evaluate available combat potential with any degree of confidence. Available combat potential is also affected by the characteristics of the physical environment in which a given force will be employed, such as the size and shape of the theater, relevant distances, military positions, terrain, weather and climate, and oceanography, to name just a few.

Combat power is a force's real combat capability generated while accomplishing a mission against a given enemy force and in a specific combat environment.[61] Once hostilities start, one's (and the enemy's) available combat potential is gradually converted into combat power. Whether that process is finalized depends mainly on the duration and intensity of combat and resulting losses. In practice, however, the entire available combat potential of a given force is rarely transformed into combat power.[62]

Many factors influence the conversion of available combat potential into combat power. Rough terrain, bad visibility, or seasonal characteristics can significantly reduce the effectiveness of weapons and troops. Combat power encompasses many unquantifiable elements that can considerably, if not decisively, affect the overall performance of a given force, service and armed forces as a whole. The characteristics of natural elements, specifically terrain, weather and climate, can be foreseen to some extent, but their effect cannot be measured with any degree of precision.

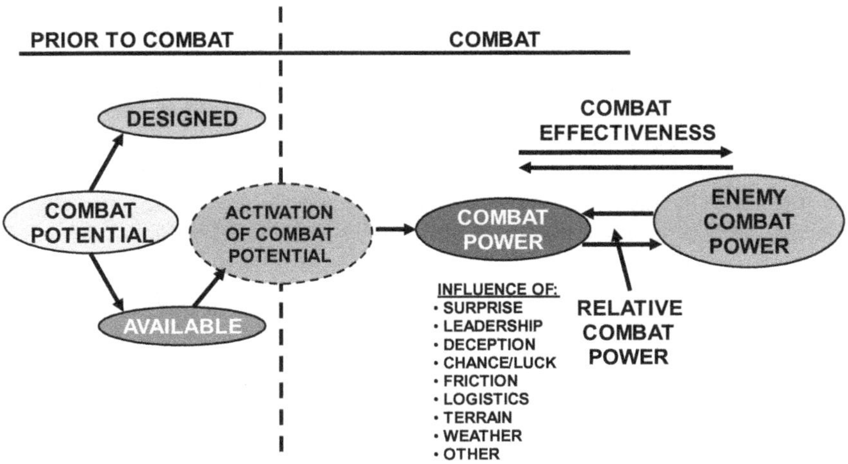

Figure 3.2 Combat Potential versus Combat Power

All these factors can greatly reduce the effective combat power activated from available combat potential.

The Clausewitzian friction and the "fog of war" is an inherent feature of warfare at any level. These factors also play a considerable role in the conversion of one's available combat potential into combat power. Friction consists of the infinite number of unforeseen things, large and small, that interfere with all activities in war.[63] It encompasses uncertainties, errors, accidents, technical difficulties and the unforeseen, and their effects on one's decisions, morale and actions.[64] It is one of the most important elements that distinguish real war from war on paper. In the Clausewitzian view, the "military machine – the army and everything related to it – is basically very simple and therefore easy to manage." Yet it is composed of many parts, and "each part is composed of individuals, every one of whom retains potential for friction." The ever-present factor of danger, combined with the physical exertions that war demands, compounds the problem. Clausewitz insisted that this "tremendous friction cannot be reduced, as in mechanics, to a few points. It is everywhere in contact with chance, and it brings about effects that cannot be measured, just because they are largely due to chance." Friction is the "force that makes the apparently easy (things) so difficult."[65]

Friction is the main reason a military action differs in its execution from the one planned. Generally, the more complex one's plan, the more friction during its execution. The principal causes of friction are the enemy's actions, human errors, fatigue, terrain, weather, inadequate or inaccurate information, and pure luck and chance. Unclear and ambiguous orders, or clear orders misinterpreted by subordinates or superiors, are also important sources of friction. Other factors that contribute to friction in combat are mistakenly selected or multiple objectives, wrongly identified centers of gravity, decisive points, inadequate logistical support and sustainment, poorly protected lines of supplies, and personal animosity between commanders, causing them to cooperate poorly or not at all. Modern information technologies can reduce but never eliminate friction. At the same time, commanders should try to master and overcome its effects.

Because combat is a clash of opposing wills, uncertainties and unknowns abound. When combined with friction, they create ambiguity, or what Clausewitz called the "fog of war," in which a commander must make his decisions. The fog of war presents both opportunities and dangers. In general, the chances of achieving surprise and deception increase as the fog of war increases. The higher the level of war, the more uncertainties the military situation entails. A force's overall effectiveness is reduced when decisions are made – as they often are – based on imperfect, incomplete, or even false information. The fog of war is the main factor that makes some commanders willing to take high risks and others extremely cautious in making their decisions.

Any factor of force is composed of a number of "physical" or tangible and so-called "abstract" or intangible (imponderable) elements. The physical elements – such as the number of personnel, weapons, and equipment; physical mobility; firepower; logistics; and quality of weapons and equipment – are for the most part measurable in some way. Nevertheless, tangible elements of combat potential or power can be either partly known or completely unknown.

In contrast, intangibles are generally hard or even impossible to quantify with any degree of precision. Intangibles pertain for the most part to human elements, such as cohesion of an alliance/coalition, strength of public support for war, quality of national or military leadership, morale and discipline, training, and command and control. Some of these elements, such as training and combat readiness, can be assessed in very broad terms: low, medium, high or excellent. Other intangible elements – such as leadership, will to fight, morale and discipline, small-unit cohesion, combat motivation and soundness of doctrine – are extremely difficult to quantify with any degree of precision or confidence.

Difficult things to evaluate are the degree and effectiveness of cooperation among services, or "jointness" ("combinedness," in a multinational employment of forces). The effectiveness of military force is also affected by the overall posture – offensive or defensive – and the effects of the physical environment, weather and seasonal factors. Because the factor of force has so many unquantifiable elements at any level, often too much emphasis is given to a simple tabulation of forces, weapons, and equipment. However, operational commanders and their staff should do everything necessary to properly and realistically evaluate all elements of the factors of force.

The tangible and intangible elements of the factor of force are usually mixed and cannot be neatly separated from each other. This is especially true in the case of forces employed at the operational and strategic levels of war. Additionally, friction and the fog of war are ever-present. Tangible factors can be properly or improperly evaluated; they can be changed over time; and they can be intentionally or unconsciously erroneously reported. They can be wrongly understood because of emotions of fear, hate, lack of confidence, fatigue, and nervous stress. Tangible elements can also be falsely evaluated. The number or size of enemy forces or weapons/equipment might be accurately observed or obtained but falsely reported or evaluated without a proper context. Information received might be accurate but is wrongly interpreted by commanders and their staff. This can occur intentionally or unintentionally. It can be caused by incompetence, lack of operations security or treason. The commander can falsely evaluate the enemy's capabilities or intentions. Misunderstandings between commanders and subordinates are frequent occurrences in combat; they cannot be predicted or quantified. The breakdown of weapons or technical equipment can occur at any time. The effects of atmospheric influences cannot usually be measured precisely. Except in some rare cases, natural events cannot be timely predicted. Hence, the unreliability of humans and technology considerably affects the performance of the factor of force on both sides in combat. The boundaries between tangible and intangible factors are in the area of chance and are fluid.[66]

The forces' numerical strength is probably most often used to evaluate the probable combat performance of forces on both sides. The size of the armies, navies or air forces employed is one of the most decisive elements in combat. It is usually a great advantage to be numerically stronger than your enemy. And nowhere is this more true than at the operational and strategic levels of war. At some point, sheer numbers are simply overwhelming, no matter what the level of skills, morale and discipline, or training and soundness of doctrine of the opposing force. History offers many examples where a numerically superior force has been a decisive factor in achieving a victory over a better-armed, trained and led but numerically smaller force. Because of the great numerical inferiority of his forces. In their war with Russia in 1904–1905, the Japanese failed to achieve decisive victories on the ground quickly because of their inferiority in numbers of troops. They were too weak in proportion to the difficulties they had to overcome and lacked sufficient forces to achieve quick victories.[67]

For all its value, the importance of *numerical superiority* should not be overstated. History shows that in many cases, superior numbers were of no avail.[68] Inferior forces have often defeated a much larger force because of the better quality of their leaders and the better training, morale, and discipline of their troops. In the Six-Day War (5–10 October 1967), the Israeli Defense Forces (IDF) defeated quickly and decisively much larger Arab armies. The Israelis mobilized 250,000 men (225,000 were in the army).[69] Their forces were smaller but much better led and trained than their Arab enemies. For the Israelis, the Sinai was the principal part of the theater. This was the reason that the Israelis concentrated some 70,000 men and 750-800 tanks on their southern front.[70] Initially, the Israelis were strategically on the defensive on the Jordanian and Syrian fronts.[71]

The main factors in increasing the combat potential of one's forces are high mobility and firepower. This fact led to a belief among many theoreticians and practitioners that new technology could lead to much smaller armies. The claims were not entirely without merit. In fact, steady advances in technology have led to a reduction in the size of military forces. In addition, it was believed that technology would result not only in smaller forces but also in less bloody combat, with fewer losses for both sides. However, experience has conclusively shown that the leading proponents of new technologies in the modern era often grossly exaggerate the impact of those technologies. Technology has never been the sole, or even the most important, factor in reducing the size of one's combat forces. Superior combat leadership, training, doctrine, unit cohesion, morale and discipline, and other intangible factors, properly integrated with the new technological advances, proved the surest way to success in combat.

In contrast to armies, navies and air forces normally maintain their full strength in peacetime, with a relatively small fraction of the total number of ships and aircraft maintained in reserve. Highly industrialized countries with excellent transportation networks generally require less time to mobilize and therefore might not need to maintain a large force in times of peace. The forward deployment of forces and prepositioning of weapons and equipment reduce the need for the initial movement of large quantities of weapons and equipment into a theater.

In combat, a force might suffer more losses than expected and in a shorter span of time than anticipated, thus having a large reduction in force quality. The quality of the factor of force is often significantly reduced in the offensive, because usually the best and most daring men are among the first casualties. Nominally increasing the numbers to replace combat losses, but sacrificing quality, rarely represents an actual increase in combat strength. On the other hand, the longer the war, the more experienced friendly forces are. This can sometimes make up for deficiencies in peacetime training.[72]

Military theoreticians often cite a *force ratio* of 3 to 1 between the attacker and the defender as a prerequisite for success in an offense. Too often these and similar force ratios have been applied to the overall strength of forces on both sides. In fact, many times an attacker has been successful despite only a slight margin of overall numerical superiority, or even numerical inferiority, with regard to the defender. In the West in May 1940, the ratio of attacker to defender was 0.7 to 1.0 or 3,740,000 Allied soldiers (including 2,240,000 French troops) facing 2,760,000 Germans. The Allies had a three-to-two superiority in artillery pieces. However, France had only three armored divisions (plus one more created during the campaign) against Germany's 10 panzer divisions.[73] In many major operations on the eastern front in World War II, the Wehrmacht was successful although the Soviets possessed a ten-to-one superiority.[74]

Experience shows that what matters most is superiority in numbers at the sector of main effort or the main point of attack (or defense), not overall superiority. Clausewitz observed that massing one's strength at the decisive point depends on the strength of the army and the skill with which this strength is employed. The aim, then, is to take the field in the greatest possible strength, either to get the upper hand or at least to make sure the enemy does not.[75]

The *number of weapons* – especially heavy weapons such as tanks and large-caliber guns, combat aircraft and large surface combatants – is another measurable element of combat potential and combat power. Generally, the larger the numbers in the main categories of weapons and equipment, the higher the chances of success – provided, of course, those other elements of power are approximately equal. However, often a major operation or campaign can be successful despite numerical inferiority in weapons, if one's forces possess superior leadership, weapons and training. In addition, sometimes it is more important to possess superiority in a selected category of weapons and inferiority in others, as was true of the Germans in their offensive in the West in May 1940. The Germans had fewer battle tanks compared to the Allies—2,754 versus 3,254 Allied tanks. However, they enjoyed marked superiority in the number of modern aircraft—3,500 versus 1,090 French aircraft – and in the strength of the air defenses of their fielded forces.[76]

The *physical mobility* of a force is one of the main prerequisites for quick action. In general, physical mobility is the capability of a force to move effectively (by foot, vehicle, ship, aircraft or any other means) before, during, and after combat, in relation to or in comparison with a hostile force. Thus, mobility is always relative.[77] Hannibal, Genghis Khan, Timur and Belisarius were successful in their wide-ranging conquests because their horse-mounted armies had a higher degree of mobility against the comparatively immobile forces of their enemies. They achieved victories in many battles in which their enemies had numerical superiority.[78]

Speed is often confused with mobility and mobility with maneuver. An ingredient of mobility is speed. However, the speed of movement alone is rarely sufficient to achieve the full advantages of mobility. To constitute mobility, speed should be combined with the direction or speed of advance relative to the enemy, surprise, timing and deception. In a maneuver, it is not as important that movement be executed at a given maximum speed as that it be faster than the movements of the enemy. It should result in bringing a preponderance of force to bear relative to the enemy's resistance. *Momentum* is a product of mass and velocity. As long as both mass and velocity are preserved, momentum is maintained; mobility usually preserves both.

A lack of tactical mobility can prevent a numerically larger force from prevailing over a smaller hostile force that is more mobile and tactically better organized.[79] In the invasion of France in May 1940, the German panzers possessed higher mobility than the French tanks. At speeds of twenty-five to thirty miles per hour with an effective range from seventy-five to 125 miles, they were about twice as fast and had twice the range of their French counterparts. A false vision of the future war had made the French emphasize firepower and protection over the speed and range of their tanks.[80]

Physical mobility directly affects a force's deployment capability. This is especially critical when the distances from the basing or staging area are large and the force must be deployed at short notice. Mobility and firepower are interrelated and significantly affect each other. The availability of highly mobile and effective means of *firepower* enhances the movement capabilities of one's forces, while at the same time it can inhibiting the movement capabilities of hostile forces.[81] The number of firing platforms and fire volume alone is insufficient to evaluate the firepower of one's forces; the circumstances and the ways it is used are also factors.[82]

Mobility without reliable and continuous *communications* throughout the chain of command means little. In 1940, the German panzer commanders controlled their units from forward-deployed command vehicles (panzers or half-tracked vehicles) specially equipped with a radio. From the top down, German command and control was flexible and positive. Wide latitude to exercise initiative was allowed and encouraged through the issuance of mission-type orders. In contrast, the French tank units could not communicate with each other by radio because their intended role was to support infantry attacks. They lacked the speed and flexibility necessary for mobile operations. The French commanders were allowed little or no leeway for flexibility and exercising initiative.[83]

Not only the number of weapons and equipment but, more importantly, their quality considerably affect the overall combat potential/combat power of a force.[84] In some major operations or campaigns, *better weapons* were one of the major contributing factors to ultimate success, while in others, success was ephemeral and ultimately the entire effort failed.

A high quality of weapons and equipment is often not enough in itself to achieve victory if other elements of combat potential are inadequate. In the initial phase of the Russo-German war in 1941, the German army's panzer and motorized infantry divisions had better tanks and assault guns than their Soviet counterparts. The German antitank guns were simple in design but of high quality.[85] The 88-mm Flak 18 gun was used in both an antiaircraft (AA) and antitank role and proved to be one of the best guns of World War II. The Soviets overproduced tanks before the war, and by the beginning of the war, most of their tanks were obsolete. By mid-June 1941, their best tanks, the famous T-34 and KV1, numbered fewer than 1,000 and 500, respectively. The Soviet guns and mortars were superior, and their small arms were comparable to their German counterparts. The German combat aircraft were of excellent design. The Germans produced fewer models of aircraft, and these models were easily modified to perform different roles. The Soviet air force, in contrast, was large, but some 80 percent of its front-line aircraft were obsolete. The Soviets lacked radar and radio direction aids.[86] The design and combat performance of the German U-boats, torpedoes and mines were far more advanced than those of their Allied counterparts, but ultimately, they were insufficient to win the victory in the Battle for the Atlantic in World War II.

The problem in estimating the actual combat performance of a force is the pervasive presence of many unquantifiable elements that fall within the scope of combat potential/combat power. The human element and its associated manifestations are often a much more crucial factor in evaluating the factor of force. The German army performed superbly in both world wars because its leadership made a conscious decision to focus on the critical and decisive importance of war fighting and leadership. Moreover, the Germans realized that the human factor is most critical for success in combat. Therefore, they built an organization around the social and psychological needs of the individual fighting men. They emphasized mutual trust, willingness to assume responsibility and the right and duty of subordinate commanders at all levels to make independent decisions and carry them out. This was the true secret of the German army's fighting power.[87]

In the Falklands/Malvinas Conflict of 1982, almost every element was weighed in the equation except for the human element. Conclusions were made based on the probable outcomes of numbers and technical capabilities: the opposing numbers of troops; the advantages of coastal defense versus amphibious landing; the quality of aircraft on both sides; the weather; and the apparently isolated nature of key terrain. However, the decisive factor in that conflict was the highly competent British leadership, the qualities of the individual British soldiers and their organization and unit cohesion. The Argentines lacked the will to prevail that is characteristic of cohesive, well-led units.[88] The British soldiers were well-trained and highly disciplined. They were more than a match for the eighteen- and nineteen-year-old Argentine draftees. Despite being outnumbered by as much as three to one, the British won the encounters. On land, they inflicted heavy casualties on the enemy while suffering low casualties.[89]

Moral elements of the factor of force include such personality traits as courage, loyalty, simplicity and the maintenance of high ethical standards. They also encompass the mental qualities of leadership in the effective employment of fighting power. It includes one's ability to think soundly, to act at the right time and to accept responsibility for one's actions. While the superiority of physical elements may actually be the most decisive factor in the eventual victory, its mere possession may not of itself ensure final success. Among other things, victory requires a skillful and judicious mix in one's factor of the force of moral, mental and physical elements that are superior to those of the enemy.[90] The psychological element of power is perhaps the most important in combat but also the most difficult to evaluate properly.

Any fighting is essentially a test of one's physical and psychological endurance. Clausewitz observed that psychological forces exert a decisive influence in a war. Superior numbers can be neutralized by the superior military and moral qualities of the numerically weaker side. *Morale* permeates war as a whole.[91] The most difficult to quantify is the moral strength of an army or a nation.[92] The morale of a force depends on many imponderable factors, which makes it extremely difficult to assess the quality of a force. These factors include small-unit cohesion, discipline, training, leadership, ideology and combat motivation.

Morale and *discipline* are opposite sides of the same coin. Instilling them in a military force depends upon the leadership's understanding of the nature of human relationships.[93] In addition to issuing training schedules, the general aim of regulations is to set an overall standard of conduct and work for all concerned. The level of discipline should at all times be gauged by what elicits the best results from the majority of dutiful individuals. Setting an example is far more effective than the written or spoken word. As the soldier sees his superiors obey, so does he usually follow their lead.[94]

Combat motivation is a combination of morale and discipline, small-unit cohesion, and training, tempered by the combat environment. At any level of command, one of the principal tasks of the commanders is to strengthen morale factors while mitigating the adverse effects of the environment. The commander should identify those elements of morale that he can influence and develop.[95] The environmental effects can be reduced by small-unit cohesion and strict and realistic training, and by informing subordinates of actions planned or taken in a timely manner. Soldiers' motivation to fight is a product of many elements. The soldiers must perceive the higher authority as legitimate; otherwise, it is highly unlikely that their motivation for combat will be very high. The soldiers are subjected to numerous rules and regulations. They also carry out numerous routine tasks that they learned during combat training.[96]

Another factor in the will to fight is fear. Although some soldiers can be paralyzed by fear or even try to flee, for many others this factor sustains their will to fight.[97] However, soldiers are principally motivated to fight through their sense of loyalty to the unit, the country and the cause.[98] They also fight well if trained well and if they have pride in their military skills. Their motivation for combat is also higher if they believe that their performance contributes to the overall success of the unit's mission and to the personal safety of their colleagues.[99]

For the armed forces as a whole and for the individual services, combat motivation is affected by several other factors, notably the character of the war, the justness of the cause, the war's legitimacy, ideology, patriotism, and a sense of national honor. The combat motivation of a force is also heavily influenced by the nature and legal aspects of the conflict. An army defending home territory is usually more initiative-taking than the invading force. However, if the regime is oppressive either to its own population or to national minorities, then combat motivation may be low or even nonexistent.

Generally, if the majority of the population perceives the war to be just and legitimate, then combat motivation is usually higher than if the war is considered unjust and illegitimate. The combat motivation of the British, U.S., and other Allied armies in World War II was generally high, both because of the sustained support from home and because of the strong and widespread belief in the justness of the Allied cause. However, sometimes the degree of demoralization and defeatism before a war is so high, as in France in 1940, that although the cause is undoubtedly just and the war legitimate, the will to fight is sorely lacking. The side that initiates a war of aggression can still motivate the armed forces to fight well through highly successful propaganda, ideology, and the survival instincts of the majority of the population. The combat motivation of the Wehrmacht as a whole was uniformly high during World War II, both in victories and in defeats. The German troops were well-led and highly disciplined. It is generally accepted that human relations were better in the Wehrmacht than in the German armies of 1870 or 1914. Despite great losses, the German U-boat crews and Luftwaffe fighter pilots retained remarkable combat motivation until the very end of World War II.[100]

Operational leadership is an almost entirely intangible element of combat potential or combat power. It is often a decisive factor in combat and can make up for greatly superior numbers of the enemy.[101] In World War II, the Soviet leadership at the highest level made many costly mistakes. However, on balance it proved to be much better than the German High Command in making sound strategic decisions and preparing detailed plans. The Germans' greatest strength was in their tactical and operational leadership, especially from the corps level downward. In contrast, the Soviet leadership at the tactical and operational levels was rather poor. The Soviet commanders lacked initiative, training and experience.[102]

The combat power of a force depends not only on combined arms tactics but also on the degree of multiservice cooperation, or *jointness* – the employment of two or more services across the full range of the operational continuum – from routine peacetime tasks to the accomplishment of the assigned military objective in combat.[103] Generally, jointness enhances unity of effort by focusing all the energy of individual services and the armed forces across the full range of military operations, at all levels of war and in every environment – peace, crisis and war – toward enhancing the effectiveness of military operations.[104] In peacetime, jointness can be achieved through a multiservice command organization that ensures joint planning and training and joint education. In a war, jointness is ensured by applying the tenets of operational art in planning, preparing and executing campaigns and major operations.

Jointness offers the operational commander multidimensional capabilities – land, sea, air, space and special forces – that are inherently more effective than single-service forces. True jointness generally greatly enhances unity of effort. The operational commanders should have a menu of capabilities from which to choose. Then they can build an organization through which to direct the operation. They should be able to assign missions to their subordinate commanders and provide the requisite support to those commanders.[105] Multiservice forces provide the operational commanders with multiple operational and tactical options in carrying out their responsibilities. An innovative operational commander uses the capabilities of land, sea, air, and special operations forces creatively. Joint forces can be used both symmetrically and asymmetrically and in synchronized ways. Employing the combat arms of several forces generates a greater impact than using those of a single service.[106] An operational commander should synchronize the multiservice capabilities to achieve the highest synergistic effect.[107]

Doctrine is a critical element of the factor of military force. Without soundly written and skillfully applied doctrine, campaigns and major operations are likely to fail. Doctrine is understood as the set of officially approved views and fundamental principles on the missions, roles, methods and procedures of employing forces and assets in both wartime and peacetime. A soundly written doctrine should allow a great deal of flexibility in its application. It should be revised as circumstances change, as new weapons are introduced or as accumulated experiences and lessons learned – in peacetime and in combat – require new solutions. Doctrine should provide a common baseline for the employment of one's forces. Its greatest value is in providing common techniques and procedures to solve military problems. In this way, it improves the training of the entire force. A sound doctrine should encourage and develop the mental agility of commanders at all levels. The commanders should display drive, energy, flexibility and willingness to seize the initiative and make quick decisions. Mental agility is determined by the intellectual flexibility of the leadership, and by the speed of the individual commander's decisions and the staff's responsiveness to those decisions.[108]

Training is another critical element of the factor of force. Doctrine and training are interrelated and affect each other in many ways. A force can be numerically larger and excellently armed and equipped but still be ineffective because of severe training deficiencies. Whether training is high, moderate or low can be easily assessed. However, the impact of training on the anticipated performance of a force in combat is difficult to measure in any meaningful way. Training also affects motivation for combat.[109] There should be no lowering of standards in times of peace, regardless of the cost. Clearly, it is better to have smaller but excellently trained forces than much larger but poorly trained forces. Normally, the more highly trained the forces, the higher their motivation for combat. Uniformly high standards of training are the chief distinction between regular forces and elite forces such as marines (or naval infantry), airborne units and special forces. Submarine forces are normally a highly trained component of any navy. The extremely high performance of the German U-boats in both world wars was due to their high standards of training.

In the interwar years, the Japanese Combined Fleet was highly trained. The most rigorous phase of the preparation for war started in 1934, when Admiral Nobumasa Suetsugu took over the command of the Combined Fleet. He emphasized training in severe weather conditions. Japanese waters have eight times as many stormy days as U.S. waters. Even on ordinary days, the waters off Japan are seldom calm. In contrast, the U.S. Navy was trained under easy natural conditions, in the cool climate of Seattle and Halifax in the summer and in the warm breezes of California and Cuba in winter. Although training in severe weather lowers fighting efficiency, the Japanese were convinced that by being trained under severe weather conditions their officers and sailors would retain higher fighting effectiveness than their U.S. counterparts. This factor also led the Japanese Navy to decide to build much bigger surface ships than any other navy of the day.[110]

Information versus Operational Factors

Information has affected the conduct of war at all levels since the first wars started some 3,500 years ago. The most successful commanders have invariably tried to obtain accurate and timely information on the enemy while denying the same to the enemy. Hence, what is today called information dominance has been an integral and increasingly important aspect of warfare, especially in the modern era. Yet it is also true that the situation today differs significantly from those that resulted from technological advances in the past. Modern information technology, electronic media and worldwide communications all contribute to what has become a veritable flood of information. Today, electronic media connected with space-based communications systems generate an enormous flow of information. Some of this information is objective, while much is subjective or entirely false. Very often an evaluation of military actions depends not only on real facts but also on subjective perceptions of those facts in the media, the public and politics.[111]

The information revolution today has affected not only cybernetic but also physical domains, simultaneously and to a significant degree; that has not been the case in the past.[112] Some military theoreticians go even further and claim that information has emerged as a fourth factor along with the three traditional operational factors.[113] They insist that an exclusive focus on the traditional factors of space, time, and force cannot be justified in light of new conditions, since nonmilitary factors will steadily influence military decision-making, planning, and the course and outcome of military actions.[114] Yet, the volume of information received cannot be regulated; nor can one compensate for a lack of information. Accuracy, timeliness and relevancy of information cannot be meaningfully measured or anticipated. Information is essentially indefinable in any meaningful way, unlike traditional operational factors. Fundamentally, the concept of operational factors rests on the operational commander's ability to bring into rough balance the definable factors of space, time and force versus a given operational or military-/theater-strategic objective. By this balancing, shortcomings or deficiencies in a factor of space, time or force are countered by advantages or surplus value in other factors so that the assigned objective can be accomplished. However, a lack of information cannot be balanced by any of the three traditional operational factors. The main problem is that the same fact is often perceived differently by different commanders because of their personal character traits, education and training.

What is indisputable is that the significance of information is growing at a rapid pace and has a decisive impact on the application of operational art.[115] It increasingly and often profoundly affects the factors of space, time and force, both individually and in combination. A proper evaluation of forces, space and time simply cannot be done without accurate information on various aspects of the military situation. The enormous advances in information technology in recent years have elevated information into a common, and increasingly critical, link among the factors of space, time and force. Information affects the morale and inner cohesion of friendly and enemy forces and population. These influences can have direct and indirect effects on operational leadership.[116]

The operational commander today can reliably and, when necessary, continuously communicate with increasingly dispersed forces in the theater. Information can be transmitted and received globally, within not only a given theater but also hundreds and thousands of miles outside the boundaries in which friendly and enemy forces are in contact. News of military actions is broadcast in real time, or near-real time, to friends, neutrals and enemies alike. The global information environment has already had a profound impact on geopolitics, economics, military activities in general and operations. Numerous sources of information, generated not only by governments but also by nongovernmental organizations and private volunteer groups, exist. The more a country relies on information, the more vulnerable its economy and society are to disruption, or even paralysis, by actions of hostile governments or groups harboring grievances or outright hostility to one's way of life.

The huge increase not only in the volume but also in the accuracy and speed of information on all aspects of the situation in a theater has great potential for policymakers and operational commanders, both in crisis prevention and during armed conflict. The top political and military leadership should have a clear, timely, comprehensive and accurate picture of the key elements of the situation in a given theater, specifically as to society, economy, infrastructure, nonmilitary information systems and transportation network and traffic.[117]

New information technologies allow operational commanders and their staff to have much more detailed and accurate information than they had in the past on all aspects of the physical space. This is a critical element in the operational commander's estimate of the situation, planning and execution of a campaign or major operation. Full knowledge of climate and weather is critical for the employment of precise ammunition. This is essential for the timing of the start of a campaign or major operation. In the past, effective reconnaissance and surveillance depended on clear weather. Today, reconnaissance and surveillance can be conducted through clouds and in darkness.[118]

The planning and conduct of campaigns and major operations are considerably affected by nonmilitary aspects of space, specifically foreign and domestic politics, diplomacy, economic activity, society and other types of human activity. Information has an increasing influence on the formation of public opinion in open societies. Oppressive governments supposedly cannot ensure control of the flow of information. This gives their domestic or external adversaries greater opportunities to undermine their control over the population. Yet in the recent years, dictatorial or authoritarian regimes were quite successful in ensuring their stay in power by controlling the flow and content of information among the populace. In many poor countries, the majority of the population, especially in the countryside, relies on radio or TV as the main source of information.

Information has great potential to transform society by affecting the forms of commercial organization and social control and by generating wealth. Information is also becoming increasingly critical to the country's economic competitiveness. Instantaneous global communications and rapid, cheap transportation are creating less hierarchical, more flexible commercial organizations. The emerging networked organizations possess significant advantages over more traditional hierarchical forms of organizations. They are generally less vulnerable to hostile disabling attacks intended to seriously disrupt their functioning or even make them useless. The lack of central control nodes in networked organizations, as well as their physical dispersion and numerous interconnections, makes these organizations very difficult to destroy or neutralize. At the same time, new communications technologies allow these organizations to coordinate actions successfully over vast distances.[119] In contrast, traditional organizations are incapable of quick, flexible responses because of their high degree of centralized decision-making.

The rapid technological advances have greatly increased the importance of the factor of time and information. The side that wins the struggle for time and has an information advantage and quicker reaction is able to surprise the enemy and obtain the initiative. *Information advantage* is the principal factor for enhancing the factor of time and thereby increasing one's freedom of action.[120] New information technologies allow a significant, and in some cases radical, reduction in the time needed to plan, prepare and execute military actions. The time to respond to aggression, to alert one's forces and move them to their deployment or concentration area, is also much compressed. Yet, what is technologically possible is not necessarily applied in practice. The political leadership may vacillate, delay and restrict the employment of one's forces so that the operational commanders cannot act quickly and decisively despite the availability of modern information technologies.

Information allows a dramatic reduction of time in making decisions and of the time required for operational planning. The time it takes to recognize and understand a situation, or a change in a situation, and to identify and evaluate various courses of action is greatly reduced today. One of the key prerequisites for achieving information advantage is to seize and maintain the initiative and react swiftly to any information the enemy might try to use to influence public opinion at home and abroad.

Only with better and more timely information can the operational commander make sound decisions in combat. Information warfare enthusiasts assert that today commanders at all levels can obtain diverse information in unprecedented quantity and quality. By using a mix of military and commercial information technologies, the operational commander can acquire an accurate and timely picture of the situation in the theater and thereby obtain a decisive advantage over his opponent. The more accurate the picture, the smaller the likelihood that the enemy can achieve surprise. The operational commander's decision timelines are increased.[121] Modern information technology has also greatly improved the operational commander's knowledge about the locations and movements of friendly and enemy forces.

Information plays an essential role in the command and control of combat forces. Once hostilities start, commanders should be able to adjust their plans, exploiting opportunities and limiting vulnerabilities.[122] The beginnings of what is called the *common operating picture* for commanders at all levels of command are becoming a reality. Wide-area communications are greatly improved; garbled or incomplete information is being replaced by standardized message sets that offer less chance for confusion or ambiguity.[123]

The main prerequisite for the application of centralized command and control (C2) is that the commander possesses complete, actual, precise and reliable information on the situation.[124] With the advent of wireless telegraph it became possible for higher headquarters to control subordinate forces deployed far away and to direct their movements, often in minute detail. Recently adopted advanced information technologies have changed the relationship between the factors of space and time. The will of the higher commanders can now be applied directly and quickly. Tactical leaders can be reached via radio, wire, fax or satellite communications. They are linked, but they are bound as well.[125]

Information is generated in the course of reducing uncertainty so that decisions can be made. Information allows a commander's decision execution cycle to function and gives direction to the forces' actions to accomplish their mission. Without accurate, relevant and timely information, it is difficult to make a sound decision. Yet despite all the technological advances at the disposal of military commanders, the danger of being ignorant of the actions of friendly and enemy forces is still considerable.

The aim of commanders at all levels should be to shorten their own decision cycle so that the enemy does not act but is always forced to react or is unable to react at all. The essential factor is speed, and higher speed is a matter not only of technology but also of the operational commander's mental agility and ability to act decisively and boldly and, above all, wisely.

Information probably has the single greatest effect on the deployment and logistical support/sustainment of combat forces. More accurate and timely information allows an operational commander to plan, prepare, and execute the phasing of his forces in a theater much more effectively than in the past. The real- or near-real-time knowledge of the status of all aspects of logistics in the theater greatly facilitates planning and executing logistical support and sustainment. Massive amounts of supplies of all kinds are required to fight a high-intensity conventional war with multiservice or multinational forces. Commercial techniques and procedures could be used for real-time knowledge of positions and destinations of critical supplies.[126] Modern information technology can provide deploying units with access to the latest intelligence and information in the theater and the logistics data for follow-on units.

In mobile warfare, it is important to deny the enemy an accurate picture of the situation in the theater. Improved information capabilities increase the operational commander's ability to know the location and movement of the enemy's force. These capabilities can provide vital clues to the enemy commander's intent. While space-based sensors can provide nearly constant coverage of many of the military activities of potential opponents, this capability is still not perfect.[127] An operational commander's knowledge about the enemy's dispositions can greatly help in making sound decisions. Similarly, knowing where the enemy will attack allows the operational commander to timely strike the enemy's forces or prepare a timely counter-concentration or counter-maneuver.

During the planning and execution of the operation, the selection of combat forces should be evaluated, in terms of not only their military suitability and purposefulness but also possible political consequences. Sometimes a tactical success accomplished with unsuitable forces in a sensitive area or at a wrong time can endanger or even doom an entire major operation.[128]

Another aspect of information is cooperation with the media. Military leaders at all levels should be aware that in the information age, all military events and actions are exposed through the media and can considerably affect both public opinion and the employment of forces.[129]

Conclusion

The main characteristic of modern warfare is the continuous expansion of space in which military movements and combat action take place. Combat actions have been extended to encompass not only a ground and sea/ocean surface but also subsurface and air/space. However, in terms of time, "space" has steadily shrunk because of the advent of aircraft, missiles and new means of reconnaissance/surveillance. The importance of the factor of space in planning, conducting and sustaining campaigns and major operations cannot be overstated. Many military enterprises ultimately failed, or the effort required too much time and too many resources to reach a successful outcome, because either the factor of space was neglected or some of its key elements were improperly analyzed and hence led to a flawed operational idea. Despite all the technological advances, physical space cannot be wished away, as some information warfare enthusiasts believe. A large force needs a larger space to move and maneuver. Space, with its distances and physical features, remains a critical factor in the employment of forces for both sides in a conflict. While in the past forces were essentially limited within the boundaries of a given theater, today many elements of the enemy's military and nonmilitary sources of power are located farther away. The significance of the "space" should not be overemphasized, because its ultimate effect depends considerably on the factors of time and force.

The factor of time is the most critical and precious factor in the conduct of warfare. It is one of those rare commodities that, once lost, cannot be recovered. The higher the level of war, the more critical the time factor becomes. At the national-strategic level, time is a vital factor for establishing and developing armed forces, a defense industrial base, and mobilization. Military/naval theoreticians and their civilian counterparts should develop as accurate a vision of the character and duration of a future war as possible. The operational commanders and their staff can most directly influence the time required for the planning, preparation and conduct of a major operation or campaign. There is a great – and sometimes decisive – advantage to acting and reacting faster than the enemy. Disadvantages of space and inferiority in forces can sometimes be remedied by acting faster and accomplishing the assigned objectives within a given time frame. In the past, the pace of warfare changed slowly; hence, there was no drastic compression of the time factor. Today, new technologies will enlarge the area of combat even more and shrink the factor of time even further.

The importance of properly evaluating the factor of force in planning for and conducting a campaign or major operation cannot be adequately emphasized. Yet no task is more difficult than correctly evaluating the capabilities of one's forces, both before and during combat. Gross numbers of personnel or weapons and equipment and their quality are inadequate indicators for realistically evaluating the factor of force. Things are more complex than they often appear. The difficulties are due to the many hard-to-measure or impossible-to-measure elements of combat potential or combat power. Human elements of power – specifically morale and discipline, small-unit cohesion, combat motivation, leadership, doctrine and training – cannot be precisely evaluated. Thus, operational commanders and their staff must make approximate judgments about the true capabilities of their forces, which means they must use very good judgment and experience in assessing their own and friendly forces. In doing so there is no place for falsified data, political judgments, or unfounded optimism. Operational commanders and their staffs should always face realities, no matter how unpleasant; otherwise, the planned campaign or major operation is bound to fail.

Information increasingly affects all three traditional operational factors. Its effect on the factor of space is growing, particularly in the nonmilitary aspects of the situation. This has already had some significant consequences on the use of one's military forces for the accomplishment of operational and strategic objectives in a theater. At the same time, the importance of information should not be overemphasized at the expense of the human factor in the conduct of war. Ultimately, information dominance is useless if the operational commander lacks intellect, wisdom and moral courage in making sound decisions and sufficient determination in executing them.

The value of information should not be underestimated, but it should also not be exaggerated. The sheer complexity and volume of information relationships are increasing and will have a decisive impact on the outcome of campaigns or major operations. Yet information should properly be considered as an aid to, not the master of, operational commanders and their staffs. As in the past, the human element will dominate the conduct of war as long as war is a clash of human wills.

Notes

1 Some Indian scholars are not sure or doubt that Kautiliya existed; however, they generally agree that whoever wrote *The Arthashastra* compiled the wisdom of the Indian civilization over many centuries. *The Arthashastra* was discovered by Dr. R. Shamasastry of Mysore in 1904 who then published them for the first time in 1909; the English translation appeared in 1915; Kautiliya, *The Arthashastra,* ed. rearranged, trans. and intro. L. N. Rangarajan (New Delhi: Penguin Books of India, 1987), 2, 18.
2 Kautiliya, *The Arthashastra*, 625.
3 Hans Jűrgen Usczeck, *Scharnhorst. Theoretiker-Reformer-Patriot* (East Berlin: Militärverlag der Deutschen Demokratischen Republik, 1979), 138–40.
4 Volker Wieker, *Operation, Operative Führung zwischen Taktik und Strategie – Bedeutung und Entwicklung des Begriffes seit Moltke d. Ae. – Bearbeitung aus der Sicht der Landstreitkräfteführung* (Hamburg: Führungsakademie der Bundeswehr, December 12, 1988), 8.
5 Usczeck, *Scharnhorst*, 139.
6 Erich H. Kőnen, *Die operativen Ideen Mansteins hinsichtlich Nützung des Raumes, Gewinnen der Initiative, Schaffen von Handlungsfreiheit und Wahl zwischen offensivem and defensivem Vorgehen. Eine Untersuchung anhand der Beispiele "Rochade" und Schlacht bei Kharkow des Winterfeldzuges 1942/43* (Hamburg: Führungsakademie der Bundeswehr, November 1988), 2; Führungsakademie der Bundeswehr, Arbeitspapier, *Operative Führung* (Hamburg: Führungsakademie der Bundeswehr, August 1992), 19.
7 Inspector of the Army, *Truppenführung*, 3rd ed. (Strausberg: Presse-und Informationszentrum des Heeres, 2018), 65.
8 Gustav Däniker, *Raum, Kraft und Zeit in der Militärischen Kriegführung* (Frauenfeld: Verlag von Huber & Co. Aktiengesellschaft, 1944), 13.
9 Lothar Rendulic, *Grundlagen militärischer Führung* (Herford/Bonn: Maximilian Verlag, 1967), 22.
10 Däniker, *Raum, Kraft und Zeit in der Militärischen Kriegführung*, 29.
11 Arbeitspapier, *Operative Führung*, 14.
12 Center for Military History, U.S. Army, CMH Pub 104-5, *Terrain Factors in the Russian Campaign*, facsimile ed. (Washington, DC: U.S. Government Printing Office, [July 1951] 1986), 1.
13 John M. Collins, *Military Geography for Professionals and the Public* (Washington, DC/London: Brassey's, 1998), 17.
14 Harmut Behrendt, *Die Handlungsfreiheit der militärischen Führung – Mőglichkeiten und Grenzen aufgrund des heutigen Kriegsbildes* (Hamburg: Führungsakademie der Bundeswehr, January 1968), 6.
15 Erich H. Kőnen, *Die operativen Ideen Mansteins hinsichtlich Nützung des Raumes, Gewinnen der Initiative, Schaffen von Handlungsfreiheit und Wahl zwischen offensivem and defensivem Vorgehen. Eine Untersuchung anhand der Beispiele 'Rochade' und Schlacht bei Kharkow des Winterfeldzuges 1942/43* (Hamburg: Führungsakademie der Bundeswehr, August 1992), 4.

16 Däniker, *Raum, Kraft und Zeit in der Militärischen Kriegführung*, 11.
17 Collins, *Military Geography for Professionals and the Public*, 17.
18 Martin Blumenson and James L. Stokesbury, *Masters of the Art of Command* (Boston, MA: Houghton Mifflin, 1975), 25–26.
19 Werner Lange, *Raum, Zeit und Mittel als Faktoren strategischer Zielsetzung. Überlegungen zu Napoleons Feldzug in Russland 1812* (Hamburg: Führungsakademie der Bundeswehr, November 1964), 16.
20 Ministry of Defence (NAVY), *War with Japan, vol. 1: Background to the War* (London: Her Majesty's Stationery Office, 1995), 1.
21 Collins, *Military Geography for Professionals and the Public*, 20.
22 Collins, *Military Geography for Professionals and the Public*, 15.
23 Däniker, *Raum, Kraft und Zeit in der Militärischen Kriegführung*, 14.
24 Lange, *Raum, Zeit und Mittel als Faktoren strategischer Zielsetzung. Überlegungen zu Napoleons Feldzug in Russland 1812*, 2.
25 Friedrich von Bernhardi, *On War of To-Day, vol. 2: Combat and Conduct of War*, trans. Karl Donat (London: Hugh Rees, 1913), 243.
26 Däniker, *Raum, Kraft und Zeit in der Militärischen Kriegführung*, 12.
27 Brereton Greenhous, "The Israeli Experience," in *Case Studies in the Development of Close Air Support*, ed. Benjamin Franklin Cooling (Washington, DC: Office of Air Force History, United States Air Force, 1990), 491.
28 Collins, *Military Geography for Professionals and the Public*, 20.
29 Daniel Frei, "Neutralisierte Zonen: Versuch einer strategisch-machtpolitischen Theorie," *Wehrwissenschaftliche Rundschau* 12 (December 1969), 666.
30 Frei, "Neutralisierte Zonen," 666–67.
31 Frei, "Neutralisierte Zonen," 671–73.
32 Mark W. Janis, "Neutrality," in *The Law of Naval Operations,* ed. Horace B. Robertson, Jr. (Newport, RI: Naval War College Press, 1991), 148.
33 *Neutral internal waters* encompass those waters of a neutral nation that are landward of the baseline from which the territorial sea is measured; a *neutral territorial sea* is like neutral land, because it cannot be used by any belligerent either as a sanctuary or as a base of operations; belligerents are also obliged to refrain from all hostile acts; belligerents are not allowed to move troops, war material, or supplies across neutral territory; troops that enter neutral territory must be disarmed and interned until the end of the conflict; neutral nations can, but are not obliged to, close their ports and roadsteads to belligerents.
34 Frei, "Neutralisierte Zonen," 671–73.
35 D. P. O'Connell, *The Influence of Law on Sea Power* (Manchester: Manchester University Press, 1975), 51.
36 O'Connell, *The Influence of Law on Sea Power*, 28, 162.
37 J. F. C. Fuller, *The Conduct of War 1789–1961* (New York, NY: Da Capo Press, 1961), 50.
38 Lange, *Raum, Zeit und Mittel als Faktoren strategischer Zielsetzung*, 1.
39 Božidar Lukić, "Suština Faktora Vremena i Prostora kao Vojnih Kategorija," *Vojno Delo* 6 (November–December 1974), 35.
40 Lange, *Raum, Zeit und Mittel als Faktoren strategischer Zielsetzung*, 2.
41 Harry Marx, *Raum-Zeit-Mittel als Faktoren strategischer Zielsetzung. Überlegungen zu Napoleons Feldzug in Russland 1812* (Hamburg: Führungsakademie der Bundeswehr, 1964), 11.
42 Friedrich von Bernhardi, *Conduct of War To-Day, vol. 2: Combat and Conduct of War* (London: Hugh Rees, 1913), 232, 242.
43 von Bernhardi, *Conduct of War To-Day*, 224, 229.
44 Jürgen Fehling, *Das Problem der Übertragbarkeit taktisch/operativer Erfahrungen aus der Kriegsgeschichte* (Hamburg: Führungsakademie der Bundeswehr, January 1988), 14.

45 Charles T. Stewart, Jr., "Time as a Concept in Military Strategy," *Military Review* 4 (April 1959), 3; Bernard Brodie, ed., "Implications for Military Policy," in *The Absolute Weapon: Atomic Power and World Order* (New Haven: Yale Institute of International Studies, 1946), 79; T. F. Walkowicz, "Strategic Concepts for the Nuclear Age," *Annals, American Academy for Political and Social Science* (May 1955), 121.
46 Charles T. Stewart, Jr., "Time as a Concept in Military Strategy," *Military Review* 4 (April 1959), 6.
47 Stig Förster, "Der deutsche Generalstab und die Illusion des kurzen Krieges, 1871–1914. Metakritik eines Mythos," *Militärgeschichtliche Mitteilungen* 54 (1995), 62.
48 Fehling, *Das Problem der Űberträgbarkeit taktisch/operativer Erfahrungen aus der Kriegsgeschichte*, 14.
49 Charles T. Crenshaw, *Distinctions Between Tactical and Operational Level of War – Are Some More Important Than Others?* (Fort Leavenworth, KS: School of Advanced Military Studies, U.S. Army Command and General Staff College, May 27, 1986), 19.
50 Wittigo von Dobschütz, "Der Faktor Zeit in der Lagebeurteilung," *Wehrkunde* 11 (November 1970), 581.
51 Hans-Joachim Schubert, "Mehr Raum, weniger Zeit und Kräfte," *Truppenpraxis* 5 (May 1995), 328.
52 Eliot A. Cohen and John Gooch, *Military Misfortunes: The Anatomy of Failure in War*, 1st ed. (New York: Vintage Books, 1991), 102.
53 Ajay Singh, "Time: The New Dimension of Warfare," *Joint Force Quarterly* (Winter 1995–96), 59.
54 Center for Military History, CMH Pub 104-5, *Terrain Factors in the Russian Campaign*, facsimile ed. (Washington, DC: Center for Military History, United States Army, [July 1951] 1986), 7.
55 Center for Military History, *Terrain Factors in the Russian Campaign*, 58.
56 Burkhart Müller-Hillebrand, *Die Improvisierung einer Operation* (*Dargestellt an der Vorbereitung der deutschen Operation gegen Jugoslawien 1941*), Project no. 21a, MS # P-030, ZA/1 1848, Studien der Historical Division Headquarters, United States Army Europe, Foreign Military Branch, BA-MA, November 1951, 70–72, 78.
57 Jőrg Beck, *Schlagkraft – ein Schlagwort? Zur Problematik der Erfassung und Bewertung militärischer Leistungsfähigkeit* (Hamburg: Führungsakademie der Bundeswehr, October 30, 1978), 1.
58 Lange, *Raum, Zeit und Mittel als Faktoren strategischer Zielsetzung*, 1.
59 Marx, *Raum-Zeit-Mittel als Faktoren strategischer Zielsetzung*, 9.
60 Edmund L. DuBois, Wayne P. Hughes, Jr., and Lawrence J. Low, *A Concise Theory of Combat*, 2nd printing (Monterey, CA: Institute for Joint Warfare Analysis, October 1998), 74.
61 DuBois, Hughes, and Low, *A Concise Theory of Combat*, 71.
62 DuBois, Hughes, and Low, *A Concise Theory of Combat*, 74.
63 Charles D. Franklin, *Time, Space, and Mass at the Operational Level of War: The Dynamics of the Culminating Point* (Fort Leavenworth, KS: School of Advanced Military Studies, U.S. Army Command and General Staff College, April 28, 1988), 9.
64 Peter Paret, "Clausewitz," in *Makers of Modern Strategy: From Machiavelli to the Nuclear Age, ed.* Gordon Craig and Felix Gilbert (Princeton, NJ: Princeton University Press, 1986), 202.
65 Paret, "Clausewitz," 202–3.
66 August Winter, "Wägbares und Unwägbares bei der Entstehung von Führungsentschlüssen (I)," *Wehrkunde* 3 (March 1965), 117.
67 von Bernhardi, *On War of To-Day*, 234.
68 von Bernhardi, *On War of To-Day*, 80, 82–83.
69 Trevor N. Dupuy, *Elusive Victories. The Arab-Israeli Wars: 1947–1973* (Garden City, NY: Military Book Club, 1978), 231.
70 Dupuy, *Elusive Victories*, 244.

71 Dupuy, *Elusive Victories*, 233.
72 von Bernhardi, *Vom Heutigen Kriege*, 89, 82.
73 Eliot A. Cohen and John Gooch, *Military Misfortunes: The Anatomy of Failure in War*, 201.
74 Maximilian Fretter-Pico, *Verlassen von den Sieges Gőttern* (Wiesbaden: Kyffhäuser-Verlag, 1969), 75; Dankmut Schinzer, "Verteidigung gegen sowjetische Angriffe," *Truppenpraxis* 1 (January 1980), 23.
75 von Clausewitz, *On War,* 195–96.
76 von Clausewitz, *On War*, 201, 206.
77 Reginald Bretnor, *Decisive Warfare* (Harrisburg, PA: Stackpole Books, 1969), 85.
78 Rajindar Singh, "Mobility as a Factor of War," *Armor* 1 (January–February 1962), 37.
79 von Bernhardi, *Vom Heutigen Kriege*, 83.
80 Historical Evaluation and Research Organization (HERO), *Developing a Methodology to Describe the Relationship of Mobility to Combat Effectiveness, Report Prepared for Research Analysis Organization under Subcontract FY 67-ARD1-1-2* (Fairfax, VA: HERO, December 27, 1966), 30, 31, 57.
81 Historical Evaluation and Research Organization (HERO), *Developing a Methodology to Describe the Relationship of Mobility to Combat Effectiveness*, 45.
82 Trevor N. Dupuy, *Numbers, Predictions and War: Using History to Evaluate Combat Factors and Predict the Outcome of Battles*, revised ed. (Fairfax, VA: Hero Books, 1985), 13.
83 Historical Evaluation and Research Organization (HERO), *Developing a Methodology to Describe the Relationship of Mobility to Combat Effectiveness*, 60.
84 Eliot A. Cohen and John Gooch, *Military Misfortunes: The Anatomy of Failure in War*, 1st ed. (New York: Vintage Books, 1991), 8.
85 Seaton, *The Russo-German War 1941–45*, 436; Earl F. Ziemke, *Stalingrad to Berlin: The German Defeat in the East* (Washington, DC: Center of Military History, United States Army, [1968] 1987), 70–76.
86 Seaton, *The Russo-German War 1941–45*, 436; Ziemke, *Stalingrad to Berlin: The German Defeat in the East*, 86–87, 90, 93.
87 Martin van Creveld, *Fighting Power: German and U.S. Army Performance, 1939–1945* (Westport, CT: Greenwood Press, 1982), 164–65.
88 William D. Headerson, *Cohesion: The Human Element in Combat. Leadership and Societal Influence in the Armies of the Soviet Union, the United States, North Vietnam, and Israel* (Washington, DC: National Defense University, 1985), 3.
89 Robert W. Duffner, "Conflict in the South Atlantic: The Impact of Air Power," *Air University Review* (March–April 1984), 80.
90 Department of Operations, *Sound Military Decision*, 15.
91 von Clausewitz, *On War*, 127, 184.
92 Marx, *Raum-Zeit-Mittel als Faktoren strategischer Zielsetzung*, 16.
93 S. L. A. Marshall, *The Officer as a Leader*, 1st ed. (Harrisburg, PA: Stackpole Books, April 1966), 139.
94 Colmar von der Goltz, *The Nation in Arms: A Treatise on Modern Military Systems and the Conduct of War,* trans. Philip A. Ashworth, 5th German ed. (London: Hugh Rees, 1906), 162–63.
95 John M. Spiszer, "Leadership and Combat Motivation: The Critical Task," *Military Review* 3 (May–June 1999), 66.
96 William L. Hauser, "The Will to Fight," in *Combat Effectiveness: Cohesion, Stress, and the Volunteer Military, ed.* Sam Sarkesian (Beverly Hills, CA/London: SAGE, 1980), 188–89.
97 Hauser, "The Will to Fight," 190.
98 Hauser, "The Will to Fight," 192.
99 Hauser, "The Will to Fight," 194.
100 Dirk W. Oetting, *Motivation und Gefechtswert. Vom Verhalten des Soldaten im Kriege* (Frankfurt am Main/Bonn: Report Verlag, 1988), 42, 55.

101 von Bernhardi, *Vom Heutigen Kriege*, 90.
102 Seaton, *The Russo-German War 1941–45*, 436; Ziemke, *Stalingrad to Berlin: The German Defeat in the East*, 85.
103 Michael C. Vitale, "Jointness by Design, Not Accident," *Joint Force Quarterly* 3 (Autumn 1995), 27.
104 Vitale, "Jointness by Design, Not Accident," 27.
105 Leonard P. Picotte, "Fighting Joint," *Proceedings* (January 1994), 42.
106 Vitale, "Jointness by Design, Not Accident," 27.
107 Vitale, "Jointness by Design, Not Accident," 28.
108 Historical Evaluation and Research Organization (HERO), *Developing a Methodology to Describe the Relationship of Mobility to Combat Effectiveness*, 44, 58.
109 Dirk W. Oetting, *Motivation und Gefechtswert. Vom Verhalten des Soldaten im Kriege* (Frankfurt am Main/Bonn: Report Verlag, 1988), 185.
110 Shigeru Fukudome, "The Air Battle off Taiwan," in *The Japanese Navy in World War II: In the Words of Former Japanese Naval Officers,* ed. David C. Evans, 2nd ed. (Annapolis, MD: Naval Institute Press, 1986), 340–41.
111 Peter Kraska, "Operative Leitlinie 1994. Die neugefassten Fűhrungsgrundsätze des Heeres," *Truppenpraxis* 5 (May 1994), 416.
112 Edward V. Rowe, *Technology's Impact on the Operational Level of War* (Fort Leavenworth, KS: School of Advanced Military Studies, U.S. Army Command and General Staff College, April 26, 2000), 7.
113 Christian Millotat, "Operative Űberlegungen fűr das deutsche Heer in der gegenwärtigen Sicherheitslage," *Ősterreichische Militärische Zeitschrift* 1 (January–February 1996), 59; Kraska, "Operative Leitlinie 1994," 416.
114 Arbeitspapier, *Operative Fűhrung* (Hamburg: Fűhrungsakademie der Bundeswehr, August 1992), 3.
115 Kraska, "Operative Leitlinie 1994," 416.
116 Christian Millotat, "Operative Űberlegungen fűr das deutsche Heer in der gegenwärtigen Sicherheitslage," 60.
117 Karl Ernst Strachwitz, "Wissen ist Macht," *Truppenpraxis/Wehrausbildung* 9 (September 1996), 581–82.
118 Edward Harschberger and David Ochmanek, "Information and Warfare: New Opportunities for U.S. Military Forces," in *Strategic Appraisal: The Changing Role of Information in Warfare, ed.* Zalmay M. Khalilzad and John P. White (Santa Monica, CA: RAND, 1999), 168.
119 Jeremy Shapiro, "Information and War: Is It a Revolution?" in *Strategic Appraisal: The Changing Role of Information in Warfare, ed.* Zalmay M. Khalilzad and John P. White (Santa Monica, CA: RAND, 1999), 119.
120 Reimar Scherz and Theodor Hodouschek, "Kampf um Zeit und Information. Die Digitalisierung des Gefechtsraumes – Der Ansatz des deutschen Heeres," *Truppenpraxis/Wehrausbildung* 11 (November 1999), 751.
121 Edward Harschberger and David Ochmanek, "Information and Warfare: New Opportunities for U.S. Military Forces," in *Strategic Appraisal: The Changing Role of Information in Warfare, ed.* Zalmay M. Khalilzad and John P. White (Santa Monica, CA: RAND, 1999), 165.
122 Manuel W. Wik, *Mobilization for a New Era* (Copenhagen: Royal Danish Defence College, Information Warfare Conference, November 16, 1998), 8.
123 Harschberger and Ochmanek, "Information and Warfare: New Opportunities for U.S. Military Forces," 167.
124 Hans Felde and Peter May, "Auftragstaktik oder Befelhstaktik? – Kriterien für die Auswahl eines Führungsprinzips," *Wehrwissenschaftliche Rundschau* 2 (February 1978), 57.
125 Dieter Stockfisch, "Im Spannungsfeld zwischen Technologiefortschritt und Führungsverständniss. Auftragstaktik," *Marineforum* 12 (December 1996), 11.

126 Harschberger and Ochmanek, "Information and Warfare: New Opportunities for U.S. Military Forces," 166–67.
127 Harschberger and Ochmanek, "Information and Warfare: New Opportunities for U.S. Military Forces," 163–64.
128 Peter Kraska, "Operative Leitlinie 1994," 416.
129 Inspector General of Bundeswehr, *Operative Leitlinie für Einsätze der Streitkräfte (OpLESK)* (Bonn: Ministry of Defense, 1999), 8.

4 Methods

One of the main responsibilities of commanders at any level is to properly determine the method of employing combat forces to accomplish the assigned military objective with the least losses for the friendly side and in the quickest possible time. The larger the scale of the military objective, the larger and more diverse the force, the more time, and the more complex the method of combat force employment. Depending on the scope of the military objective, a distinction is made today among tactical actions, major operations and campaigns. Operational and military-/theater-strategic objectives are principally accomplished by conducting major operations and campaigns, respectively. The methods of combat force employment are the result of the long evolution of warfare. The most radical changes in these methods occurred over the past two centuries, due almost exclusively to the effects of technological advances.

Major Operations

In the western militaries, the term "operation" is imprecise and is widely abused. It can mean many things, such as a training, administrative action, or tactical, operational and strategic mission.[1] This was the reason that the U.S. Army adopted in the early 1980s, termed "major operation" so as to clearly define employment of combat forces to accomplish an operational objective based on the tenets of operational art. Yet there is no consensus in Western militaries on what constitutes a major operation. In the U.S. military, a major operation is currently defined as "a series of tactical actions (battles, engagements, strikes) conducted by combat forces, coordinated in time and place, to achieve strategic or operational objectives in an operational area."[2] For some reason this term is inappropriately applied for non-combat operations as well.[3] A major problem with its definition is that it confused the issue. A campaign, not a major operation, is aimed at accomplishing a strategic objective. Generically, a major operation can be described as consisting of *a series of related major and minor tactical actions aimed to accomplish a single operational objective. It is normally planned and conducted by a single commander and in accordance with an operational idea (scheme).*

Major operations are conducted by each service of one's armed forces. Hence, major land (or ground), air, and naval operations, respectively, can be differentiated.

DOI: 10.4324/9781003333685-4

In some cases, special forces can conduct a major operation. In respect to their main purpose, a major operation can be offensive or defensive. An offensive major operation would normally aim at the destruction or neutralization of the enemy forces. In contrast, a major defensive operation is aimed to prevent the enemy from accomplishing his objective by weakening his forces over time.

Depending on the degree of service participation, a major operation can be planned and conducted by a single service, two or more services (joint) and services of several nations (combined). Single-service major operations are, as the term implies, conducted exclusively or predominantly with combat arms/branches of a specific service. Examples of such major operations are those aimed at capturing an operationally significant position/area, large city/capital, or an economic area on land, destroying or neutralizing an enemy fleet at sea or at its bases, destroying enemy fighters in the air or suppressing enemy air defenses, and defending and protecting political-military centers or industrial areas against enemy attack from the air.

In modern warfare the ultimate success was achieved by close cooperation of two or more services. Although advanced technologies provide unprecedented capabilities, no single weapon or force reaches its full potential unless employed with the complementary capabilities of combat arms/branches of other services – hence the critical need to employ multiservice or joint forces. A major joint/combined operation *consists of a series of major and minor tactical actions conducted by two or more services of a single or several nations and aimed to accomplish the main operational objective. It is planned and conducted by a single joint/combined commander in accordance with an operational idea (scheme).* Normally, major land (or ground) operations are conducted with the participation of air forces. In a case where such operations are conducted in a coastal area, naval forces also take part. All amphibious and anti-amphibious landing operations are inherently joint and often combined.

Principal advantages of major joint/combined operations are complementary capabilities, greater flexibility and, hence, a greater number of options in the employment of friendly forces. A joint force allows the operational commander to pose multidimensional threats to the enemy. The enemy would have a much greater problem countering the capabilities of multiservice rather than single-service forces. One of the principal advantages of having multiservice forces is the operational commander's flexibility in employing these forces asymmetrically, e.g., land versus air, or sea versus land. Hence, it is possible to use strengths against the enemy's weaknesses more effectively or to prevent the enemy from exploiting one's weaknesses. In contrast, symmetric employment of forces often requires substantial numerical superiority and/or much more advanced weapons to achieve success and minimize friendly losses or casualties. In contrast, the employment of dissimilar forces can be extremely lethal.[4]

Among the disadvantages of employing multiservice forces are sometimes considerable differences in service ways of warfare, parochialism, service doctrine, and lack of interoperability and commonly agreed operational terms. Another major problem is often the lack of common logistical support and sustainment. The planning, preparation, and execution of joint operations is more complex than major operations conducted with a single service.[5]

The operational commanders must have full knowledge and understanding of the capabilities of their subordinate forces. They must carefully orchestrate quick, decisive actions; otherwise, success will be elusive.[6] The need for the closest cooperation among services should not lead to the elimination of their cultural differences and way of warfare. The need for conformity should not be imposed at the expense of uniqueness. In short, service cultures should not be suppressed or eliminated but preserved and exploited.[7]

Major operations with multinational forces are more difficult to organize and execute than those with national forces. The reasons are the great and sometimes incompatible differences in national and/or service ways of warfare, doctrine, training, logistical support and sustainment, and rules of engagement, and the difficulties in communicating due to the lack of a common language and common vocabulary. Moreover, operations with nontraditional allies can greatly increase the risk to one's forces. Coalition partners have often very different views about the value of human life. Hence, their treatment of prisoners of war (POWs) and avoidance of collateral damage might differ considerably from one's own. These disadvantages are often compensated for by some substantial advantages. The participation of services of other nations considerably enhances the political, diplomatic, psychological, and informational status of both stronger and weaker partners. By including the forces of other nations, the stronger partner gets access to bases, installations and host-nation support. Perhaps the greatest benefit of major combined operations for the stronger coalition partner is that they considerably enhance the legitimacy of using military forces in the world's public opinion.

The *main* (or principal) and *supporting* major operations can be differentiated. A major land operation is invariably aimed at accomplishing the principal operational objectives of a campaign. The air and naval forces would be employed to accomplish the secondary operational objectives. In the Normandy invasion (Operation Neptune) in June 1944, the amphibious landing was the main major joint/combined operation, while Allied air forces and naval forces conducted major joint/combined supporting operations, respectively. In the Leyte operation in October 1944, the main major operation was the amphibious landing in the Leyte Gulf, while supporting major operations were the Third Fleet's actions in isolating the northwest approaches to the Philippines, conducted between 9 and 17 October, and its actions as distant cover and support from 17 to 26 October, respectively. Likewise, the Far East Allied Forces (FEAF) conducted supporting major air operations by bombing a great number of targets on land and enemy shipping in the northeastern part of the Netherlands East Indies (NEI) prior to the main landing on Leyte and in its aftermath.

In terms of their timing, a major operation can be initial or successive (consecutive). The *initial* major operation is aimed to accomplish the first and most important operational objective. It is then usually followed by a *successive* major operation aimed at accomplishing another operational objective by destroying the remnants of the enemy forces. This is usually the case in capturing a large island or archipelago. In the Leyte operation, the Allied amphibious landing on 17–20 October and until the end of the amphibious phase on 24 October was the initial major operation. This was followed by another major operation aimed to capture the entire island of Leyte. That operation ended on 25 December 1944.

One of the principal features of all major operations is to defeat the enemy quickly and with the fewest losses of one's forces and materiel. The aim is to seize and then retain the initiative and thereby ensure the commander's freedom of action. A major operation is usually characterized by high-intensity actions; skillful maneuvering of one's forces, extensive use of cover/concealment and deception; and all-encompassing operational support.

A major operation is normally planned and conducted by a service or functional component commander. The decision-making process for a major operation is based on the commander's estimate of the situation of a much larger scope. In contrast to the planning of a tactical action, major operations are planned by using a "regressive" (or backward) planning process in which the ultimate objective is determined first and is followed by the determination of several intermediate, usually major tactical objectives. These intermediate objectives are then accomplished sequentially and/or simultaneously.

A major operation requires the employment of two or more combat arms of a single or, in many cases, several services. For a major operation, the closest cooperation between two or more services – jointness – is a key prerequisite for success. In a major operation, the command and control (C2) process is more complex than in the planning and execution of a tactical action. The two most critical elements for the successful outcome of a major operation are the initial deployment of one's forces into a theater and their logistical support and sustainment. Logistics must be properly synchronized with operations; otherwise, success in combat will be wanting.

A major operation is conducted in an area of operations. Yet, the deployment of combat forces can encompass a large part of a theater of operations. In contrast to a tactical action, the *duration* of a major operation is measured in terms of several days or weeks. The operation starts with the beginning of an operational deployment and ends after the assigned operational objective is accomplished the duration of a major operation depends primarily on the correlation of force on both sides, the intensity and effectiveness of the enemy's resistance, and characteristics of the operating area.

A major operation encompasses three major phases, arbitrarily called the precombat, combat, and postcombat phases, respectively. The *precombat phase* consists of pre-deployment and deployment sub-phases. In a major operation, combat forces conduct operational and, rarely, strategic deployment. Deployment is normally conducted from home or forward bases to the area of operational concentration or dispersal. A *combat phase*, especially in case of a major land operation, may consist of two or overlapping sub-phases; hence, there is no clear break in fighting. The time between consecutive phases varies. Postcombat phase is normally used to regenerate the combat potential of friendly forces.

A major operation comprises a series of related major and minor tactical actions that collectively lead to the accomplishment of the assigned operational objective. Each type of major operation is aimed at accomplishing an operational objective with forces that operate in very different physical environments. It is difficult to generalize how each major operation would unfold and what would be its various elements. After the ultimate objective of a major operation is accomplished, operational success is consolidated through an energetic pursuit aimed at destroying or at least neutralizing the remaining enemy forces. Redeployment is conducted during the postcombat phase of a major operation. Then, friendly forces can be redeployed to another area of operations, in either the same or an adjacent theater of operations, or they might be directed to return to their home base. In the postcombat phase there might be a need to reconstitute one's forces heavily weakened in combat. This is not often the case with air or naval forces. Yet reconstitution of these forces might be necessary if losses in ships, aircraft and personnel are so high that units are not capable of conducting their assigned combat mission.

Major Land Operations

A major land operation is the most important method of employing ground forces to accomplish *a single operational objective*. They consist of *a series of related tactical actions conducted by branches of ground forces aimed to accomplish a single operational objective. They are planned and executed by a single ground commander and in accordance to an operational idea (scheme).*

Major land operations are differentiated according to their main purpose and timing. A major land operation can be primarily offensive or defensive in its purpose.

The offensive is a decisive form of war. Even when on defense, one's forces should conduct offensive actions to obtain and maintain the initiative. *Major offensive land operations* should be planned in a situation in which a favorable correlation of forces exists. The objective of a major offensive land operation can be destroying a major part of the enemy army; preventing enemy mobilization; or capturing an operationally important geographic or economic area, the enemy capital, or some other large urban center. They can be planned to preemptively disrupt the enemy's planned offensive. The principal aim is to seize and retain the initiative and inflict a decisive defeat upon the enemy. One should attack the enemy force to throw the enemy off balance and overwhelm him, thereby ensuring his defeat or destruction. The offensive ends when one's forces accomplish their assigned operational objective or risk overreaching and passing the point of culmination. Sometimes inflicting physical damage on the enemy is incidental to the success of a major offensive land operation. Large gains can be achieved by destroying the coherence of the enemy defense, isolating the enemy forces and advancing into the depth of the enemy's defense to secure decisive objectives.[8]

The main prerequisites of a major offensive land operation include at least superiority in combat potential at the point of the main attack. This, in turn, means that the operational commander must accept higher risks elsewhere. Once the operation starts, it should proceed to the utmost to prevent the enemy from recovering from the initial shock, regaining his balance and reconstituting a coherent defense or mounting a counterattack.[9]

A soundly planned major offensive land operation is characterized by the covert movement of one's forces to their assigned assembly area, speed of movement, surprise, high tempo of actions and boldness. The operational commander maneuvers his/her forces to an advantageous position before establishing contact with the enemy force. At the same time, he should organize all-encompassing combat support and maintain high operations security. The enemy should be denied the ability to acquire accurate, timely and relevant information on friendly forces.[10]

One of the key prerequisites for success is to surprise the enemy. This is achieved, among other things, by acting at a place and/or time or in a way that is new to the enemy. Surprise delays the enemy's reaction to one's actions. It overloads and confuses the enemy's commanders and the staff. It can radically, although only temporarily, reduce the enemy's combat power. It can allow the attacker to succeed with fewer forces than he might otherwise require. Generally, it is difficult to surprise the enemy once hostilities start. However, by operating in a manner contrary to the enemy's expectations, one can still achieve surprise by radically altering the structure or tempo of the battle.[11]

The scope of major offensive land operations is usually expressed in terms of width, depth, tempo and duration.[12] Rates of advance are much higher for smaller units. They are generally higher for lightly engaged forces and lower for longer operations. The longer the operation lasts, the slower its average advance rate. The rate of advance is determined by force ratios. The force ratio is the most important single factor governing advance rates. The force's advance rate is dictated primarily by the speed of its governing mode of transportation.[13]

Normally, a *major defensive land operation* is planned and executed in the course of a strategic defensive. It can be conducted as part of the defense of a sector of a continuous front; as an attempt to break out from operational encirclement; and in defense of a large city.[14] It should be prepared in the course of an offensive operation when a continuation of one's attack becomes impossible or not recommended because the enemy has obtained temporary superiority of forces.[15]

The main purposes of a major defensive land operation can be to repulse the enemy offensive, retain currently held positions, buy time and heavily attrite enemy forces, thereby creating prerequisites for going over onto the offensive. During a major defensive land operation, friendly forces might await the enemy's attack by conducting offensive tactical actions to weaken the enemy's forces prior to their major attack. The enemy cannot be defeated by being on the defensive. Ultimate success can be achieved only by eventually going on the offensive.[16]

Defense on land can be used to gain time for reinforcements to arrive or to economize forces in one sector while concentrating forces in another. In either case, a defense or a delay may achieve these purposes. Friendly forces might be compelled to be on the defensive because they are unable to go on the offensive. In that case, they can use the advantages of a favorable position and superior knowledge of the terrain to weaken the stronger enemy forces by counterattacks in combination with fires. Sometimes part of friendly forces may be required to retain key or decisive terrain. Even in offensive operations, an air assault or amphibious force may be needed to defend a deep objective until they linkup with a larger force on the ground. The immediate challenge of any defensive operation is to regain the initiative and thus create the opportunity to shift to the offensive.[17]

In a land campaign, major land operation is the main operation; all others are secondary or supporting operations. In terms of timing, a major land operation can be initial or successive. In the past, major land operations were conducted by army groups, armies, and in many cases army corps or their equivalents. In World War I, major operations were conducted primarily by armies. Major battles were fought by divisions and army corps. The armies ranged in size from 80,000 to 200,000 men. The size of an army group varied from 600,000 to 800,000 men.[18] On the Eastern Front in 1941–1945, major land operations were conducted by several army groups (by the Germans) or fronts (by the Soviets). The army corps operations were part of army operations by the U.S., British, and German armies.[19] Since the end of World War II, the enormous increase in the mobility and lethality of ground forces led gradually to a reduction in the size of forces capable of accomplishing an operational objective. Today, even a single division can accomplish an operational objective, as the example of the 3rd Infantry Division (ID) and 101st Airborne divisions in the war in Iraq in 2003 (Operation Iraqi Freedom) shows.

A major land operation can be conducted from an interior or exterior position. Hence, one's forces operate from either interior and diverging or exterior and converging lines of operation. Each position offers certain advantages but also some disadvantages. In operating along *interior lines*, the attacker can shift the weight of the main effort against enemy forces attacking in different sectors and beat them separately. However, the area available for deploying one's forces can be very limited.[20]

By operating from interior lines, the attacker can strike at widely separated enemy forces while keeping his forces concentrated. This gives the attacker an opportunity to isolate and defeat the separated enemy forces before they can join to form a stronger defense. Interior lines allow fixing or slowing down an enemy force and then massing numerically superior forces at the decisive point in time and space to defeat the enemy force in detail. At the operational level, interior lines allow for sequencing major operations in a manner that takes advantage of an enemy who may be overextended and may have difficulty in enforcing unity of time and objective within the various elements of his command.[21] A major land operation on interior lines would evolve into a penetration when the aim is to defeat one part of the enemy force while keeping the other away. If that is unsuccessful, then there is a danger that the attacker can be attacked from the rear.[22]

A prerequisite for successful operation from interior lines is that the operational commander fully understands the operational features of the terrain and how they affect the movement and combat employment of his forces. He must also be able to visualize the actions of the enemy forces using the same terrain. The operational commanders and their staff can maximize the advantages of a central position and interior lines.[23]

Greater operational mobility is a definite advantage when operating on interior lines. Boldness, resolution and willingness to take high but calculated risks are requirements for success for the operational commander using interior lines. The commander must show great determination and resolution to move units rapidly and anticipate enemy movements to prevent the concentration of the enemy's forces against friendly forces. The greater distances involved require the operational commander to look farther and make decisions that will set the conditions for tactical engagements days and weeks later.[24]

A major land operation conducted along *exterior lines* consists of concentric advances from several directions toward a given physical objective. Because of steady improvements in communications, enemy forces operating from exterior lines can coordinate their actions much better than in the past. They can attack from different sectors and eventually encircle the forces operating from the central position. Operations from interior lines are successful if the enemy cannot properly coordinate the employment of his forces.[25]

A force operating along exterior lines moves at a higher speed than its opponent. However, its force elements must keep within mutual supporting distance to avoid being defeated in detail. Such a major operation also requires sound synchronization of the movement and actions of friendly forces. This is the most demanding thing to do, especially when the distance between one's base of operations and the physical objective is long.

The German theoreticians paid great attention to planning and executing operations from interior versus exterior lines. The term "exterior line" was used for the first time in the German military literature after 1866.[26] Moltke, Sr, differentiated between concentration on the battlefield and off the battlefield. On the day of a major battle, one's forces should advance from separate points and then concentrate on the battlefield. An operation had to be so conducted that one's army needed to traverse only a short distance before striking the front and the enemy's flanks.[27] Moltke, Sr, believed that there was an indisputable advantage to interior lines that pertained if one had sufficient space to meet the enemy in a few marches and time to defeat one part of an enemy force and then turn against the other parts enemy force. Otherwise, there would be a danger of having to deal with the other enemy forces at the same time. One would obtain an advantage by falling in the enemy's rear or flanks. Sometimes, the advantage of interior lines of operation could become a tactical disadvantage in the battle.[28] This happened to Napoleon I in the Battle of Leipzig in October 1813 and to the Austrian General Ludwig Benedek in the battle of Königgrätz (Sadowa) in July 1866.

A major land operation comprises many diverse major and minor tactical actions. Major tactical actions in land warfare are attacks, strikes, raids, engagements and battles. Minor tactical actions include reconnaissance, surveillance patrolling, ambushes, feints and demonstrations. In land warfare, the term attack is used to describe actions aimed at accomplishing a wide range of objectives, from minor tactical to operational. The term also refers to the overall posture of the ground troops along an entire front or a major part of it. Attacks are normally integral parts of engagements or strikes. They might be aimed at capturing key terrain or securing vital lines of communication. Attacks might also be conducted to secure critical war-supporting resources. In other cases, an attack might be conducted to force the enemy to reveal his strengths, dispositions or intentions. A reconnaissance in force may develop into a major attack if initial probes disclose an exploitable weakness in the enemy's defenses. Attacks may also be intended to deceive or distract the enemy. Such diversionary attacks can delay the enemy's identification of and reaction to one's main effort and may induce him to shift forces away from critical areas. Alternatively, an attack may be designed to fix an enemy force in position, thereby preventing interference with friendly maneuvers elsewhere.[29]

A *strike* consists of a series of related swift attacks, carried out simultaneously or almost simultaneously, to accomplish a tactical objective. In the past, a strike was considered part of an engagement or battle; today, however, its importance can approximate that of a major battle or engagement. A *spoiling attack* is a form of attack that preempts or seriously impairs an enemy attack while the enemy is in the process of planning or preparing it. The objective is to disrupt the enemy's offensive capabilities and timelines while destroying his personnel and equipment, not to secure terrain and other physical objectives. A commander conducts a spoiling attack whenever possible during friendly defensive operations, to strike the enemy while he is in an assembly area or attack position preparing an offensive operation or while he is temporarily stopped. Attack can be used by a defending force to disrupt an expected enemy attack and upsetting the enemy's preparation, thereby buying time. Such a spoiling attack may develop into a major offensive operation if the attack reveals an exploitable enemy weakness.[30]

A *raid* is a special variant of a spoiling attack designed to destroy installations or facilities critical to the enemy. It is a form of attack that is usually small-scale, involving a swift entry into hostile territory to secure information, confuse the enemy or destroy installations. It ends with a planned withdrawal from the objective area after the completion of the mission. A raid can also be used to support actions aimed at rescuing/recovering individuals and equipment in danger of capture. Raids might be mounted prior to or in conjunction with other offensive operations to confuse the enemy or divert his attention. They are characterized by surprise, concentration, speed, flexibility, and audacity.[31] A large-scale raid is designed to achieve an operational objective and may be used in combination with a standoff, long-range attack in concert with the direct action of forces operating in the objective area and employing close combat to complete the mission.[32]

An *engagement* comprises a series of related attacks, strikes, and smaller tactical actions, by a single or several combat arms, aimed to achieve a major tactical objective. It is normally an integral part of a battle. A special case of an engagement is a *meeting engagement*, which occurs when one's force engages the enemy force at an unexpected place and time and is itself unprepared for the engagement. Such an engagement is usually the result of faulty reconnaissance and inadequate intelligence. A meeting engagement can also occur when each of the opposing forces is aware of the other's general presence but lacks sufficient information on the precise location, so both make the decision to initiate combat immediately. The key to success in a meeting engagement is speed of action. The movement of one's forces must be combined with both direct and indirect fires. The operational commander must do everything possible to obtain and maintain the freedom to act and thereby exercise the initiative.

A *land battle* consists of a series of related engagements, strikes and attacks, conducted by several combat branches and coordinated in time and place, to accomplish a major tactical or sometimes even operational objective in a given area of operations. A battle is normally planned, but it can be fought without a plan having been drawn in advance. A major battle is an integral part of a major land operation.[33]

An *ambush* is a form of attack by fire or other destructive means from concealed positions against a moving or temporarily halted enemy. A distinction is made between hasty and deliberate ambushes. A hasty ambush is an immediate reaction to an unexpected opportunity. A deliberate ambush is planned as a specific action against a specific target.

A *feint* is a limited-objective attack that involves contact with the enemy. It is conducted for the purpose of deceiving the enemy as to the location and/or time of the actual main offensive action. A feint can vary in size from a raid to a supporting attack. It can be conducted before, during or after the main attack. It may be independent of the main effort. It can be used to cause the enemy to react in one of three predictable ways: employing his reserves improperly, shifting his supporting fires or revealing his defensive fires.

A *demonstration* is an attack or show of force on a front where a decision is not sought. It is aimed to deceive the enemy. A demonstration differs from a feint in that there is no intent to establish contact with the enemy.[34]

Major Air Operations

Major air operations emerged during World War II as the main method of combat employment of air forces to accomplish an operational objective. Several major air operations have also been conducted during regional wars since the end of World War II. Yet the theory of major air operations was largely neglected in the U.S.A. and other Western militaries. Hence, it is not surprising that there is no commonly accepted definition of what a major air operation is in the U.S. military. The term "air operation" and its variations are used extensively but are generally poorly defined or not defined at all in proper operational terms. In contrast, the Soviets and, today, the Russians pay more attention to precisely defining various operational terms, including what they call an "air operation." In one of the Soviet definitions, an air operation was described as

> the entirety of sequenced and synchronized related combat actions according to the objective, place, and time conducted according to the common idea and common plan and within limited time. An air operation is aimed at destroying a major part of the enemy air forces and air defenses.[35]

In generic terms, a major air operation can be described as *a series of related tactical actions conducted by several combat arms of an air force and/or air defense aimed to accomplish a single operational objective. It is planned and conducted by a single air commander and in accordance to an operational idea (scheme).* Normally, a major air operation is an integral part of a land or maritime campaign. In some cases, a major air operation can be planned and executed independently.

A major air operation is normally aimed to obtain and maintain air superiority or air supremacy or to deny the same to the enemy in a given part of a theater. Control of the air is aimed at creating preconditions for the success of the air force's other missions and the missions of land and maritime forces, respectively. Obtaining and maintaining control of the air is one of the supporting objectives in a land/maritime campaign. A major air operation is planned and executed by a service or functional air component commander. This would ensure unity of effort through unity of command. Experience shows that the most effective use of air forces is possible only through a high degree of centralization in planning and mission command in execution.

In terms of its main purpose, an air operation can be offensive or defensive. A *major offensive air operation* is primarily aimed at obtaining control of the air in a given part of the theater. This is accomplished through the destruction or serious weakening of a major part of the enemy air force and ground-based air defenses. The focus is usually on destroying a major part of the enemy fighter strength in the air and on the ground. The weakening of the major part of the enemy air force can in some cases be achieved by destroying enemy air installations and runways.[36]

A major offensive air operation as part of a campaign can be aimed to cut off enemy reinforcements and material to a part of the theater where a new campaign will be conducted. Efforts to weaken the enemy's military-economic potential would consist of many tactical actions throughout the entire war. Such an effort would also include some major air operations aimed at achieving a specific operational objective. A *major defensive air operation* consists of a series of related defensive actions and measures aimed to detect, identify, intercept, and destroy or negate enemy air and missile forces trying to attack or penetrate friendly territory.

The main *characteristics* of a major air operation are decisiveness of actions, employment of all available aircraft and often forces of other services, diverse methods of combat actions and limited duration.[37] One of the greatest advantages of a major air operation is that the commanders can obtain the initiative and therefore enlarge their freedom of action. A major air operation is conducted within a specific time frame, determined during its planning. By conducting a major air operation, more costly and time-consuming attrition warfare at the operational level would be avoided.

The main features of any major air operation are its spatial extent expressed in terms of its width and depth, its duration, and its operational tempo. The width and depth of a major offensive air operation depend primarily on the size of the theater and the number and location of the enemy's and friendly air bases and ground-based defenses. A defensive major air operation would normally be limited to the size and extent of friendly territory and hence would be smaller in scope than a major offensive air operation.

A major air operation can last from several days to several weeks. The Israelis' initial major air operation on 5 October 1967 in the Six-Day War (operation Moked) lasted only a few hours. Sometimes a single initial and several successive major air operations take place over several months to accomplish the ultimate objective, as was the case in the Battle for Britain in 1940 (10 July–31 October).

A major air operation consists of a large number of air-to-air, air-to-ground and ground-to-air tactical actions. The principal tactical actions are air battles, strikes, raids and attacks. An air battle consists of related strikes and attacks conducted by several air combat arms. It is aimed to accomplish a major tactical objective by destroying or neutralizing a major part of the enemy's air forces in the air and/or on the ground. Several combat arms of air forces, air defenses and, in some cases, also land or naval forces can take part in an air battle.

Today, the most prevalent methods of tactical employment of aircraft are attacks and strikes. An *air attack* is conducted by two or more aircraft aimed at achieving a minor tactical objective. A *strike* consists of a series of related attacks conducted simultaneously or sequentially by several air groups of a single air combat arm and aimed to accomplish a minor or, in some cases, major tactical objective. *Successive strikes* are planned when there is not sufficient time for preparing forces for conducting simultaneous strikes. They can be used when one's aircraft are based at several widely dispersed bases or crews are inadequately trained to conduct simultaneous strikes.[38] Successive strikes are most effective against targets with weak air defenses. They are highly effective against targets requiring repeated strikes such as airfields and railway hubs.[39] Successive strikes are simple to prepare and execute. Conditions for the distribution of targets are improved. They offer the advantage of striking the newly detected targets. Each successive tactical group can utilize the effects of the previous group against a weakened enemy. The tactical radius of action of individual strike groups is extended. The conditions for target assignment are improved and the precision of missiles is enhanced. Each aircraft group requires the employment of a larger number of supporting aircraft. The disadvantage is that the aircraft leader has little time to prepare to execute the strike. In conducting successive strikes against the targets on the ground, aircraft must spend more time over enemy territory. Another disadvantage is that each strike group requires the employment of a large number of supporting aircraft.[40]

Simultaneous air strikes are aimed to destroy or neutralize large and important targets.[41] To achieve high effectiveness, all groups must cooperate closely. Disorganized and untimely strikes would lead to high losses. Strikes would last only for a few minutes.[42] The main advantage of simultaneous strikes is their higher effectiveness and small losses for one's forces.[43] Their main disadvantage is the longer time for planning and accomplishing the mission. The enemy would have more time to react. They also require supporting aircraft, but not as many as in the event of successive strikes.[44]

In Western terms, the main methods used to destroy enemy air force are massive strikes, fighter sweeps, escort missions, and the destruction or neutralization of the enemy defenses. *Massive strikes* against enemy aircraft in the air and on the ground and air-related ground facilities/installations were the principal method used in the past. With the ever-increasing range and lethality of air weapons, the number of aircraft and sorties required to destroy the targets in the air and on the ground has been steadily reduced. Today, far fewer aircraft are required for destroying or neutralizing a major part of enemy air force, compared with the numbers required in past wars. Massive strikes should be conducted by using diverse forces and weapons against selected targets. They can be conducted simultaneously or sequentially or in combination. They should be conducted with the largest available force and from different directions. Most critical to the success of the entire operation are the strikes at the beginning of hostilities.

A *fighter sweep* is aimed to seek out and destroy enemy aircraft in an assigned air area of operations. It can be autonomous but is most effective when linked to air-to-surface action. The air-to-surface attack force acts as bait, drawing the enemy fighters into the air where they can be destroyed. The aim is not only to destroy enemy aircraft but also to kill aircrews.[45] Generally, two main methods of fighter sweep differentiated are an area sweep and a route sweep. A fighter sweep is aimed to seek out and destroy enemy aircraft in an air area of operations. It can be autonomous but is most effective when linked to air-to-surface action. The air-to-surface attack force acts as bait, drawing the enemy fighters into the air where they can be destroyed. The aim is not only to destroy enemy aircraft but also to kill aircrews.[46] Fighter sweep can accompany or not accompany a strike package.[47] The following two main methods of fighter sweep are differentiated: an area sweep and a route sweep. The *fighter area sweep* can be conducted independently, or in support of another attack force by decoying or destroying enemy aircraft that pose a direct threat. A *route fighter sweep* is aimed to provide direct support to an attacking force of friendly aircraft by clearing the attacking aircraft's planned route of the presence of enemy aircraft.[48]

Escort involves the use of fighter aircraft to protect other friendly aircraft during a mission. Escort aircraft are used to provide an air-to-surface attack force and thereby force the enemy to break through the escort before he can engage the attacking force. They can also be used for protecting friendly airborne surveillance or tanker aircraft.[49]

Destruction or neutralization of enemy air defenses is one of the most important but also most complicated and time-consuming tasks in a major offensive air operation. Air superiority can be considerably limited unless the enemy air defenses are sufficiently weakened. To obtain air supremacy, the enemy air defenses must be destroyed to such a degree that friendly aircraft can operate without undue losses or interruption.[50]

The targets can be engaged by conducting independent searches, free searches, on call and ambush missions. The *independent search* is aimed to detect and destroy important but weakly defended targets whose location is unknown but must be destroyed within a short time.[51] It is characterized by high speed and decisiveness of action.[52] Its major advantage is free selection of placed and time of strike. By combining reconnaissance and immediate destruction there is a higher chance of achieving surprise. Disadvantages of an independent search are high requirements for training of the flight crews and great difficulties in overcoming the enemy air defenses.[53]

Single fighters, fighter-bombers, or helicopters can be employed for conducting a *free search* (or "roving" patrol). The pilots are allowed to select the targets and method of accomplishing their mission. The aim is to detect and localize important air, ground or sea targets in the assigned area of independent search.[54] Success depends on surprise, speed and boldness.[55]

Fixed-wing aircraft or helicopters kept in a state of high combat readiness may conduct *on-call* missions. This method is applied in the case of the newly detected targets. Aircraft are called into action from alert either on the airfield or airborne. Due to time constraints, only a rough estimate of the required number/type of aircraft is conducted. Other elements of the mission are determined shortly before takeoff or after the aircraft are airborne. This method is highly effective in conducting close air support missions.[56]

An *ambush* is conducted by a single fixed-wing aircraft or helicopter or a small group of helicopters flying in an area in which the enemy cannot observe either visually or by radar. The aim is to destroy the detected targets in a surprise attack. The aircraft follow the route of their enemy counterparts, the movement of enemy tanks or other vehicles and the actions of troops.[57]

Major Naval Operations

There is no consensus on a definition of what constitutes a major naval operation. The term "naval operation," so often used by the U.S. Navy and other Western navies, is too broad and imprecise. In comparison, in the early 1930s and through the 1980s the former Soviet navy devoted substantial and consistent efforts to developing a theory of what it called "naval operations." Naval operations constituted the very heart of the Soviet naval operational art. In one of the many Soviet definitions of the term, a naval operation (*morskaya operatsiya*) was described as a series of naval battles, engagements and strikes coordinated in terms of objective, place and time and conducted in an assigned area (zone) of an oceanic or sea theater of military action (TV; theater of war, in Western terms). It was conducted by a fleet commander either independently or in cooperation with formations, forces or troop units of other services, according to a common idea and plan for accomplishing a single operational or operational-tactical task.[58]

In generic terms, a major naval operation can be described as *a series of related tactical actions aimed to accomplish a single operational objective. They are planned and conducted by a single naval/maritime commander and in accordance to an operational idea (scheme).* Major naval operations are normally an integral part of a maritime campaign. Sometimes they can be conducted independently. They are normally aimed to achieve a decisive result quickly and at the least losses for one's forces.[59]

A major naval operation is aimed to destroy/neutralize a major part of the enemy's combat forces. If fully successful, the result of a major naval operation would be a radical or drastic change of the situation in a given maritime theater of operations. Yet if a major naval operation is only partially successful, the operational situation is likely to remain as it had been, as the Battle of Jutland on 31 May–1 June 1916 illustrates. A major naval operation can also have a strategic effect as well. The operational objective of the surprise attack on Pearl Harbor on 7 December 1941 was to destroy enough of the U.S. Pacific Fleet to prevent it from interfering with the Japanese invasion of the Philippines. However, the unintended and highly negative (for Japan) effect was to shift American public opinion from isolationism to unequivocal support for complete victory over Japan.

Major naval operations can be conducted on the open ocean or in the littorals (waters contiguous to the continental landmass, and peripheral, enclosed, and semi-enclosed seas, respectively). Most major naval operations have taken place fairly close to continents, large islands or oceanic archipelagoes. Many have taken place in enclosed or semi-enclosed seas (collectively called "narrow seas"). Today, because of enormous advances in the range and lethality of weapons, even coastal navies composed of small surface combatants, submarines, and land-based aircraft can conduct major naval operations. Major naval operations in the littorals are generally more complicated to prepare and execute than those on the open ocean, because they involve diverse forces of all services of the country's armed forces. They also differ from major naval operations on the open ocean because of the much smaller operational space involved and the smaller forces used.

A major naval operation can be offensive or defensive in terms of its purpose. A *major offensive naval operation* is usually planned and conducted in the initial phase of the war to obtain sea control. An offensive major naval operation can also be planned as part of a defensive maritime campaign, as was the Japanese commitment of a major part of the Combined Fleet in defense of the Philippines in October 1944. Major naval operations against enemy maritime trade and amphibious landing operations are inherently offensive in their purpose.

A *major defensive naval operation* is usually planned and executed when one side is forced onto the defensive at sea by its permanent or temporary weakness. Then, the weaker side may conduct major defensive naval operations to oppose enemy attacks on its naval bases, amphibious landings and attacks on coastal installations or facilities, and to carry out major evacuation of friendly troops and civilians. Both sides in a war at sea will occasionally mount major naval/joint operations in defense and protection of maritime trade.

Based on the degree of participation of services, naval, joint (multiservice) and combined (multinational) major naval operations can be differentiated. A major naval operation, as such, is conducted predominantly by a navy, as the examples of the battles of Cape Matapan (March 1941), Coral Sea (May 1942), Midway (June 1942), and the Philippine Sea (June 1944) illustrate.

A *major naval/joint operation* is planned and conducted by forces of the navy but with substantial participation by other services. In a maritime theater encompassing a large ocean or sea area, major naval operations would involve land-based aircraft. In contrast, major naval operations in littoral waters are likely to require the participation of all three services. All major amphibious landing operations are inherently joint as are major operations against enemy maritime trade or in defense of maritime trade.

A *major naval/combined operation* is conducted with two or more navies or services of the armed forces of two or more countries. Because of the downsizing of most navies, such operations might well become routine even in a regional war. Examples of major naval/joint operations are major amphibious landings such as the invasion of Sicily (Operation Husky) in July 1943, the landing in Normandy (Operation Neptune) in June 1944 and the UN amphibious landing at Inchon in September 1950 (Operation Chromite).

In terms of their timing, initial, successive, and preliminary major naval operations are differentiated. The *initial* major naval operation is planned and conducted to accomplish the first, and most operational, objective in a maritime The Allied invasion of Guadalcanal on 7 August 1942 (operation Watchtower) was the initial major naval/joint operation in the projected Solomons Campaign (August 1942–December 1943).

A *successive* (or *consecutive*) major operation normally starts during or shortly after the initial major naval operation. Its purpose is to consolidate or expand the operational success of the preceding major naval operations. The U.S. invasions of the Marshalls (Operation Flintlock and Operation Catchpole) and the Marianas (Operation Forager) are examples.

A *preliminary* major naval operation is usually planned to isolate a maritime area of operations in which a new campaign or major naval operation would be conducted. The purpose is often to prevent the arrival of enemy reinforcements in troops, aircraft, or materiel. Another purpose might be to deceive the enemy as to the intended sector of the sector of main effort. Task Force (TF)-38's series of air strikes and other actions against the Ryukyus, Formosa and Luzon between 10 and 17 October 1944 constituted a preliminary major naval operation – that is, with respect to the invasion of Leyte.

In generic terms, the ultimate objective of a major naval operation today in the case of a high-intensity conventional war at sea could be as follows:

- Destroy the enemy fleet at sea and/or its bases.
- Capturing part of the enemy coast/island/strait.
- Defense of the coast/island/strait.
- Destroy enemy coastal installations/facilities.
- Destroy the enemy maritime trade.
- Defend and protect maritime trade.
- Destroy enemy sea-based strategic nuclear forces.
- Defend and protect one's sea-based strategic nuclear forces.

A major naval operation aimed at *destroying an enemy fleet at sea and/or in its base* is the quickest and most effective – but most difficult – way to establish sea control in a given maritime area of operations. Such operations can be conducted on the open ocean or in an enclosed/semi-enclosed sea (popularly called "narrow sea"). Most fleet-on-fleet encounters in World War II took place when one fleet provided distant cover and support of a major amphibious landing, as happened at Leyte in October 1944, or when the stronger fleet used the landing to lure an enemy fleet into a "decisive" battle, as the Japanese Combined Fleet attempted in Midway–Aleutians Operation, which led to the battle of Midway in June 1942.

A major naval/joint operation to destroy an enemy fleet might be necessary when the attacker provides support of the flank to his forces operating in the coastal area or when the enemy fleet is forced to operate in a certain sea or ocean area. If the stronger side at sea attempts to isolate certain sea/ocean areas, then the objective could be the destruction of the major parts of the enemy fleet forces. Likewise, though a major operation to destroy the enemy fleet at sea is difficult for a numerically weaker side, it might decide to plan one in order to weaken or lift a blockade. In a typical narrow sea, such an operational objective could be accomplished by massive strikes by surface combatants, submarines and land-based aircraft. Preparation time would be short because of rapid changes in the situation at sea and in the air. Such an operation would most likely encompass a single phase, with one or several synchronized strikes by naval forces, aircraft, and coastal missile or artillery batteries.[60]

Today, nuclear-powered attack submarines (SSNs), modern conventionally powered attack submarines (SSKs), land- and carrier-based aircraft and surface ships armed with long-range cruise missiles are the most effective platforms for destroying enemy ships in their bases. Attack submarines armed with antiship cruise missiles (ASCMs) or land-attack cruise missiles (LACMs) can strike enemy naval bases from several hundred nautical miles away.

In the initial phase of war in an enclosed sea theater, the principal objective of a major fleet versus fleet operation would be to obtain a desired degree of sea control, thereby creating favorable conditions for supporting friendly ground forces on the coast.[61] This would be accomplished by destroying the enemy's surface forces and submarines; destroying enemy aircraft on the ground; enemy antisubmarine forces at sea and in their bases; and suppressing or destroying enemy air defenses.[62]

The objective of a *major amphibious operation* is capturing an operationally significant position on the opposed shore such as large island, small archipelago, one or both shores of a major strait, or capturing a major naval base/port. It can also be aimed at capturing a large part of the enemy's coast, thereby creating preconditions to seize some objectives deep into the enemy's littoral area. They are one of the most complex undertakings to plan and execute. They are inherently joint in character, regardless of their purpose or the size of the forces involved. Amphibious landings within a typical narrow sea are conducted across much shorter distances and are normally smaller than those mounted against coasts on open oceans. Sometimes, a major landing in the littorals can have a strategic effect, in case of the opening of a new front as was true of the Allied Normandy invasion of June 1944, or causing a radical change in a strategic situation, as was in the case of the UN landing at Inchon in September 1950.

A *major anti-amphibious operation* would have just the opposite objective of a major amphibious operation. Ground forces and land-based aircraft would have the most critical role in such an operation. The mission of naval forces prior to the enemy landing would include reconnaissance/surveillance and detection/tracking of the enemy amphibious task force at sea, conducting strikes aimed to destroy enemy force of distant cover and supporting a screen of the amphibious force and amphibious force itself, and conducting defensive mining of the approaches. After the landing and battle ashore naval forces efforts would be focused on destroying enemy forces on the coast and providing surface fire support to defending friendly troops.

A major naval operation aimed to *destroy enemy coastal installations/facilities* can be conducted as a preliminary to an amphibious landing or as an integral part of a naval blockade. Such major operations can be conducted either with naval or air forces or with both jointly. The primary objective can be destruction or neutralization of the enemy naval or commercial ports, airfields/airports, shipbuilding/ship-repair facilities, rail and road traffic, and shore/offshore oil and gas production, and refining industries.

Today, longer-range and more capable carrier- and land-based aircraft, armed with cruise missiles and "smart" bombs, make it possible to attack the enemy coastal installation and facilities along a large part of the enemy coast and far into the depth of his defenses. Major navies today have a much greater ability than in the past to attack a variety of targets far in the enemy's operational and even strategic depths.

Major naval operations on the open ocean and in littoral waters are conducted to destroy/weaken the enemy maritime trade or to defend and protect friendly maritime trade. The emerging task is destruction or defense of undersea infrastructure (internet, gas/oil pipelines, mineral deposits, etc.). A major naval operation aimed to *destroy enemy maritime trade* would focus on attacking the enemy military and commercial shipping at sea/ports plus selected shore-based elements of maritime trade. It would be normally conducted when sea control is in dispute, or the enemy has slight superiority. Such an operation would consist of series of massive missile, torpedo, air and artillery strikes, plus offensive use of mines. Major naval/joint operations will be conducted only occasionally. Efforts to destroy enemy military-economic potential at sea cannot be effective unless conducted systematically over a relatively large sea/ocean area. The most important targets will be enemy convoys, major ports and cargo loading/offloading facilities; rail/road junctions in the littoral area; shipyards/ship-repair facilities; shipping-related industries and destroying or neutralizing critically important digital support structure.

One of the most effective methods of preserving maritime component of friendly military-economic potential is by conducting major naval operations aimed to *defend and protect maritime trade*. Specifically, such operations would be aimed at preventing interruption or cutoff of shipping carrying troops, weapons, military equipment, or strategic raw materials. In some cases, such an operation can be aimed to protect friendly shipping in coastal waters.[63] Ideally, all elements of friendly military-economic potential at sea should be defended and protected. However, this is not possible because one's resources are limited. Hence, a major naval/joint operation should be aimed to defend and protect selected elements of friendly maritime trade. Duration of such an operation would depend on the length of the shipping routes, how long ships stay in ports and anchorages, and convoy-assembly time.[64]

Today, only large navies have the capability to mount major naval operations aimed at *destroying the enemy's sea-based strategic nuclear forces* – that is, ballistic-missile submarines (SSBNs) and their supporting elements. Such major operations might be focused on destroying enemy SSBNs either in their basing areas, in their patrol zones or in transit between them. Likewise, a major naval operation might be conducted to *protect one's sea-based strategic nuclear forces*. Such a major operation would most likely be carried out predominantly by naval forces, specifically SSNs in the open ocean and patrol aircraft overhead. However, in areas closer to friendly or the enemy's coast, such operations will probably be more joint, because air forces will be expected to take part.

A major naval operation consists of a series of naval tactical actions with and without the use of weapons (reconnaissance/surveillance, search and patrolling). Naval tactical actions can be planned or unplanned. They can be conducted at any time regardless of the ratio of forces in a given theater. Naval tactical actions are much simpler to plan and execute than major naval/joint operations or maritime campaigns. They are conducted in a sea/ocean area varying in size from a combat zone/sector to a maritime area of operations.

There are several methods used for the employment of naval forces aimed to accomplish tactical objectives. Yet there is lack of common agreement both in regard to terms and to their true meaning. Generally, the principal methods of tactical employment of naval forces with the use of weapons are attacks, strikes, raids, engagements and battles. The oldest and the simplest tactical action of naval forces is a *naval attack – a combination of a maneuver and decisive use of weapons/sensors aimed to accomplish a minor tactical objective*. It is normally a part of a strike or an engagement/naval battle. It could be also conducted independently.[65]

A naval attack can be conducted by a single or several types of ships, submarines, or aircraft. It is intended to destroy or neutralize a single enemy platform or a tactical group or a target ashore. An attack can be conducted by using missiles, torpedoes, guns and bombs singly or in combination. A gunnery attack is the use of shipboard or coastal guns against targets at sea or on land. It could be conducted either during a platform's movement or from a stationary position. Surface ships, submarines, aircraft or their groups can carry out a missile attack against enemy ships or coastal installations from one or more sectors. A torpedo attack is conducted by surface ships, submarines or aircraft or their groups against hostile warships or merchant vessels. An air attack is conducted with a single fixed-wing aircraft or helicopters or their groups against sea or ground targets, independently or in cooperation with other forces.

A *naval strike* consists of *a series of successive or simultaneous attacks coordinated in time and place, to accomplish a major tactical objective*. It is today a dominant method of combat employment for naval combat aircraft and missile-armed submarines and surface combatants. It is characterized by the rapidity of action (often measured in minutes), swift maneuvering of platforms and quick firing of weapons.[66] The main features of a well-planned strike are surprise, purposeful tactical related actions of the participating forces, concentrated and relatively simultaneous employment of weapons against selected targets, skillful maneuver and exploitation of the weaknesses in the enemy defenses.[67] The success of a strike can be anticipated with a higher degree of certainty than can the outcome of a naval battle. Strikes allow destruction of the targets in the shortest possible time. They can possibly destroy an enemy force that in the past required the fighting of a naval battle.

Regarding the number of platforms, sequencing, and number of targets, individual, group, concentrated and massed strikes are differentiated. As the term implies, an *individual strike* is conducted by a single platform or group of platforms using a single type of weapon against a single target. A *group strike* is simultaneous use of several groups against several targets with the use of single-type or multitype weapons. A *concentrated strike* consists of a series of successive strikes of several groups against a single target by using single-type or multi-type weapons. A *massed strike* is defined as the simultaneous employment of a large number of the same type of or multitype weapons by one or several groups against the main target in a naval battle. In theory, a strike could be carried out against an unprepared enemy, or an enemy who is ready to fire his weapons, or an enemy after he has fired his weapons. In terms of the relative positions between the attacking platforms and the target, flanking and frontal strikes are differentiated. A distinction is also made among preliminary, main and secondary strikes in regard to their importance and timing.

A strike is usually conducted by a small number of platforms of a single type of force – for example, missile craft, submarines or attack aircraft. A strike can be conducted using conventional weapons or weapons of mass destruction (WMD). A conventional strike is carried out using missiles, torpedoes, guns, bombs or a combination of these. In a combined strike, the longer-range weapons are usually used first, to weaken the enemy's defenses, followed by firing the short-range weapons to finish off the target. Like a naval attack, a naval strike can be conducted from a single or two, three or more attack directions or sectors.

A *naval raid* can be described as *a covert penetration of the enemy-held coast, aimed to accomplish a tactical objective followed by a swift withdrawal.* It is a broader variant of naval strike. It could be conducted independently or as an integral part of a major operation. A naval raid is usually conducted when the enemy has general sea control but not local control. Naval raids are normally conducted by a weaker side at sea. A stronger side might conduct raids to divert the enemy's attention or force the weaker enemy to react in the sector of secondary effort.

When an integral part of a major naval/joint operation, a raid is usually part of the deception effort. If conducted independently, the objective could range from destruction/neutralization or harassment to collection of combat information, capture/killing of some important enemy commanders or political leaders, demonstration, recovery or rescue of friendly personnel. In contrast to attacks and strikes, a naval raid invariably aims to achieve a psychological effect (boosting one's morale and diminishing the enemy's). Forces assigned for a naval raid might include fast and advanced combat craft or slow and old and expendable ships, surface/subsurface craft, helicopters and a variety of unmanned vehicles (UAVs/USVs/UUVs).

Raids could be intended to destroy some critically important enemy naval/military or economic facility/installation or any target that requires high precision and is not susceptible to destruction by other means; testing enemy beach defenses prior to a larger landing at some future date; prevent the enemy from further use of some important naval base or anchorage; capture or decapitate some high civilian or military/naval leaders. Any naval raid usually has a positive psychological effect for one's force/population and negative effect on the enemy's forces/population. In some cases, the primary objective of a naval raid might be to raise the morale of one's forces and demoralize the enemy forces. Such raids could be conducted at the beginning of the hostilities or after a string of setbacks suffered by one's forces. Some naval raids might even generate large psychological effects.

The term *naval engagement* is widely used, but usually its meaning is not well defined or understood. In U.S. military parlance, it is used interchangeably with the terms "encounter" and "attack." In the proper understanding of the term, a *naval engagement* is a major clash of opposing forces that is larger than an attack/strike and smaller than a battle. It consists of *a series of related strikes/counterstrikes and attacks/counterattacks by a single or several naval combat arms, aimed to accomplish a major part of the objective in a naval battle.* Offensive and defensive, air, antiair, naval and combined arms engagements are differentiated. The attacks and strikes of the U.S. and Japanese aircraft against the opposing fast carrier groups during the Port Moresby–Solomons Operation (the Battle of the Coral Sea for the Allies) on 8–11 May 1942 and in the Midway–Aleutians Operation (the Battle of Midway for the Allies) on 4 June 1942 are examples of naval engagements that decided the outcomes of the respective major naval operations.

A *naval battle* is the highest and most complex method of tactical combat employment of naval forces. Until the advent of submarine and aircraft, it was fought only on the sea surface. Since then, a naval battle has encompassed all three physical mediums or dimensions. A naval battle can be defined as *a series of mutually related attacks/counterattacks, strikes/counterstrikes, and defensive measures, coordinated in time and place to accomplish a major tactical objective.* It can be fought exclusively on the surface, or it can encompass subsurface and airspace. It can be conducted on the open sea/ocean or in the littoral waters. A naval battle could be fought when one's forces possess local sea control, when the enemy has sea control, and in a situation when a state of uncommanded sea prevails.

A naval battle is normally fought with the participation of two or more naval combat arms, and often combat arms of other services (air force, troops defending the coast). In the littorals, ground forces and coastal missile/gun batteries can take part in a naval battle. In a naval battle, the main forces would be tasked to destroy the enemy's main forces and expand combat success by completing the destruction of the remaining enemy forces. Supporting forces would conduct strikes/attacks against the screen of the enemy's main forces and defend friendly main forces against enemy strikes/attacks.

The main purpose of a naval battle could be offensive or defensive. An offensive naval battle could be to destroy (1) a major tactical force at sea; (2) an amphibious task force at sea; (3) a major convoy; or (4) forces defending a major naval base/port. A defensive naval battle is conducted (1) in defense of a major tactical force at sea; (2) against an amphibious landing; (3) in defense of a major convoy; or (4) in defense of a major naval base/port.

A naval battle can be fought as part of a major naval operation or independently. The battles in the Sibuyan Sea, Surigao Strait, off Samar, and off Cape Engaño fought on 24–25 October 1944 were all part of the Leyte operation. In contrast, most of the naval battles fought during the Solomons campaign were fought within a strategic framework but were not an integral part of major naval/joint operations.

Between 9 August 1942 and 25 November 1943, fifteen major surface actions were fought in the waters around Guadalcanal, New Georgia, and Bougainville. All battles but three were fought in the night. No fewer than seven naval battles and engagements were fought in the struggle for Guadalcanal alone. The Japanese were much better than the Allies in night fighting and the use of guns and torpedoes in combination. They won or achieved a draw in ten naval battles.[68] A great majority of naval battles took place during the Japanese attempts to bring in fresh reinforcements and supplies or to bombard the Allied positions on the coast. In the five-month-long struggle for Guadalcanal, the Japanese lost two battleships, three cruisers, twelve destroyers, sixteen transports and more than one hundred aircraft.[69]

The main features of a modern naval battle are decisiveness and high intensity of actions, sudden and radical changes of the situation, and the use of diverse tactical methods and procedures. A modern naval battle would have a large spatial extent because of the participation of two or more naval combat arms.[70] It would be characterized by the use of long-range and highly precise and lethal weapons, such as ASCMs/LACMs, and advanced torpedoes. In contrast to a major naval operation, during a naval battle there would be no clear break in fighting.

The common characteristics of a major naval operation are the complexity of its planning, preparation, and execution; the involvement of diverse naval combat arms and often combat arms or branches of other services. Today, a major naval operation would be characterized by the decisiveness and high intensity of actions, relatively large spatial extent, sudden and drastic changes of the situation. The objective of a major naval operation should be accomplished within relatively short time. Hence, there is an absolute need to obtain and then maintain the initiative.[71]

A major naval operation on the open ocean will be characterized by its large spatial extent and participation of SSNs and aircraft carriers, heavy surface combatants, and long-range land-based aircraft. The long distances between the home bases and the operating area would require long transit times. The critical task would be to overcome the enemy air defenses and ASW defenses.[72] In contrast, a major naval operation in a typical narrow sea would differ considerably from one on the open ocean because of the much more complex physical environment and short distances. The combat will take place in the entire theater; on the surface, subsurface, on the coast and in the air. The encounters will be sudden, short and violent. Combat actions will occur largely at night or in bad visibility. That necessarily means fighting at close range. Covertness of action and the opportunities for achieving surprise would be considerably enhanced because of short distances and use of long-range cruise missiles and heavy torpedoes.

Campaigns

For most of recorded military history, wars were won or lost by conducting campaigns. In the past, a war consisted of a single campaign, which, in turn, comprised several major battles. However, upon the advent of mass armies and until relatively recently, most wars consisted of several campaigns. The apparent trend today is to conduct a single campaign that would end hostilities. In contrast to tactical actions and many major operations, campaigns are conducted with the participation of all services and, in some cases, multinational forces.

In generic terms, a campaign can be defined as *a series of related major operations (land, air, naval) and numerous tactical actions aimed to accomplish a single military-/theater-strategic objective. It is planned and conducted by a single operational commander and according to an operational idea (scheme).* Campaigns are invariably conducted with participation of two or more services of the armed forces and very often with multinational forces. In a general war, a campaign would represent a distinctive phase of the war.

In the era of the Napoleonic Wars (1803–1815), the employment of army divisions and army corps provided a more flexible organizational structure, while theaterwide logistics gave large formations more mobility. A campaign consisted of a single or several decisive battles and a series of related smaller tactical actions. A particular area of the study of tactics was the decisive battle – a series of related attacks and engagements fought in a relatively limited area over several days. The objective was to destroy a major part of the enemy's army and thereby decisively influence the outcome of a campaign or even the war. Strategy was then primarily concerned with planning and conducting land campaigns. The armies played the principal role, while the navies were employed in a supporting role (amphibious landings, bombardment of forts, support for the army on the coast, etc.).

Depending on the predominant characteristics of the physical space (or "medium") in which combat actions take place, land (continental) and maritime campaigns are differentiated. In a conventional high-intensity conventional war, a *land campaign* consists of a several related land (ground) and air/space major operations, while a *maritime campaign* consists of several naval and air/space major operations. In both cases, many tactical actions, some planned and some unplanned, will be conducted as well. A blue-water navy or coastal navy bordering an enclosed or semi enclosed sea would conduct major operations or tactical actions as part of a land campaign; such was the case with the German and Soviet navies in the Baltic and Black seas in World War II. Allied navies took part in a land campaign fought in Italy in 1943–1945. In both land and maritime campaigns, the third dimension – air/space – and, increasingly, the fourth dimension – cyberspace – are integral and indelible parts of the land or maritime campaign. The air elements are the most mobile and flexible instruments, and flexibility and mobility are essential to successful campaigning.[73]

Depending on the nature of the military or theater-strategic objective, a campaign can be *offensive* or *defensive*. In a high-intensity conventional war, a land campaign can be aimed either to defend national or captured territory or to gain control of the enemy's territory. The Germans conducted an offensive campaign in their invasions of Poland in September–October 1939, Denmark/Norway in April–June 1940, the Benelux countries/France in May–June 1940, Yugoslavia/Greece in April–May 1941 and Soviet Russia in June 1941. In each of these examples, the defenders conducted a defensive land campaign.

A maritime campaign can be aimed to capture an enemy-held large island or archipelago chain; or attacking the enemy's or protecting and defending one's maritime trade. The Japanese conquest of the British Malaya/Singapore, the Philippines and the Netherlands East Indies (NEI) in 1941–1942 were each an example of an offensive maritime campaign. The Allied capture of the New Guinea Papua in 1942–1945, the Solomons Archipelago in 1942–1943, the Japanese held island strongpoints in the central Pacific in 1943–1944, and the Philippine Campaign in 1944–1945 were each offensive maritime campaign. The Germans conducted an offensive campaign aimed at cutting off Entente's maritime shipping in 1914–1918 and the Allied maritime trade in 1939–1945. At the same time, the Entente's Powers conducted defensive maritime campaign as did the Allies in World War II. The defense of Alaska and the Aleutians in 1941–1943 was a defensive maritime campaign. The Germans conducted an offensive maritime campaign aimed at cutting off Allied trade in the Atlantic in 1939–1945, while the Allies conducted a defensive campaign.

Accomplishment of military or theater-strategic objective would require planning and executing a single campaign in a given theater of operations. In a general war, two or more campaigns can be waged sequentially or simultaneously in a theater of war. The Allies started the Central Pacific campaign in November 1943, when the Solomons campaign was almost completed in the Southern Pacific Area (SOPAC). A campaign requires more time to complete than a major operation does. Its duration depends principally on the scope and complexity of the strategic objective, the factors of space–time–force and the enemy's resistance. A campaign on land in a developed theater normally lasts longer than one in an undeveloped theater. Initial campaigns usually last longer than those in the final phases of a war. The Allied Papua New Guinea campaign lasted about 27 months (June 1942–September 1944), while the Philippine campaign lasted only about 10 months (October 1944–August 1945). The defensive and offensive phases of the Allied North African campaign lasted almost three years (June 1940–May 1943). The German campaign in Norway in 1940 lasted only three months (April–June), and the one in the West slightly more than one month (10 May–17 June).

The lengths of the defensive and offensive phases of a campaign can vary considerably, depending on whether the ultimate strategic objective is limited or unlimited, the correlation of forces between the opposing sides, the size and geomorphological characteristics of the theater, demography and the degree of urbanization, the level of enemy resistance and weather/climate. The defensive phase of the Gulf War of 1991 (Operation Desert Shield) lasted about five and a half months (3 August 1990–17 January 1991), while the offensive phase (Operation Desert Storm) lasted only about six weeks (17 January–28 February 1991). The major combat phase of the war in Iraq in 2003 (Operation Iraqi Freedom) was planned to last 125 days, but it lasted only 40 days (20 March–1 May).[74] Maritime campaigns to attack the enemy's or protect friendly trade are conducted for the duration of the war; hence, they can be quite long. The German attack on Allied trade in the Atlantic and the Allies' protection of their trade lasted more than five and a half years (September 1939–May 1945).

The strategic framework for a campaign comprises three main phases: pre hostilities, hostilities, and post hostilities. The corresponding phases for the operational commander are the pre – major combat, major combat, and post – major combat phases, respectively. The major combat phase can encompass an initial defensive phase followed by an offensive phase. For the U.S.-led coalition, the Gulf War of 1990–1991 consisted of the defensive phase (3 August 1990–16 January 1991), followed by the offensive phase (17 January–28 February 1991). The Allied campaign plan for the invasion of France, Benelux and Germany (plan Overlord) in 1944 was overambitious, because the planners aimed to achieve two theater-strategic objectives in a single campaign. The better solution was to plan two successive but related land campaigns; one for liberation of France and the Benelux countries and another aimed at the occupation of western part of Germany.

In the case of a global war, several campaigns might be required to accomplish the ultimate national or alliance/coalition strategic objectives in a given theater of war. Then, only the last campaign would have a post hostilities phase; all others would end with the post – major combat phase. In the case of a major regional war, the phase of a campaign that ends hostilities would have a post hostilities phase. In both cases, the post hostilities phase will end with a transition phase – the handover of authority to the legitimate indigenous government or an international governing body.

Normally, all land or maritime campaigns should be planned. In planning a campaign, it is necessary to ensure that each combat action is linked to each other and is a part of a broader operational framework. This is ensured by developing the common basic idea.[75] Sometimes the national or alliance/coalition strategy might suffer from lack of clarity and poorly determined military strategic objectives. In such a case, a campaign might be seen as such only in retrospect. A classic case of ad hoc strategy was the Allied performance in the Italian theater in World War II. After the defeat of the Axis forces in North Africa, the invasion of Sicily was the next Allied objective. As events proved, the Allies had high and unrealistic expectations that capturing Sicily would accomplish the main strategic objective of the war in the Mediterranean – the overthrow of the Fascist regime and Italy's change of sides in the war. As it happened, the Allied invasion of Sicily led only to an operational, not a strategic, success. Italy's dictator, Benito Mussolini, was overthrown, and Italy sided with the Allies, but that did not end the war in the Mediterranean. Instead, German troops poured into Italy. The invasion of Sicily was followed by the Allied landing in Calabria (Operation Baytown) and at Salerno (Operation Avalanche) in September 1943. However, the German forces offered stubborn resistance and did not capitulate until May 1945. The main reason for this course of events was that the Allies never had a campaign plan for the Italian theater; they improvised a series of operational objectives until the theater strategic objective was accomplished in May 1945.

Maritime campaigns aimed to defend or capture an enemy-held large island archipelago are usually planned. The Allies planned campaigns to capture the Solomons, Papua New Guinea, the central Pacific, and the Philippines in World War II in the Pacific. However, a maritime campaign can also be seen as such only in retrospect. This usually happens when a stronger sea power declares a naval blockade to which the weaker opponent responds with a counterblockade; if the resulting hostility at sea becomes protracted, the actions of both sides can in retrospect be seen as a maritime campaign. An example of such a campaign was the German effort to cut off Allied maritime trade in the Atlantic (and adjacent sea areas) in 1939–1945. Meanwhile the Allies conducted a defensive maritime campaign aimed at protecting their maritime trade. Both sides conducted numerous offensive and defensive tactical actions of varying intensity throughout the theater; a relatively small number of major operations were conducted. The most notable exceptions were defense of several large Allied convoys to northern Russia in 1942 and the large convoy operations in 1940–1942 to relieve the British-held island of Malta in the Mediterranean. The Germans also conducted several major operations with their heavy surface ships and U-boat screens in the north and central Atlantic.

Conclusion

For most of the nineteenth century, the European and Russian military theoreticians considered the movements of an army, from its base of operations toward the areas in which it would fight – what the Germans called the major battle, and the Russians called the general engagement – as an operation. The battle itself was considered something separate and purely the domain of tactics, while the operation was a matter of strategy. It was only in the late nineteenth century that several German and Russian theoreticians began to consider the movement of armies and the major battle/general engagement as an inseparable entity and to call it an "operation." That major change was due to the effects of the major technological advances, and concurrent changes in the character of warfare in Europe, in the late nineteenth century. Collectively they led to new methods of combat force employment.

Modern major land operations are the result of a long and complex evolution of war on land. Any of the many decisive battles on land in the past contained the rudiments of today's major land operations in terms of main purpose and the need to properly synchronize the movements of one's large forces on the battlefield. However, not until the Napoleonic Wars did it become necessary to plan such major battles based on a common idea and as part of a larger operational framework. A large number of major land operations have been fought in many wars of the modern era. In both world wars, major land operations involved the movement of large forces on both sides. They also encompassed a relatively large part of the theater. Since the end of World War II, new technological advances have led to the smaller size of modern armies. Today, even a single division can conduct a major operation.

In the future, the importance of major land operations to the success of a campaign will not be diminished. As in the past, the size of one's land forces and methods in planning and conducting major offensive and defensive land operations will continue to evolve. In the future, major land operations will most likely be conducted with much smaller but more capable forces than they are today.

Major air operations are the principal method of accomplishing operational objectives in the air. As important as they are, these objectives are subordinate to the operational objectives assigned to friendly land forces. Hence, they are not air campaigns, as often claimed by airpower advocates. The service or functional air component commander should have full responsibility and authority to plan, prepare and execute major air operations as part of a land or maritime campaign. Major offensive air operations are planned to obtain and then maintain the desired degree of air superiority or air supremacy. Major defensive air operations are intended to deny the enemy the ability to use the air over friendly territory for his military and commercial aircraft.

As in war at sea or on land, it is not always possible to conduct major air operations. To be successful, they require relatively large forces. Major air operations are normally planned only when it is necessary to drastically change the situation in the air and thereby create prerequisites for one's air forces and other services to carry out other missions. The main advantage of major air operations is that the operational objective is accomplished decisively and relatively quickly, and with the fewest losses for friendly forces. Also, by planning and executing a major air operation, the operational commander chooses the place and time of achieving the objectives in the air and imposes his will on the enemy. The successful outcome of a major offensive air operation should enlarge the operational commander's freedom to act.

In the future, major air operations will remain the main method of employing one's air forces to accomplish operational objectives in the air. The methods of tactical combat employment in such an operation will steadily evolve. Major air battles are already largely a thing of the past. The importance of air strikes and attacks will increase. As in war on land and in the sea, major air operations will most likely be conducted with relatively smaller but highly capable forces.

A major naval operation is the principal method of accomplishing operational objectives in a maritime theater. It can be the most effective way to achieve decisive results within a given time frame, results in bringing about a drastic change in the situation in a theater. The only alternative is attrition warfare, which should be avoided even when an operational commander enjoys quantitative superiority. Nonetheless, Western naval theoreticians and planners have generally neglected major naval operations as an area of study in operational art. Too much emphasis is given instead to the tactical employment of naval forces. The real danger is creating the impression that no major naval operations will be conducted in the future – an impression that could not be more wrong. Yet in either a regional or global war, a navy would need to plan and execute major naval/joint operations as an integral part of its struggle for sea control. Threats to national interests at sea tend to come with little warning. Fleets are built and maintained primarily to wage war, whether regional or global.

Because of the enormous increase in the combat potential of platforms, major naval operations of the future are likely to be highly intense. This will be especially true in a littoral environment, because of the small area and correspondingly short distances. For that reason, in major naval operations today extremely high consumption rates of fuel and ammunition should be expected.

The focus of the theory should be on major naval operations of the types most likely to be conducted. However, this does not mean that the other types of major naval operations should be neglected. Because of the absence at present of major naval opponents, major naval operations to destroy enemy fleets at sea are not likely in the near future; nevertheless, it would be very wrong for theoreticians to ignore them. Likewise, no blue-water navy, focusing on major naval operations in littoral waters, should neglect meanwhile the possibility of encounters on the open ocean.

The term campaign is used today all too often without regard for its true meaning. Sometimes any large military effort is defined as a campaign, regardless of whether it might be more appropriately styled a major operation or even a major battle or engagement. Yet the term campaign should be reserved for those major efforts aimed at accomplishing a single military or theater-strategic objective. The term "land" or "maritime" campaign has nothing to do with services but the predominant features of the physical environment in which the campaign is conducted. Campaigns are the most complex of all the methods of combat force employment. A campaign encompasses the entire spectrum of methods of combat force employment, ranging from the smallest tactical action, such as an ambush or a raid, to a major joint/combined operation. These actions take place on land, at sea and in the air. A campaign usually requires the employment of multiservice and often multinational forces. Because of the large scope and complexity of strategic objectives in a mature theater, no single service can accomplish campaign objectives by its efforts alone. This is not to say that each service makes an equal contribution to the conduct of a campaign.

Notes

1 *DOD Dictionary of Military and Associated Terms*, November 2021, 159.
2 *DOD Dictionary of Military and Associated Terms*, 119.
3 *DOD Dictionary of Military and Associated Terms*, 119.
4 Michael C. Vitale, "Jointness by Design, Not Accident," *Joint Force Quarterly* 3 (Autumn 1995), 27.
5 Vojnopomorski Školski Centar, *Skripta Iz Operatike RM, V Lekcija (Projekt) Zajedničke Operacije Vidova* (Split, 1970), 46.
6 Frederick R. Strain, "The New Joint Warfare," *Joint Force Quarterly* (Autumn 1993), 23.
7 Bernard E. Trainor, "Jointness, Service Culture, and the Gulf War," *Joint Force Quarterly* (Winter 1993–94), 74.
8 Headquarters, Department of the Army, *FM 100-5: Operations* (Washington, DC: U.S. Government Printing Office, May 1986), 94.
9 Headquarters, Department of the Army, *FM 100-5: Operations* (1986), 94.
10 Headquarters, Department of the Army, *FM 3-0: Operations* (Washington, DC: U.S. Government Printing Office, June 2001), 7–4.
11 Headquarters, Department of the Army, *FM 100-5: Operations* (1986), 95.
12 Jovo Ninković, *Nastanak i Razvoj Ratne Operacije* (Belgrade: Vojno-Izdavački Zavod, 1984), 103.
13 Robert L. Helmbold, *Rates of Advance in Historical Land Combat Operations* (Bethesda, MD: U.S. Army Concepts Analysis Agency, June 1990), 4–1.
14 Ninković, *Nastanak i Razvoj Ratne Operacije*, 113.
15 *Grundlagen der modernen operativen Kunst* (Dresden: Military Academy Friedrich Engels, 1962), 43.
16 Headquarters, Department of the Army, *FM 3-0: Operations*, 8–1; Headquarters, Department of the Army, *FM 100-5: Operations* (1986), 129.
17 Headquarters, Department of the Army, *FM 100-5: Operations* (1986), 131.

18 Ninković, *Nastanak i Razvoj Ratne Operacije*, 40, 43.
19 Ninković, *Nastanak i Razvoj Ratne Operacije*, 99–100.
20 Aleksander A. Svechin, *Strategy*, ed. Kent D. Lee (Minneapolis, MN: East View, 1992), 297.
21 Joseph O. Rodriguez, *How to Maximize the Advantages of Interior Lines at the Operational Level* (Fort Leavenworth, KS: School of Advanced Military Studies, May 1, 1987), 1–2.
22 Wilhelm von Blume, *Strategie. Ihre Aufgaben und Mittel* (Berlin: E. S. Mittler & Sohn, 1912), 360.
23 Rodriguez, *How to Maximize the Advantages of Interior Lines at the Operational Level*, 29.
24 Rodriguez, *How to Maximize the Advantages of Interior Lines at the Operational Level*, 37.
25 Svechin, *Strategy*, 296.
26 Heinrich Aschenbrandt, *Űbersicht űber die deutsche Militärliteratur im 19. und 20. Jahrhundert,* P-150, ZA/1 2165, BA-MA, November 1951, 8.
27 Dieter Brand, "Freie Operationen (1)," *Truppenpraxis/Wehrausbildung* 10 (October 1997), 581.
28 Helmuth von Moltke, Sr., *Ausgewählte Werke, vol. 1: Feldherr und Kriegslehrmeister* (Berlin: Verlag von Reimar Hobbing, 1925), 211.
29 Headquarters, Department of the Army, *FM 100-5: Operations* (1986), 95.
30 Headquarters, Department of the Army, *FM 100-5: Operations* (1986), 95.
31 Headquarters, Department of the Army, *FM 100-5: Operations* (1986), 95.
32 Harry E. Mornston, *Raids at the Operational Level – To What End?* (Fort Leavenworth, KS: School of Advanced Military Studies, United States Army Command and General Staff College, May 15, 1992), 9.
33 Adolf Feil, ed., *Grundlagen der Theorie der operativen Kunst der SSK* (Translated from the *Russian Operativnoye iskusstvo voyenno-morskogo flota*) (Dresden: Military Academy Friedrich Engels, March 13, 1989), 21.
34 MCWP 3-40.4, *Marine Air-Ground Task Force Information Operations* (Washington, DC: U.S. Government Printing Office, July 2003), 3–2.
35 Lecture, *Die Luftoperation* (Dresden: Military Academy Friedrich Engels, 1981), 5.
36 *Die Operative Kunst der Luftstreitkräfte*, prepared by F. Beer et al. (Dresden: Military Academy Friedrich Engels, 1974); from the Russian original *Kurs operativnogo iskusstvo voyenno-vozdushnykh sil* (Moscow: Monino, 1974), 119–20.
37 Lecture, *Die Luftoperation*, 6.
38 Translated from the Russian by Manfred Arnold et al., *Taktik der Marinefliegerkräfte (Original: Taktika aviatsii Voyenno-Morskoga Flota)*, Teil I (Dresden: Military Academy Friedrich Engels, [1977] 1982), 63.
39 Arnold et al., *Taktik der Marinefliegerkräfte (Original: Taktika aviatsii Voyenno-Morskoga Flota)*, 63.
40 Arnold et al., *Taktik der Marinefliegerkräfte*, 61, 63.
41 Arnold et al., *Taktik der Marinefliegerkräfte*, 61.
42 Arnold et al., *Taktik der Marinefliegerkräfte*, 60.
43 Arnold et al., *Taktik der Marinefliegerkräfte*, 158.
44 Arnold et al., *Taktik der Marinefliegerkräfte*, 160.
45 Andrew G. B. Vallance, *Doctrine of Air Power Strategy and Operational Art* (New York, NY: St. Martin's Press, 1996), 72–73.
46 Vallance, *Doctrine of Air Power Strategy and Operational Art*, 72–73.
47 Flight Training Instruction, *Basic Fighter Maneuvering (BFM) Advanced Strike/Fighter NFO Training System T-45C/VMTS (P-826)* (NAS Corpus Christi, TX: Naval Training Command, CNATRA P-826, June 18, 2018), 17–10.
48 Vallance, *Doctrine of Air Power Strategy and Operational Art*, 73.

49 Vallance, *Doctrine of Air Power Strategy and Operational Art*, 73.
50 Vallance, *Doctrine of Air Power Strategy and Operational Art*, 72.
51 Arnold et al., *Taktik der Marinefliegerkräfte*, 64.
52 Arnold et al., *Taktik der Marinefliegerkräfte*, 161.
53 Arnold et al., *Taktik der Marinefliegerkräfte*, 64.
54 "Samostoyatel'nyy poisk," *Voyenno Entsklopedicheskiy Slovar*, 1st ed. (Moscow: Voyenizdat, 1983), 653; "Slobodni lov," *Vojni Leksikon* (Belgrade: Vojno Izdavački Zavod, 1981), 563.
55 "Samostoyatel'nyy poisk," *Voyenno Entsklopedicheskiy Slovar*, 653; *Vojni Leksikon*, 563; F. Heinig, "Methods of Combat Actions of Fighter Aviation," 34.
56 "Deystviya aviatsii po vyzovu," *Voyenno Entsklopedicheskiy Slovar*, (Moscow: Voyenizdat, 1st ed. 1983), 226.
57 "Zasada," *Voyenno Entsklopedicheskiy Slovar*, 1st ed. (Moscow: Voyenizdat, 1983), 269.
58 Adolf Feil, ed., *Grundlagen der Theorie der operativen Kunst der SSK*, trans. Dieter Förster and Bernd Kulbe (Dresden: Military Academy Friedrich Engels, March 1989), 24. (Originally published in Russian as *Operativnoye iskusstvo voyenno-morskogo flota.*)
59 Heinz Engelmann, "Die Sicherstellung von Seeoperationen," *Militärwesen (East Berlin)* 3 (March 1980), 69; S. Filonov, "Morskaya Operatsiya," *Morskoy Sbornik* 10 (October 1977), 24.
60 Tomislav Bolfek, "Rasprava o podeli i osnovnim karakteristikama operacija na JPV," *Mornarički Glasnik (Belgrade)* 6 (November–December 1978), 992.
61 *Upotreba Strategijskih Grupacija (Privremeno Uputstvo)* (Belgrade: Savezni Sekretarijat Narodne Odbrane, 1990), 97.
62 Gerhard Regner et al., trans., *Operative Kunst der Seestreitkräfte, Teil II/1: Seeoperationen und gemeinsame Handlungen der Flottenkräfte und Truppen der Front in Küstenrichtung* (Dresden: Military Academy "Friedrich Engels" 1981); originally published as *Operativnoye iskusstvo voyenno-morskogo flota* (Leningrad: A. S. Grechko Naval Academy, 1975, 4–5).
63 Delimir Kolec, "Operacije na jadranskom pomorskom vojištu," *Mornarički Glasnik* (Belgrade) 5 (September–October 1985), 757.
64 Tomislav Bolfek, "Rasprava o podeli i osnovnim karakteristikama operacija na JPV," 993.
65 Horst Metschke, "Entwicklung der Hauptformen taktischer Kampfhandlungen zur Bekämpfung leichtere und mittlerer Überwasserkräfte," *Militärwesen (East Berlin)* 6 (June), 1984, 55.
66 Hans Tietze, "Zu einigen Kategorien der Theorie der Taktik der Seestreitkräfte," *Militärwesen (East Berlin)* 3 (March), 1983, 48.
67 Horst Metschke, "Entwicklung der Hauptformen taktischer Kampfhandlungen zur Bekämpfung leichtere und mittlerer Überwasserkraefte," 52.
68 Paul S. Dull, *A Battle History of the Imperial Japanese Navy (1941–1945)* (Annapolis, MD: Naval Institute Press, 1978), 295.
69 Raizo Tanaka, "The Struggle for Guadalcanal," in *The Japanese Navy in World War II: In the Words of Former Japanese Naval Officers,* ed. David C. Evans, (Annapolis, MD: Naval Institute Press, 2nd ed., 1986), 211.
70 Hans Tietze, "Charakteristische Merkmale taktischer Kampfhandlungen der Seestreitkräfte," *Militärwesen (East Berlin)* 3 (March), 1982, 54.
71 Translated from Russian by Dieter Förster and Bernd Kolbe, *Grundlagen der Theorie der operativen Kunst der SSK (Operativno iskusstvo voyenno-morskogo flota)* (Dresden: Military Academy "Friedrich Engels," 1989), 31–32.
72 Gerhard Regner, *Operative Kunst der Seestreitkräfte*, translated from the Russian by Heinz Engelmann et al., Teil I/2 (Dresden: Military Academy "Friedrich Engels," 1981), 141.

73 Irwin, *The Levels of War, Operational Art and Campaign Planning*, 8.
74 Kevin C. M. Benson, *Expectation of Valor. Planning for the Iraq War* (Havertown, PA: Casemate Publishers, 2024), 16.
75 Wilhelm Stanger, *Grundzűge der Lehre von der Strategie. Studienbehelf fűr did K.K. Kriegschule, vol. 1, Theorie mit Kűrzeren Bespielen* (Vienna: Verlag der K.K. Kriegschule, 1884), 56.

5 Elements

Employment of one's forces in combat encompasses a relatively large number of constituent parts, arbitrarily called *elements*. The main elements of combat force employment deployment/redeployment, concentration/dispersal, center of gravity, maneuver/counter maneuver, deception/counter deception, the culmination point, reserve, combat success, pursuit, withdrawal/retreat and regeneration of combat potential. The time–space–force factors affect the elements of warfare both qualitatively and quantitatively. Hence, there are considerable differences in the scale of these elements in a tactical action, major operation and campaign. The higher the scale of the objective, the more critical are elements of combat employment. Because of the lack of space, only selected elements of major operations and campaigns will be explained here.

Deployment

In general, deployment is understood as a combination of movement and mobility aimed to bring one's forces into the area where a combat action will be conducted. Rapid and timely deployment of combat forces and their logistical support and sustainment are the keys to the successful execution of a plan. The results of initial major operations and tactical actions often have a decisive effect on the outcome of an entire campaign or major operation. Because of the size of the forces involved, errors in the deployment of one's forces cannot be corrected easily, if at all, once hostilities start.

In war on land, the troops and their weapons and equipment are moved by railroads, roads, or internal waterways. When forces are deployed from bases outside a given theater of operations, troops are normally moved by air while the weapons and equipment are moved by surface transport, mostly by sea. A deploying force is usually moved in two groups: troops and weapons/equipment. The operational commanders retain command but do not exercise control over parts of units moving by different transport modes. Various force elements of the deploying forces arriving in the theater should reach specific locations and reassemble into tactical units before force commanders can reestablish control.[1]

Regarding its scale, deployment can be tactical, operational, and strategic. *Tactical deployment* pertains to the movement of tactical-size forces from their initial

DOI: 10.4324/9781003333685-5

or new combat zone (sector) to another zone (sector) within a given area of operations. The purpose is usually to concentrate one's forces or obtain a position for conducting a tactical maneuver.

Operational (or *intratheater*) *deployment* is aimed at moving a military force from its current area of operations to another area of operations (usually within the boundaries of a given theater of operations) to start an operational maneuver. In the nineteenth century, such a deployment in land warfare was defined as strategic – the movement of the armed forces of a state to a threatened border area.[2]

Strategic (or *intertheater*) *deployment* is aimed at moving one's combat force from its home bases in one theater (of war) to another theater (of war). It is usually conducted over long distances and entails the movement of large and multiservice forces. Strategic deployment is the most complex and critical of all operational undertakings and is related directly to strategic mobility. An example of strategic deployment was the movement of the German forces in the spring of 1941 from France, Germany, Yugoslavia and Bulgaria to their assembly areas in occupied Poland, Czechoslovakia and Romania in preparation for the invasion of Soviet Russia.[3]

The most important prerequisites for the successful deployment of one's combat forces for a campaign/major operation are a favorable geostrategic position, sufficient size of the deployment area, adequate theater transportation structure, diversity of transport modes and relatively short distances between the home bases and the deployment area.

The geostrategic position from which a force deploys has a considerable influence on the time, transport modes, and sequence of deployment. Among other things, a central geostrategic or operational position offers shorter lines of communication. In general, less time is required to deploy one's forces from one part of the theater to another or between theaters. This was one of the great advantages for the Germans in both world wars. However, a central position is usually not an advantage in deploying combat forces if the resulting physical space is too small or transportation means are inadequate. An exterior geostrategic position offers deployment advantages and disadvantages opposite to those of a central position.

The deployment area should be large enough to accommodate the movement and arrival of large forces. Its size should be proportionate to the size of the force to be deployed. It should facilitate access to the enemy's critically important areas, have direct links with one's base of operations, offer the use of the best and most direct line of operations and have sufficient communications for the rapid movement and supply of troops.[4] Having many parallel lines of communication, especially railroads and highways, complemented with sea communications and inland waterways, offers many advantages. The more numerous the routes for movement, the better. In the offensive, it is easier to deceive and surprise the enemy when using multiple lines of communication for the deployment of combat forces. A deployment area containing a single line or few lines of movement is usually at a great disadvantage, because the enemy is never in doubt as to one's forces' intentions. Moreover, one's forces would be in great danger of being cut off from their base of operations and then possibly defeated.

The theater transportation infrastructure should be sufficient to support the deployment of one's combat forces. In U.S. terms, the transportation infrastructure consists of organizational structure, physical capabilities and physical infrastructure. *Organizational structure* deals with theaterwide elements of the transportation system and logistics that support deployment. *Physical capabilities* are ports, airports, the rail system, highway and road networks, inland waterways and pipelines. *Physical infrastructure* consists of the transportation nodes and available modes of transportation. A *transportation node* is a location in a mobility system where a movement requirement originates, is processed for onward movement or terminates. Nodes include airports, seaports, river terminals, in-stream offload, railheads and staging areas.[5] Each node consists of several *terminals* – any facility, regardless of size and complexity, at which cargo or personnel are loaded and handled in transit between elements of various transportation modes. Terminals are established at points of origin, destinations and in-transit transfer points. Planners should give close attention to the availability of seaports and airports in a given theater of operations and to the transportation networks that link them. The throughput capacities of seaports and airports considerably affect the type of forces and units to be deployed through them, as well as the speed and sequence of deployment.[6]

The principal *transport modes* for the deployment of combat forces are air and surface. Air transport today is one of the most essential elements of the military transportation system. It provides rapid and flexible movement of cargo, passengers and equipment without regard to terrain restrictions.[7] It can move troops to prepositioned equipment and carry other cargo and supplies to threatened regions. It provides the ability to respond rapidly with minimal warning. Air transport can move directly to deep inland locations, even if ground and sea lines of communications are not available.[8] Airlift is highly suitable for the transport of light early-entry forces or for the movement of troops relying on prepositioned stocks or equipment transported by sea.

The disadvantages of air transport are considerable. Among other things, airlift is highly expensive per unit of cargo.[9] Air transport's capacity is limited as to the size and weight of cargo, such as fuel. It requires secure airfields large enough to handle diverse aircraft, as well as secure air lines of communication. Airlift also requires specialized material-handling equipment that must be brought to wherever the aircraft will be loading or offloading.[10] Bad weather can severely restrict air transport.

Surface transport is conducted by sea, rail, highways, inland waterways, coastal waterways and pipelines. Sea transport can be oceangoing or coastal. Sealift is the most economical and flexible means of moving bulk goods and heavy equipment; ships can be redirected to widely separated ports of embarkation or debarkation at short notice. However, it is much slower than airlift.[11] Its greatest advantage is its ability to move outsize and heavy volumes of cargo over great distances.

Rail is an excellent mode of transport for heavy bulk supplies such as ammunition. It allows deployment over a large space and in a short time.[12] It enables the fast movement of troops and materiel over long distances. The military importance of railroads depends on their capability to move trains per day, their capacity in tons or troops transported, and the number of tracks. For a side operating along interior lines, railroads are especially valuable for the rapid redeployment of troops from one front to another. They are also used for evacuating wounded, prisoners of war, and civilians. Railroads are more easily destroyed but more easily repaired than roads. A small crew can transport hundreds of tons of supplies, thus freeing manpower for other functions. In addition, rail cars serve as temporary storage platforms for parked equipment and supplies awaiting delivery to the combat zone.[13]

Combat force deployment can be accelerated considerably by having prepositioned heavy weapons and equipment and supplies ashore or afloat before the actual crisis or combat. Prepositioning allows heavy equipment to be kept in-theater, near the area where it will be needed. The advantage of prepositioning is a quick response to a crisis and troops can be flown in to marry up with the equipment. Prepositioning reduces reliance on strategic sealift and airlift. It also sends a strong political message about one's commitment to defend the country's interests in the respective theater. However, prepositioning is less flexible than using airlift or sealift. Removing prepositioned equipment to a different region tends to undermine the confidence of the host country, because that equipment is a visible sign of commitment. Another disadvantage is that it may require duplicate sets of unit equipment. The ability to preposition equipment depends on a favorable political climate in the wider deployment area.[14]

Prepositioning ashore is expensive to maintain. It also requires host-nation full cooperation and may generate, not reduce, tensions in the area of possible conflict. Prepositioned weapons, equipment and supplies are also difficult to protect from various terrorist threats. In contrast, prepositioning afloat provides critical sustained combat power in a theater lacking a forward presence or prepositioned stockpiles ashore. It also offers more flexibility, because ships with prepositioned equipment and supplies can be moved to other crisis or conflict areas in an emergency. Prepositioning afloat makes it easier to obtain host-nation support (HNS). It also is less likely to create tensions in a region. In addition, prepositioned ships are relatively easier to protect from terrorist acts. However, prepositioning afloat is expensive and entails a loss of capability during periodic maintenance of prepositioning ships. Hence, prepositioning ashore and afloat should be combined to provide optimal capabilities and ensure flexibility; together, they can significantly reduce the need for heavy-lift assets during the most critical early-entry phase of power projection.[15]

Center of Gravity

The soundness of a plan for a campaign or major operation essentially hinges on the proper determination of a center of gravity for both the enemy and friendly sides. Yet despite its importance, many commanders and their staff pay inadequate attention to the theoretical underpinnings of the concept of center of gravity. Perhaps no military term is more discussed, commented on and misunderstood than "center of gravity." One reason for this is the mistranslation of the uniquely German term

Schwerpunkt as center of gravity, instead of the "weight of main effort," in the English version of Carl von Clausewitz's seminal work *On War*.[16]

The concept of center of gravity is simple, provided one accepts the notion that to accomplish an objective, a major part of one's combat potential, efforts and time should be directed to defeat or neutralize the enemy's greatest critical strength. At the same time, the source of one's greatest critical strength must be defended and protected to prevent the enemy from accomplishing his objective. For a weaker side, a major part of force, efforts and time should be focused on degrading, avoiding or deceiving the enemy's center of gravity.

In generic terms, center of gravity[17] can be defined as *a source of massed strength – physical or moral – or a source of leverage whose destruction/ neutralization or serious degradation, and dislocation would have the most decisive impact on the enemy's or one's ability to accomplish a given military objective.* In military terms, "mass" should not be understood literally. What matters most is the "massed effect," not whether available combat potential (prior to combat) or combat power (generated in combat) is physically concentrated in a certain place. Because of modern weapons' long range, lethality and accuracy, centers of gravity in air warfare or, to a lesser extent, naval warfare do not necessarily have to be physically massed in a specific area; they may be dispersed throughout a rather large part of a theater. In contrast, ground forces are usually massed in a relatively small physical space. However, even in land warfare, modern armies are much smaller and are deployed in a smaller part of the theater than was the case in World War II.

The principal utility of the concept of center of gravity is in greatly enhancing the chances that one's sources of power are used in the quickest and most effective way for accomplishing a given military objective and with the fewest losses in personnel and materiel. The key to success is to identify the friendly center of gravity and protect it, and to identify the enemy's center of gravity and then attack it with the requisite determination.[18] The enemy is not completely defeated unless his center of gravity – his relevant mass of power – is destroyed or neutralized. One may capture the enemy's capital, attack his logistics and convince much of his population that further resistance is useless, but victory is not assured unless the enemy's forces are defeated in the field.[19]

In land warfare, a maneuver is less likely to achieve its very purpose unless it is aimed at obtaining a favorable position regarding the enemy's center of gravity. Deception and counter-deception efforts cannot be successful if the focus of one's efforts is directed against the wrong enemy center of gravity. An attack on enemy vulnerabilities will usually not cause the desired deteriorating effect unless it directly or indirectly affects his center of gravity.[20]

A center of gravity cannot be considered in isolation from the corresponding military objective. In a war, one's actions should invariably be related to some larger whole. One should always bear in mind that it is the objective and the situation that determines a center of gravity, not the other way around. A change in the military objectives would require a partial or complete reevaluation of friendly and enemy situations. This might result in determining a new center of gravity. In a joint employment of forces, the objective determines the relative importance of the corresponding center of gravity. The objectives also limit the number of centers of gravity.

The single greatest advantage of linking the objective and the corresponding center of gravity is that it ensures that one's efforts are firmly *focused on the objective* and not the effects to be generated. This should also enhance the possibility that factors of space, time and force are not mismatched or disconnected. Without that linkage, it is quite possible that a major part of one's efforts would be misdirected, and ultimately fail. Therefore, it is a serious mistake to consider or, even worse, apply the concept of center of gravity with little or no regard for the objective as the proponents of the systems approach do.

In some cases, the military objective can be accomplished without destroying or neutralizing the enemy's center of gravity. Yet this would result in a disproportionate loss of one's forces, and waste of efforts and time. In other cases, failure to defeat the enemy's center of gravity might considerably delay or even preclude one's consolidation of operational or strategic success. If the enemy is allowed to recover from his losses, he may turn the initial defeat or setback into a victory.

The number of potential centers of gravity is directly related to the number of military objectives. At the national-strategic or alliance/coalition level a single military-strategic objective and therefore single military-strategic center of gravity exists. For each military- or theater-strategic objective, two or more operational objectives must be accomplished. And for each of them a corresponding center of gravity must be identified. The sequencing of the objectives determines the sequence in which corresponding centers of gravity must be defeated or protected.

At the tactical level of war, center of gravity has a major importance only when the accomplishment of a major tactical objective requires the employment of forces of two or more combat arms/branches. At the lowest tactical level, forces are reduced to a single-type force facing the same or a similar enemy force. Hence, centers of gravity on both sides are given single-type forces.

A military center of gravity consists of what can be arbitrarily called the "inner core" and the "outer core." It is the inner core where the strongest enemy's or friendly strength resides. It is composed of several elements that collectively provide the requisite power to accomplish a given military objective (firepower, mobility, leadership, will to fight, information, computer networks, etc.). The outer core consists of the elements designed to integrate, protect and support or sustain the inner core. The outer and inner cores represent a whole. The inner core cannot function properly without their respective outer core. The integrators are the command structure/C2 and automated decision aids, service culture and way of warfare, and common doctrine/training. Protection elements include air/space defense; force protection; protection of information systems; defense against weapons of mass destruction (WMD); and cover/concealment/camouflage. Support elements are intelligence, fires, information warfare, and logistics. Any center of gravity is composed of a combination of tangible and intangible sources of power. The higher the level of war, the more intangibles fall within the scope of a given center of gravity.

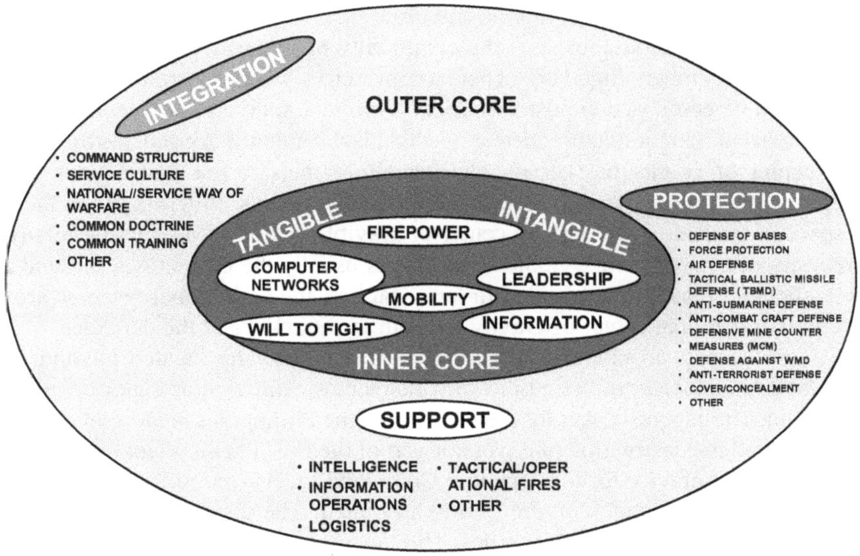

Figure 5.1 Composition of a Center of Gravity

In a campaign conducted in two or more physical mediums (land, sea, and air), not all operational objectives are equally important. The objectives at sea and in the air invariably support operational objectives on land. They are the main requirements for accomplishing the principal operational objective on land. That center of gravity must be defeated by the efforts of two or more services.

In land warfare, centers of gravity are usually found in the most mobile and lethal forces, such as armor or mechanized forces. In a war at sea, centers of gravity can range from the single carrier group, surface action (strike) group and submarine group to carrier forces, surface forces and submarine forces of a numbered or theater fleet. Likewise, in air warfare, a center of gravity can be the element of a force of combat aircraft that has the highest combat potential/power, such as a fighter/bomber squadron/wing or the entire force of fighters and ground-based air defenses or bombers and numbered/theater air fleets.

In highly netted forces, computer networks might comprise a major part of the enemy or friendly center of gravity. Yet as the centers of gravity, computer networks are fundamentally different from conventional centers of gravity, because they lack attributes of firepower and mobility. In contrast to traditional centers of gravity, the key components of a computer network are potentially highly vulnerable to an enemy's disabling attack.

Depending on the level of war, the intangible elements of a center of gravity include such unquantifiable or hard-to-measure elements as the quality of the political-military leadership, the skill and effectiveness of the operational/tactical commanders and their staff, soundness of doctrine, combat motivation, unit cohesion, the effectiveness of jointness, morale and discipline. In an alliance or coalition, the center of gravity might consist of the community of interests or common desires that hold the members together.[21] There are instances when the strategic center of gravity can be composed almost entirely of a military source of power. This situation may occur in an immature theater, one that lacks nonmilitary sources of power.

A center of gravity has certain *characteristics* that are not found among the military objectives, geographic location, or decisive points. In terms of the factor of space, a tactical or operational center of gravity is often located in the relative proximity of a given military objective. This is usually the case in war on land at the tactical and operational levels. In land warfare, the operational reserves are a potential operational center of gravity for both the attacker and the defender.

In some cases, an operational center of gravity may be located beyond the boundaries of a theater of operations in which a new campaign or major operation is to begin. The Japanese plan for the invasion of the Philippines in December 1941 contemplated the destruction of the major part of the U.S. Pacific Fleet – the operational center of gravity for the Japanese Combined Fleet – based in Pearl Harbor, Hawaii, prior to the attack on the Philippines itself. The distance from Honolulu to Manila is some 4,770 nautical miles. The Japanese objective was to prevent the U.S. Pacific Fleet from interfering with the landings in the Philippines. The Japanese attacked Pearl Harbor in the early hours of 7 December and the U.S. air and naval bases in the Philippines one day later.

An important characteristic of the operational or tactical enemy center of gravity is its ability to endanger the friendly center of gravity. Any tangible element of an enemy strategic center of gravity represents a potential threat to a friendly strategic center of gravity. Generally, the higher the level of war, the more fixed in terms of its content a given center of gravity is. A strategic center of gravity could possibly shift if the leadership is changed or removed from power or if a major

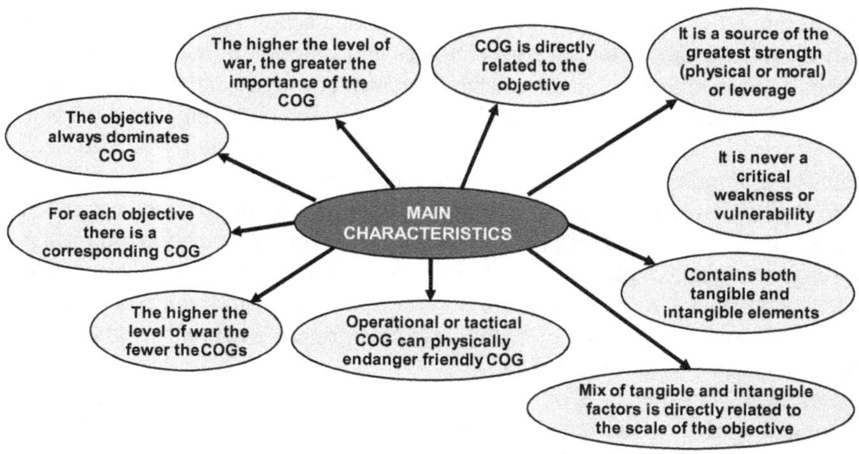

Figure 5.2 Characteristics of Centers of Gravity

part of the national/coalition military capabilities are drastically reduced in their effectiveness. The strategic center of gravity might shift if one or more members of an alliance/coalition cease hostilities or join the other side.

Prior to the start of a ground offensive in a land campaign, air or naval forces would be directed to accomplish operational objectives in their respective medium. The initial objective of air forces would be to obtain and maintain control of the air. This would require the destruction or neutralization of the enemy's fighter aircraft strength – usually his operational center of gravity. The initial operational objective for naval forces would be to obtain and maintain sea control. For them the enemy's operational center of gravity would be the striking force of the enemy fleet.

In a case where the enemy's strongest source of military power in a part of the theater is physically concentrated, it is relatively easy to identify the enemy's center of gravity. However, sometimes combat forces are either widely dispersed throughout the theater or are employed to predominantly accomplish tactical objectives. Then multiple tactical centers of gravity must be defeated or neutralized over time to ultimately defeat the enemy's strategic center of gravity. The protracted character of maritime trade warfare is largely a consequence of the fact that each side would avoid massing forces at the operational level. Hence, in an attack on the enemy's maritime trade or defense and protection of friendly maritime trade, the principal methods of combat employment of naval forces are major and minor tactical actions. The operational and ultimately strategic objectives are then accomplished over time. A similar situation exists in conducting theaterwide air defenses.

A shift in the center of gravity could occur because the intermediate objective can change from phase to phase of a major operation or campaign. There might be also a significant shift in the capabilities of friendly or enemy forces due to the introduction of some highly advanced weapons or some new major force into the theater.[22] A force different from the one that had been the enemy center of gravity at the beginning of the hostilities or military action can emerge as the center of gravity because of one's successes during the war. In the Leyte operation, the commander of the U.S. Third Fleet, Admiral William F. Halsey, apparently falsely believed that the most serious threat – the enemy operational center of gravity – was posed by Mobile Force, Main Body (fast carrier force), not battleships and heavy cruisers of the First Diversionary Attack Force. One can contend that Halsey's judgment was clouded by his obsession with fighting a decisive naval battle against enemy aircraft carriers, coupled with purposely vague orders from his superior, Admiral Chester Nimitz, CINCPOA/CINCPac. Nevertheless, it seems clear that Halsey did not sufficiently account for the declining performance of Japanese pilots after the battle of Midway in June 1942 and the very large losses in Japanese carrier aircraft in the battle of the Philippine Sea in June 1944. By October 1944, the Japanese carriers did not represent as large a threat to U.S. forces at Leyte as did the heavy surface force.

Higher-than-expected attrition, a poor state of morale and training, and a general inability to regenerate combat potential might also lead to a shift of the enemy's center of gravity to another type of force. Thus, once the plan is executed, the situation should be closely monitored and reassessed to detect potential changes or shifts in the enemy and friendly centers of gravity.[23]

During a major operation, the change in the physical medium in which combat force moves or in which combat takes place might lead to a shift in the center of gravity, while the ultimate objective remains unchanged. This is usually the case in the execution of a major amphibious or airborne operation (historic). During the sea transit phase, the initial and main objective is the defense and protection of the amphibious task force. Hence, the operational center of gravity for attacking forces in the sea transit phase is the force of distant cover and support. In many cases, such a force is a carrier group or heavy surface ship task force or land-based attack aircraft/bombers. Once the amphibious force starts landing ashore, the attacker's center of gravity would shift to that part of the landing force that has the highest combat potential, usually armor and/or mechanized forces.

Very often it is contended that the enemy amphibious task force is the proper enemy center of gravity for the defender – if that force is destroyed, then the attacker cannot conduct the planned amphibious assault. Yet troops embarked on ships of the amphibious task force cannot move and cannot use their weapons. Their combat potential is zero. Normally, the attacker would have a striking force, usually a carrier group or land-based attack aircraft/bombers to provide distant cover and support to the amphibious force. It is that force which defender must destroy, neutralize or significantly weaken, as the U.S. Navy did in the Battle of the Coral Sea and the Battle of Midway in May and June 1942, respectively. The force of distant cover and support and the amphibious task force would be normally deployed within mutually supporting distance. An amphibious task force would be also protected by a direct screen. In both cases, the respective Japanese admirals intuitively knew that the planned amphibious landing could not proceed, because even if the amphibious force landed, no naval force that remained available would be sufficiently strong to defend and sustain the landing force ashore. All these considerations at the center of gravity pertain to the situation *prior* to the landing of troops ashore. Clearly, if the attacker fails for any reason to keep his forces within mutually supporting distance, the defender should fully exploit such an opportunity and try to destroy the amphibious task force during its sea transit.

A center of gravity could shift from one force to another during a major operation because of the change in the threat and geography of the area. At the same time, the operational objective remained unchanged. This is sometimes the case in a major naval/joint operation. In the Allied resupply convoy operation to Malta on 10–15 August 1942 (Operation Pedestal), from the Allied perspective, the enemy's strongest force – the operational center of gravity – in the second phase of the operation (from Gibraltar to the Sicilian Narrows) were clearly the German heavy bombers and dive-bombers based on Sicily and Sardinia. However, in the third phase, south of Sardinia, the enemy operational center of gravity would shift to the Italian heavy surface forces (if they sortied from their bases). For the Axis, the strongest enemy force – the operational center of gravity – in the second phase was Force Z (three large aircraft carriers, with their fighter aircraft). Once the convoy entered the Sicilian Narrows, Force X (the convoy's direct screen) became the enemy operational center of gravity. Finally, the operational center of gravity would shift to the enemy fighter aircraft on Malta, once the convoy was within their effective radius. The reasons for the three successive operational centers of gravity were the flanking position occupied by the enemy land-based aircraft, the restricted operating area and the high threat posed by the enemy aircraft and submarines.

Operational Maneuver

A maneuver is one of the most important elements of a major operation or campaign. It is the heart of any operation plan. It can be described as the movement of combat forces aimed at obtaining a positional advantage relative to the enemy's force. Skillful maneuvering is one of the main prerequisites for the successful employment of forces in combat. It is usually through a maneuver that one's force obtains a position to attack the enemy's tangible center of gravity directly or indirectly. In contrast to deployment, combat forces move along lines of operation, while in a deployment, they move along lines of communication. Normally, a maneuver follows the deployment of combat forces. It is usually supported by ground and/or air fires. This is usually not the case during the deployment of combat forces. A maneuver is predominantly used by the attacker to gain the initiative and set the terms for subsequent combat. The side on defense can conduct a *counter maneuver* aimed to counteract the attacker's maneuver that is underway or planned.

In terms of their scale and potential impact, distinctions are made among tactical, operational, and strategic, respectively. A *tactical maneuver* seeks to gain a decisive advantage through a more favorable position for conducting a tactical action. It may be planned to attack selected critical vulnerabilities in the enemy's tactical depth, thereby forcing him to react tactically and sometimes even operationally.[24] A tactical maneuver at sea or in the air can be conducted using fires only. Then individual platforms or groups of platforms shift their fire from one target or group of targets to another, but with little or no change in their physical location.

An *operational maneuver* is aimed at having a decisive impact on the outcome of the initial major operations in a land campaign, either by securing the advantage of a position before attacking the enemy's operational center of gravity or by striking at some critical vulnerability, such as logistical supply lines in the operational depth of the enemy defenses. In either case, the intent is to force the enemy to react operationally or even strategically. An operational maneuver on land is supported by tactical fires and operational fires. In land warfare, an operational maneuver is normally conducted in the part of the theater that one's forces largely control, while such a maneuver at sea or in the air usually takes place within a space that is either disputed or controlled by enemy forces. A ground force conducting the maneuver moves from its base of operations along several lines of operation, via selected decisive points, to obtain a favorable position or area from which to attack the enemy's center of gravity or some critical vulnerability. An operational maneuver on land consists of a series of successive tactical maneuvers, with some breaks in the movement of one's forces.

An operational maneuver at sea or in the air is conducted almost uninterruptedly and with little or no enemy opposition until a position or area is reached from which the final assault can be launched. The latter phase involves tactical maneuvers. An operational maneuver is usually an integral part of a major naval operation aimed at destroying an enemy fleet at sea and/or its bases, an amphibious landing operation, and for major operations to interdict or cut off enemy maritime trade.

In theory, a *strategic maneuver* is aimed to engage and destroy a major part of the enemy forces in the strategic depth of the enemy's defenses, thereby greatly increasing chances of ending hostilities on one's terms. The best known example of strategic maneuver was the German execution of the Schlieffen Plan (actually a memo to his successor) of 1905 (named after the German chief of general staff, Field Marshal Alfred von Schlieffen) was a strategic maneuver. Schlieffen contemplated the deployment of 96 divisions in the west. Out of that total, seven armies with 82 divisions would serve as a mobile force. They would be deployed on the right flank of the German forces in the area between Aachen and Metz. At the same time, two armies with eight divisions acting as a fixing force would be deployed on the left flank in Alsace-Lorraine.[25] The aim was for the German right wing to conduct a gigantic movement through Belgium to a position west of Paris, to be traversed in about 41 days.[26] Afterward, the German armies would swing south of Paris and continue their advance northeast, pushing the major part of the French army against the German fortified positions in Alsace-Lorraine or driving the French army into Switzerland and internment.[27]

Regardless of the objectives to be accomplished and the forces involved, there are three basic forms of maneuver: penetration, envelopment and encirclement. Each can be conducted in several variations. In a *penetration maneuver*, the attacker seeks to break up or penetrate a selected sector of the defender's main line or position and move into his rear, thereby preventing the defender's reserves or reinforcements from moving to the main defensive line or position. The attacker then operates along interior lines of operation.[28] Secondary attacks are conducted along the defender's entire front. The most powerful coordinated assaults are directed against a section of the front offering the most favorable conditions for success. Once the penetration is made, the gap should be widened to defeat the shattered elements of the enemy force in detail.[29]

A penetration maneuver is used when the enemy's flanks are not assailable. The aim then is to mass sufficient combat potential at the point of penetration to overwhelm the enemy forces. The point of attack is usually selected where the enemy's forces are least densely deployed.[30] When penetrations develop from frontal assaults, they are called *penetrations of opportunity*. A penetration maneuver generally requires a stronger force than a flanking maneuver. If successful, it allows the attacker not only to envelop the defender's broken front. It seriously threatens the defender's lines of supply. It may compel the defender to change his front to oppose the attacker's penetrating force. The change of front in such a situation is usually very difficult.[31] The defender could in theory counteract the attacker's attempt at penetration by quickly moving forces from other parts of the front to the area of penetration, closing the breach or enveloping the attacking forces; abandoning the existing front; or withdrawing to a new defensive position.

In World War I, several attempts were made to achieve an operational penetration, but only two were successful. The Austro-German penetration of the Polish-Galician salient at Gorlice-Tarnów on 2–9 May 1915 resulted in considerable operational success. After a four-hour heavy bombardment with about 700 guns, the German 11th Army and Austro-Hungarian 2nd Army, with six army corps, broke through the enemy lines along the twenty-eight-mile front to a depth of 12.5 miles. miles.[32] This initial tactical penetration was expanded to fifty miles.[33] It was afterward followed by a successive major operation in pursuit of the Russian forces all the way to Vilnius, Lithuania. It ended in September 1915.[34] The entire Russian front collapsed. The Russians lost 210,000 (including 140,000 captured) out of 250,000 men in the area of penetration, along with vast quantities of weapons and equipment.[35] Another successful operational penetration in World War I was achieved by Austro-Hungarian and German troops at Tolmein (Tolmino)–Flitch on the Austro-Italian front on 24 October–19 November 1917. The newly formed Austro-Hungarian 14th Army, under German command, achieved operational penetration in the Tolmein–Caporetto–Plezza sector during the battle of Caporetto (the twelfth battle of Isonzo) between 24 October and 12 November. Two other Austro-Hungarian armies supported the main effort. By the time the Austro-German offensive ended, the Italians had suffered a catastrophic defeat – 40,000 men killed and wounded, 275,000 men captured, and some 2,500 guns and huge quantities of ammunition lost – while their opponents had lost about 20,000 men killed or wounded.[36]

When an envelopment maneuver cannot be conducted because of terrain or the military situation in general, then one way to reach the decision is the *breakthrough*, a logical extension of penetration. While the penetration maneuver disrupts the continuity of the defender's front, the breakthrough seeks to envelop the defender from within. In general, the broader the penetration, the deeper the penetration wedge that can be driven. And the deeper the attacker penetrates the defender's front, the more effectively he can envelop and frustrate the defender's attempts to close his front by withdrawing into the rear. When the penetration is exploited to the maximum, a breakthrough follows. The aim is then to penetrate the fragmented defender's front and, with deep thrusts diverting to the sides, open the way into the enemy flanks and rear.[37] The breakthrough is not fully successful until the attacker overcomes the defender's artillery positions and his reserves "roll up" the defender from the newly created fronts. Napoleon I at Austerlitz in December 1805 and the Germans at Sedan in May 1940 used penetration to create interior flanks, which were then enveloped.[38]

The *envelopment maneuver* (or *flanking attack*) is directed against the enemy's flanks or rear to outflank, oust, or trap enemy forces – possibly against some geographic feature.[39] Schlieffen firmly believed the flank attack was the "gist of the entire history of war."[40] Envelopment is a stronger form of maneuver than penetration, because it threatens the enemy's flanks and rear. The envelopment maneuver is aimed at overlapping one of the defender's flanks, thereby completely changing his front. Moreover, it potentially threatens the defender's ability to use his lines of retreat. The attacker can concentrate his forces and bring them simultaneously into action, while the defender has no option but to employ his forces sequentially. An envelopment maneuver is mostly conducted by operating from exterior lines.[41]

In land warfare, a flank that can be enveloped is called an *assailable* or *tactical flank.* When a force has positioned its flank against an obstruction, such as an unfordable river or an impassable swamp, and has thus protected itself from significant assault, that flank is *unassailable*, and the force is said to have *refused a flank.* A force conducting an envelopment maneuver would compel the opposing force to fight in two or more directions to meet the converging efforts of the attack.

An advantage of operational envelopment is that the attacker, by striking at the flank and rear, could interdict or even cut off the defender's lines of supplies and lines of retreat, forcing him to abandon his main defensive positions. One of the disadvantages is that if the attacker fails, his lines of supply will be open to attack; alternatively, he could be altogether driven away from his own original line of advance.[42]

The success of the envelopment maneuver largely depends on achieving surprise. The greater the effect of surprise, the less time is left for the defender to prepare and conduct a counter-envelopment maneuver.[43] Surprise is primarily achieved by secrecy in planning, high tactical mobility and covertness of movement, deception, containment of the defender's forces at the front and the attacker's superior combat potential.[44] Superior mobility of the attacking force greatly increases the chances of success.[45] This form of maneuver can achieve both a psychological and a physical effect by preventing the arrival of enemy reinforcements and resupply and by cutting off enemy lines of retreat. The chances of success are greatly enhanced if the attacker deceives the defender as to his point of the main attack.

An important prerequisite for success in envelopment is determining the distribution of forces between the mobile force and the fixing force. Doing that properly is an art of leadership.[46] In the conduct of envelopment, a supporting attack is usually necessary to fix the defender's force so that he cannot maneuver. Such an attack can also be used to deceive the opposing force regarding the form of maneuver being employed.[47] To enhance the element of surprise, it is sometimes advantageous for the fixing force to attack first, followed by the thrust of the mobile forces into the defender's flanks. A fixing force can greatly complicate and even endanger the success of an envelopment maneuver if, instead of tying down the defender's forces, it acts too aggressively, penetrates the enemy line and even starts advancing into the enemy's depth. The deeper the fixing force advances, the greater its negative impact on the movement of the mobile forces.

In land warfare, an envelopment maneuver is usually advantageous provided that the base of operations of the enveloping force can be transferred to its change of front. The envelopment maneuver is an advantage if a lateral base of operations is used. If the attacker is forced to redeploy his forces from the defense of the front to take part in the envelopment maneuver, then his front might, in turn, be penetrated by an agile and aggressive defender. The longer the attacker's front line, the greater the danger of this happening. The weaker the effects of the envelopment on any given sector of the front, the less compelling it becomes for the defender to send reinforcements to that threatened point.[48]

The defender's options to counteract the envelopment include conducting a counter-envelopment maneuver, extracting or diverting the attacker's strength from another part of the front, attacking the envelopment's axis and bringing additional reinforcements and supplies by air.[49] The defender also could bend the threatened wing back to form a new defensive flank with reserves, thereby creating a fresh front to the enemy envelopment.[50] An envelopment maneuver can be conducted by concentrating one's forces either off or on the battlefield. The former method was used in the German attack on France in May 1940 and in the failed German offensive in the same area in December 1944 that led to the "Battle of the Bulge."

An envelopment maneuver can be conducted along one or two axes; hence, single- and double-sided envelopment maneuvers are differentiated. In the *single-sided envelopment* on land, the attacker uses a single axis and usually limits the defender's opportunity to avoid maneuver, either by placing strong holding or fixing forces on one or both flanks or by restricting the defender's options by trying to reach some objective on the enemy's coast, as the Germans did in Flanders in May 1940. Such a maneuver is designed to act as a "revolving door" against the enemy force. It has a *pivot*, usually a strong point, around which the forces taking part revolve, either clockwise or counterclockwise. The German *sickle cut* maneuver (as British Prime Minister Winston S. Churchill called it) in May 1940 envisaged a fixing force on both German wings (Army Group B, with two armies in the north, and Army Group C, with two armies in the south) and an envelopment maneuver following operational penetration with the Panzer Group von Kleist (two panzer corps and one motorized infantry corps) of Army Group A (four armies) between Sedan and Dinant, with a subsequent advance to the French Channel coast.[51] The pivot of the maneuver was east of Luxembourg. The farther the Allied forces advanced into Belgium in accordance with their Dyle Plan, the easier it was for the German panzer forces to thrust all the way to the French Channel coast. This 215-mile-long maneuver, conducted in a counterclockwise direction, took eight days for the German panzer columns to reach Abbeville on the French Channel coast.[52]

Single-sided envelopment maneuvers are also conducted in naval and air warfare. In contrast to such maneuvers on land, there is no pivot and usually no clearly recognizable fixing force (the enemy's forces are usually tied to a specific area by the threat of one's action or through deception). Amphibious landings aimed to accomplish objectives within the enemy's operational depth constitute a form of single-sided envelopment maneuver. An amphibious feint can be used as part of a fixing force during a single-sided envelopment by ground forces.

A *two-sided* (or *double*) *envelopment* (also called a *pincer movement*) is an offensive maneuver along two axes, in which the main attack passes around or over the enemy's primary defensive positions to secure objectives to the enemy rear.[53] One method of effecting double envelopment is for the maneuvering force to bypass both of the defender's flanks. In another method, the maneuvering force conducts a supporting attack in the center to fix the defender, while the wings attack the flanks and rear to envelop the defender.[54]

A double envelopment is aimed to force the defender either to abandon his objectives or to execute a general withdrawal. It is usually a prelude to encirclement and destruction of the trapped enemy forces. In defense, such a maneuver can be conducted by having the forces defending the central part of a front fall back to create a pocket while the wings remain static (as in the Battle of Cannae in 216 BC). As the attacker continues his attack, the static wings envelop the flanks of the advancing force, resulting in a double envelopment.

In both the single and double envelopment on land, it is essential that a secondary frontal attack contain as much of the defender's force as possible, to hinder his counter reaction. The appearance of the attacking force in the defender's rear would usually prevent the defender from reinforcing his forward defense line, and his lines of supply could be interdicted or even cut off. The losses for the attacking forces are reduced because their attacks are directed against less prepared and less capable second-echelon forces.[55] Because the double envelopment aims to encircle the enemy forces, this maneuver is potentially more decisive than a single envelopment.[56]

The German and Soviet experiences in World War II showed that a single-sided envelopment alone did not allow for the encirclement and destruction of large enemy forces on the open terrain in western Russia. Hence, it was often necessary to plan and execute two-sided envelopments by using armored and mechanized forces as a spearhead. German forces conducted several highly successful operational double-envelopment maneuvers in the initial phase of the invasion of Soviet Russia in 1941. The German double envelopment of the Soviet Second and Sixth armies at Kyiv ended on 26 September 1941, after only a week of fighting, with the destruction of four Soviet armies (45 divisions), the partial destruction of two others, and the taking of 650,000 prisoners.[57]

A *wide envelopment* (or *turning maneuver*) is aimed to pass around the enemy's main forces to strike at some critical point deep in the rear, preferably beyond the operational depth of the enemy's defenses. Its purpose is to indirectly attack the enemy's operational center of gravity by threatening or cutting off his lines of supplies in the operational depth. Normally, the defender could relatively easily avoid the attacker's envelopment, but not the turning movement. This is especially true if the defender is forced to fight with a reverse front and his lines of supply are shifted.[58] Among other things, the success of a wide envelopment depends on the speed of movement of one's forces and the strength of the enemy flank to be turned.[59]

In a turning maneuver, the enemy may be fixed by a supporting attack, as in envelopment. However, the mobile and fixing forces would then operate beyond mutual supporting distance.[60] Thus, it is easier for the defender to avoid envelopment than a turning maneuver, especially if he is forced to fight with a reverse front and his lines of supply are cut off.[61] Deception, secrecy and high mobility are critical to the successful execution of a turning movement.[62]

An example of the highly successful turning movement which also had strategic effects was the landing of the U.S. X Corps of some 70,000 men (including 9,000 South Koreans) at Inchon (Incheon) on 15 September 1950 (Operation Chromite). This highly risky maneuver turned the North Korean position on the

Pusan perimeter.[63] The landing took place approximately 150 miles behind the main enemy lines. It forced the North Koreans to fight from a reverse front. Their lines of supply toward the north were cut off. Inchon was the principal port on the western coast of Korea, some 18 miles from the capital of Seoul, which served as the hub of the North Korean lines of communication toward the south. The capture of Inchon and Seoul would result in the disruption of the North Korean forces' rear area and at the same time provide UN forces with a valuable staging and supply point for further offensive operations.[64]

Sometimes the best way to obtain a quick decision is to conduct a penetration maneuver followed by a single-sided envelopment. One of the prerequisites for a *penetration-envelopment maneuver* is to obtain timely, accurate and relevant intelligence on the defender's forces at the front and then select a point of the main attack in the sector where the defender has the least capable forces. The attacker should also try to deceive the defender as to his sector of main effort (thrust) and point of the main attack. This could be accomplished through an elaborate operational or strategic deception. Deception, if successful, should allow the attacker to covertly move and concentrate his forces to the point of the main attack.

In the first phase of their campaign in the west in May 1940, the Germans used combined penetration and single-sided envelopment maneuvers to cut off and destroy a major part of the Allied forces deployed in northern France and Belgium. It was General Erich von Manstein's idea to combine single-sided envelopment with penetration. He selected the Sedan–Dinant sector to be the point for a frontal attack followed by a penetration maneuver. This would be followed by a single-sided envelopment maneuver all the way to the French Channel coast. The aim was to avoid a frontal attack on the Allied forces as they moved into Belgium, but to cut them off in the rear of the Somme River.[65] The Germans assigned their newly created Panzer Group von Kleist to spearhead the attack in the west. The German offensive in the west opened on 10 May, and three days later, to the great surprise of the Allies, the German XIX Panzer Corps began to cross the Meuse River at Sedan. A few days later, other divisions of Panzer Group von Kleist and the XIV Panzer Corps (with two motorized divisions) crossed the Meuse River on a wide front, from Sedan in the south to Dinant in the north and began their drive toward the French Channel coast. The subsequent envelopment maneuver ended on 20 May at Abbeville on the French Channel coast. The lines of retreat to the south were cut off for about 1.5 million Anglo-French troops.[66]

An *encirclement maneuver* is a logical extension of double envelopment. In fact, the envelopment alone does not decide anything; it must be a preliminary phase of the subsequent encirclement.[67] Encirclement usually takes place when one's forces conducting a double envelopment meet in the rear of the enemy forces. In encirclement, the attacking force must engage every section of the perimeter, preferably simultaneously. The aim then is to capture a large part of the enemy force. It can also result from penetrations, turning movements, infiltrations, and single-sided envelopment.[68] In conducting this form of operational maneuver, the attacker risks finding his forces being enveloped by the defender, who might decide to counterattack or even to counter encircle.

Single- and double-encirclement maneuvers are differentiated. *Single encirclement* is the most common variant, wherein only an *inner front*, facing the enemy's troops, is created. In this case, if the encircled forces are strong, they may be able to break out; if they do, encircling forces could find themselves attacked from the rear. To prevent this, the attacker might establish a second, wider encirclement to prevent other enemy forces from reaching the encircled forces, thus creating a *double encirclement* with both an *inner front*, facing the encircled enemy, and an *outer front*.[69] If the encircled force is unable to withdraw and no reinforcements and supplies can be brought in, then the defender is "completely encircled." In that case, the attacker does not need to surround the defending force with an unbroken cordon of troops. Cutting off the defender's lines of retreat and supply is sufficient to finish off the encircled force.

The best-known example of double encirclement in modern military history was the Soviet destruction of some 22 Axis divisions, including six panzer and motorized divisions, with about 284,000 troops at Stalingrad in late November 1942.[70] This included the entire Sixth German Army, about one-third of the 4th Panzer Army, about half of the Romanian forces on the Don, and about one-fifth of the Romanian forces deployed south of Stalingrad. The battle for Stalingrad started on 9 September; three days later, the first German troops entered the city. However, the stiffening Soviet resistance prevented the Germans from consolidating their gains in the city until late November. In the meantime, the Soviets successfully massed the forces of three fronts (Stalingrad, Don and Southwest)—1,015,000 men and 92½ divisions along a 530-mile-long line.[71] The first Soviet counteroffensive conducted between 18 and 23 November (Operation Uranus), trapped the Sixth Army and major parts of the 4th Panzer Army in the thirty-seven- by twenty-eight-mile Stalingrad pocket.[72]

The new Soviet offensive, with the forces of two fronts, started on 16 December (Operation Maly Saturn – Little Saturn). Its principal aim was to destroy the Italian Eighth Army protecting the left flank of Army Group Don.[73] The Italian Eighth Army and the Romanian Third Army offered little resistance, and after only four days of fighting, the Soviets streamed southward through an open 100 -mile-wide front.[74] The massive Soviet attack on the Stalingrad pocket started on 10 January with the forces of the Don Front and massive bombardment from the air and artillery.[75] The German forces in the Stalingrad pocket surrendered on 31 January. The Soviets claimed that 91,000 Germans were taken prisoner and 147,000 died on the battlefield.[76] Soviet losses are estimated to be four to eight times larger than those suffered by the Axis forces. Losses of Soviet civilians are not known.[77] Some unofficial Russian sources claim that the Soviet losses were heavy; 324,000 dead and missing and 320,000 wounded in the defensive phase of the battle for Stalingrad. The corresponding numbers for the battle's offensive phase were 155,000 and 330,000, respectively.[78]

Point of Culmination

There is probably no aspect of operational art as complex as the concept of point of culmination. The critical importance of this concept has been shown repeatedly in military history. Many campaigns and wars ended unsuccessfully when the concept of the point of culmination was not properly applied or, even worse, ignored. Strategic culmination eventually doomed Napoleon I's invasion of Russia in 1812. The German general staff in the conduct of its campaign in the west in 1914 neglected Clausewitzian teaching on the culminating point and friction.[79]

General Carl von Clausewitz was the first military theoretician who explained in some detail the concept of culmination. He defined the culminating point (*Kulminationspunkt*) as "the point where the remaining strength is just enough to maintain a defense and wait for peace." He implied that once the culminating point is passed, the chance of victory is foreclosed unless the enemy yields from fear without engaging in decisive combat. If the enemy were to choose to fight it out, he would prevail.[80]

In generic terms, a culmination can be understood as *a point in terms of time and space reached by the attacker or the defender, after which his stated objectives cannot be accomplished and continued effort to reach them will significantly increase the chances of failure or even defeat.* The concept of the culminating point is often misunderstood. It is sometimes believed that any defeat of friendly or enemy forces is attributed to overshooting a point of culmination. This is not necessarily the case, because in many instances the attacker has won a decisive victory, or the defender has decisively beaten the attacker, without approaching or arriving at their respective culminating points because of his higher skills in maneuvering and combat.

A culminating point is not a "point," but rather an area of uncertainty or nonrecognition in terms of space and time. The higher the level of war, the larger the area of uncertainty and the more difficult it is to anticipate or sense the culminating

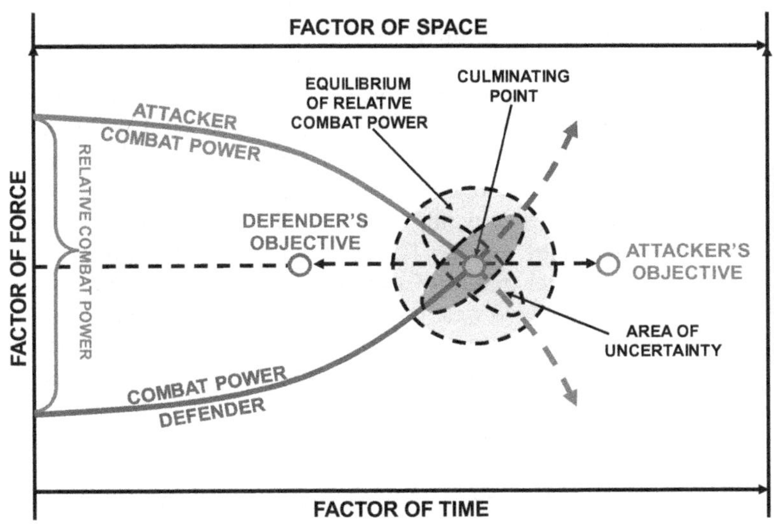

Figure 5.3 The Culminating Point: Theoretical Construct

point's arrival. One reason for this is the extreme complexity of the change in the balance of combat power. The mutual relationship of elements of combat power is highly dynamic and hard to quantify with any degree of certainty. This area is at the heart of what Clausewitz called the equilibrium of the combat power of both sides in a war. Clausewitz also noted that because so many factors contribute to an equation of force, it is difficult to determine which side has the upper hand "often it is entirely a matter of imagination."[81] It is a point at which the attacker can still revert to defense to protect his gains and regenerate his combat potential. When equilibrium exists, theoretically a determined and agile defender can take the advantage and go on the counterattack or counteroffensive.

A culminating point in combat is reached when the relative combat power drastically shifts between two sides. It is the point beyond which the difference in relative combat power begins to decrease rapidly.[82] Theory does not say that if a commander continues beyond the culminating point, he will face a defeat. It only warns that the risks of setback and even defeat will be high if he does so. In practice, a commander might well gamble and continue his attack beyond the point of culmination, after weighing all the pertinent elements in the situation, and be ultimately successful in accomplishing the assigned objective. Clearly, if a commander goes beyond the culminating point and suffers unacceptable losses and even defeat, then he has gambled too much.

The concept of the culminating point is applicable to war on land, at sea and in the air. Yet its full complexity can be demonstrated largely in land warfare. The concept of culmination applies to both the attacker and the defender. Clausewitz in *On War* focused predominantly on what he called the "culminating point of attack." He insisted that the defense is intrinsically stronger than the offense because it is easier to hold ground than to capture it. Therefore, it follows that defense is easier than attack, assuming both sides have equal means. The superiority of defense rests on the utilization of terrain; the possession of an organized theater of war; and often, but not always, the support of the population. Defense has a passive purpose – preservation; and attack has an active (or positive) one – conquest. Clausewitz believed that defense should be used only so long as weakness compels. It should be abandoned as soon as the defender is strong enough to pursue a positive object. He also observed that pure defense would be contrary to the very idea of war, because then only one side would be waging it. The principal advantage of defense over attack is that of waiting. The defender enjoys all the benefits of any error or mistake the attacker makes. Yet the defender should not be passive; he should do everything possible to wear down the attacker by conducting counterattacks. Victory normally results from the superiority of one side – that is, from a greater aggregate of physical and psychological strength. As a war unfolds, armies are constantly faced with some factors that increase their strength and with others that reduce it. In general, every reduction in strength on one side can be considered an increase on the other.[83]

In terms of time, the culminating point is usually not the same for the attacker as for the defender. An attacker who can no longer continue with his attack or offensive is culminated, but the defender may not be. The opposite is also true.

Because of the fog of war and friction, the arrival time of the culminating point is extremely difficult to determine precisely. It usually takes some time before one or both sides realize the shift in relative combat power and the approach of the point of culmination. Sometimes culmination is a matter of the commander's judgment or imagination, in which case it may not actually be reached or exceeded.

A culminating point can shift over time through the attacker's or the defender's actions. To prevent the culminating point from arriving too early, the attacker should maintain higher combat power over the defender in both current and future combat actions. At the same time, friendly forces should be fully protected. In land warfare, this can be achieved through properly assigning forces to echelons to maintain operational momentum, and through adequate sustainment.[84]

The attacker's culmination is the point in time and space at which the attacker's combat power no longer exceeds that of the defender. The relative reduction of the attacker's strength over time and space is the fundamental cause of his arrival at the point of culmination.[85] Among other things, the attacker's culmination could be reached because momentum is being lost or because of losses the defender has inflicted. Culmination occurs when and where the attacker must stop and defend his gains if he wishes to avoid losing them.

Clausewitz observed that the force of attack diminishes over time, while that of defense increases. The attacker must not only overcome the effects of friction but also move against an enemy who can use terrain and can strike when and where the attacking force has exposed itself. The attacker must act with greater strength and urgency than the defender, who, as Clausewitz noted, gains relative strength with the passage of time. Therefore, the art of attack at all levels of war is to accomplish the assigned objective before reaching one's point of culmination.[86]

In land warfare, the strength of the attacker is progressively diminished as the attacker advances deeper into the enemy's territory. His lines of supply become progressively longer and more vulnerable to enemy counterattacks unless he assigns adequate forces to protect them. The attacker must detach relatively large forces to protect these links with his base of operations. Other things being equal, the longer the distance from his base of operations, the longer the time the attacker needs for bringing up supplies and reinforcements. Either additional forces or part of the existing forces must be assigned to protect these lines of supply. The attacker also needs to destroy or neutralize bypassed enemy forces and strong points. The attacker's strength can be further reduced if his policies lead to an open mass uprising of the enemy's population.

The defender's ability to resist the attacker's advance might increase if he avoids major combat and his rear area is relatively secure from enemy attack. For the attacker, any pause in fighting to regenerate his combat potential might result in an advantage for the defender. Moreover, it might happen that the defender can regenerate his combat power faster than the attacker can.

If the attacker approaches the point of culmination before accomplishing his objective, the best he can do is to revert to defense before he suffers a major setback or is forced to retreat and surrender his gains. Clausewitz was certain that at the culminating point of attack, the initial superiority of an attack is expended and the ratio of strength between the two sides is steadily reduced, leaving the attacker with inadequate strength to conduct a successful defense.[87]

The defender should be alerted to perceive an enemy's error in continuing to attack after having reached his culminating point and then to quickly exploit the situation to the fullest. If the defender does not sense that the enemy has reached his culminating point, he will miss the opportunity to defeat him. An operational commander's inability to foresee or anticipate the arrival or overshooting of his point of culmination has been a major reason for setbacks or even the failure of an entire major operation. In the first battle of El Alamein, 2–18 July 1942, the Afrika Korps reached its operational culminating point, thereby paving the way for Allied counteroffensive in November 1942 and the Allies' eventual victory. Despite the presence of indicators, neither Rommel nor the British Eighth Army commander, General Claude Auchinleck, foresaw or acknowledged the arrival at the point of culmination.[88] Initially, Rommel apparently sensed that the tactical initiative had passed to the British. He had only 26 operational tanks, stretched supply lines, and exhausted troops, and he faced increasingly stiff British resistance. By the end of the day on 3 July, Rommel admitted defeat and gave the order to dig in. For the next few days, the Afrika Korps repulsed repeated British attacks, but by the slightest of margins. However, Rommel still did not acknowledge the passing of his operational culmination. After several days' rest, he attacked and was again repulsed. The men and materiel he unwisely expended would be critically needed at the second battle of El Alamein in November 1942.[89] In contrast, Rommel's nominal superior, Field Marshal Albert Kesselring, CINC South headquartered in Rome, correctly anticipated the arrival of the operational culmination for the Afrika Korps. He predicted on 26 June (after the capture of Tobruk) that the Axis forces could not obtain a decisive victory – not only due to logistical difficulties, but also because the British would certainly bring in reinforcements from the Middle East to defend Egypt. In contrast, both the Italian chief of the general staff, Marshal Ugo Cavallero, and Rommel believed that the Axis forces could be in Cairo within 10 days.[90]

An example of the attacker overshooting his point of culmination and suffering a significant defeat at the hands of numerically inferior but much better-led and trained forces is the Soviet failure in the major offensive in southern Russia in January–March 1943. After the end of the Battle of Stalingrad, a 200-mile gap opened in the German front lines because of the disintegration of Romanian, Italian and Hungarian formations.[91] The Soviet operational objective was to cut off and destroy two German army groups with about seventy-five divisions then deployed in southern Russia and the Caucasus.[92]

The German counteroffensive started on 19 February. Two days later, the right flank of the Soviet 6th Army was cut off, and the Soviet center began to crumble. By 23 February the Germans had cut off the escape route of the Soviet forces. Then they counterattacked and defeated the widely dispersed Soviet thrusts in detail. One day later, the Soviets suspended all operations and reverted to defense. However, by then the Soviet forces had already overshot their culminating point. The Soviet offensive ended after 21 days. The Soviets paid a heavy price in personnel and materiel.[93]

By 24 March the Germans had recaptured Kharkiv. They had destroyed Mobile Group Popov (X Tank Corps and IV Guard Tank Corps, with 55,000 men and 212 tanks). In addition, four armies of the Southwestern Front (65th, 1st Guards, 3rd Guards, and 5th Tank armies, with 325,000 men and 362 tanks) had suffered such losses that they were incapable of further combat.[94] This significant operational success was accomplished by only 160,000 German troops and about 100 tanks.[95] Manstein not only saved the German forces from being cut off in the Caucasus but also stabilized the entire southern part of the German front. The success of Manstein's counteroffensive in southern Russia probably also saved the Germans from losing the war on the eastern front in 1943 or 1944.

In land warfare, a defender reaches culmination when he can no longer go on the counteroffensive or defend successfully.[96] However, if the defender's objective is to go on the offensive, then the culminating point is where the defender should revert to a holding action and await reinforcements. If the defender's objective is to retain terrain, the culminating point is where the defender must withdraw, delay or retreat.[97] The art of defense is to speed up the attacker to his point of culmination and then quickly shift to the offensive.[98]

The attacker would suffer greater losses than the defender in taking on a prepared defense. As the attacker advances deeper into the enemy's territory and moves away from his base of operations, his combat power is steadily reduced. Thus, the defender's strength increases relative to the attacker's.[99] As long as the attacker's strength is declining faster than the defender's, it is to the defender's advantage to wait. The point of culmination will necessarily be reached when the defender must decide to act, after the advantages of waiting have been completely exhausted.[100] The defender's best chances for a successful counterattack occur when the attacker has reached or exceeded his culminating point. To speed the attacker's arrival at the culminating point, an agile and offensive-minded defender should always be on the tactical offensive. The defender must wait to revert to a counteroffensive or major counterattack until the attacker reaches and possibly overshoots his culminating point. Then the defender's aim is no longer to attrite the attacker, but to seize and maintain the initiative and accomplish objectives beyond destroying the enemy or regaining the lost space.[101]

A point of culmination exists at each level of war: national strategic, military strategic, theater-strategic, operational and tactical. The higher the level of war, the more serious the consequences of overshooting the point of culmination. At the tactical level, the commander can relatively easily determine when the actual culminating point has been reached or passed. In contrast, a culminating point at the operational or strategic level is extremely difficult to anticipate or sense because so many factors affect its occurrence. To complicate the matter, the indicators of approaching operational and especially strategic culmination are long in coming and are difficult to discern, even for a very experienced operational commander.

Today, Clausewitz's terms *culminating point of attack* and *culminating point of victory* are inadequate to address the concept at the tactical and operational levels of war. The term *culminating point of attack* pertained to what we understand today as the operational and tactical levels, while *culminating point of victory* dealt with a campaign or strategic level of war in Clausewitzian times. To make a better and more precise distinction among points of culmination at the three basic levels of war, it is important to use different terms. Therefore, the term *culminating point of attack* (or *defense*) will be used here when referring to the tactical level of war and *culminating point of offensive* (or *defensive*) for the operational level of war. For the national-strategic level of war, the Clausewitzian term *culminating point of victory* still seems to be the most appropriate.

Tactical culmination occurs in the course of a major battle or engagement when one's combat power is reduced if it is not regenerated in time. Then one's force must either stop its actions or continue fighting and accept the risk of failure or even defeat. If a force can prevent or postpone the culmination but opts not to, then it contributes to its own culmination.

Operational culmination may be reached during a major operation or campaign. A culmination in a major operation usually happens only once, while in a campaign several operational culminating points can occur, in sequence or almost simultaneously. The German Army Group Center overshot its point of culmination during the first phase of the German campaign against Soviet Russia. This army group, directed in early October 1941 to capture Moscow (Operation Taifun-Typhoon), in retrospect clearly did not have sufficient combat strength to do so. The early onset of the muddy season in Russia turned the ground into a sea of mud, considerably slowing the movement of German mechanized forces.[102] These enormously complicated German efforts to complete the operation quickly.[103] The combat effectiveness of the German infantry divisions dropped by 35 percent, and panzer divisions by 40 to 50 percent; the panzer attrition rate varied from 65 to 75 percent. There was also a high percentage of officer losses in combat. The flow of German supplies was disrupted, and only the most urgently needed rations, ammunition, and gasoline could be moved forward – and then only with difficulty. There was practically no winter clothing, and equipment for army units would not become available before early January 1942.[104] The Soviets sensed that the Germans had reached their point of culmination and prepared a counteroffensive.[105] The German offensive on Moscow finally stopped on 4 December, after its momentum was lost. Hitler ordered a shift to defensive in his directive no. 39, issued on 8 December. Army Group Center had to defend a front line of 620 miles with 67 heavily weakened divisions and no reserves.[106]

The *strategic culmination point* pertains to the theater-strategic and national or alliance/coalition levels of war. It occurs only once during a war fought in a single or multiple theaters. The strategic culmination is a point in space and time at which the combination of military and nonmilitary sources of power is reduced to such an extent that the chances of winning a war are foreclosed and one must go on the defensive or risk ultimate defeat. In terms of sources of military power, arrival at a strategic point of culmination is caused, among other things, by a lack of sufficient strategic reserves in manpower and weapons/equipment, inadequate mobilization, and one's inability to field adequately trained forces in the longer term.[107]

In referring to what is today commonly understood as strategic culmination, Clausewitz insisted that as the attacker's strength is reduced, a moment comes when both forces are roughly in balance, and "the remaining force is just enough to maintain a defense and wait for peace." If the attacker proceeds beyond this moment, then the defender's moral and physical strength becomes superior to the attacker's. Clausewitz contended that there are times when the superiority of the attacker is not sufficient to defeat the opposing army. In such instances, the attacker conducts war to acquire certain advantages over the defender. These advantages may even be psychological. The key consideration in such wars is that the attacker does not have the strength to destroy the opposing army but could achieve and maintain political advantage and equilibrium. This moment of equilibrium is called the *culminating point of victory*.[108] If the attacker attempts to go beyond that point, it is not merely a useless effort that cannot add to success; it is a damaging one that will lead to a reaction, and experience shows that such reactions usually have a completely disproportionate effect.[109] To expand the Clausewitzian concept further, the *culmination point of defeat* also exists for a nation, alliance, or coalition on the strategic defensive, dealing with military and nonmilitary resources available over long periods, as well as being closely linked to military strategy and national or alliance/coalition policy and strategy.[110]

Arrival at the point of culmination can occur because of many diverse factors. At the tactical level, an attack or defense can be temporarily overextended by the commander's and his forces' sheer fatigue or exhaustion, or by depletion of supplies of all kinds – a problem that could be resolved by simply replacing the commander or entire units, sending reinforcements, or rushing in supplies.[111] One of the common causes of culmination at the operational level is the pursuit of *multiple objectives* without properly balancing the factors of space, time, and force versus each objective.

The characteristics of the *terrain* can considerably affect culmination. The defender can use terrain to weaken advancing enemy forces. One way the attacker can overcome the factor of terrain is to operate at a high tempo. Operating at a higher tempo might create other effects that degrade attacking forces. The factor of *distance* poses a major challenge in both offense and defense. The width and depth of the theater or area of operations in which combat takes place necessarily cause the dispersal of combat power, which is an especially great problem for the attacker.

Time is another major factor in culmination. Clausewitz observed that both sides need time, and the question is only which of the two is in a situation to derive special advantage from it.[112] Failure to use available time is unforgiving; once lost, time can never be recovered.[113] The defender seeks to delay the decision. He can use the time to increase the relative advantage. In contrast, the attacker must move to hasten the decision since the passage of time accrues to the benefit of the defender.[114]

Very often one of the major contributing factors for culmination is the great reduction of combat potential through weakening of the enemy forces over time or attrition. The combat power of the attacker usually decreases faster than that of the defender and can become equal to, or lower than, that of the defender. Decreases in combat power can be due to fatigue, unfavorable position, difficult terrain or heavy weather. The arrival at, or overshooting of, the culminating point is affected by the inability to concentrate forces in the critical area to gain the advantage of surprise and psychological shock and momentum. Another factor is the inability to protect friendly forces. If one's superior *mass* is dissipated prior to attaining the assigned objective, then one's ability to go on the offensive is essentially foreclosed. It is not numerical superiority itself that is one of the main prerequisites for success, but rather the application of superior combat potential/power at the decisive place and time.[115] Sometimes the attacker mistakenly believes that his forces have reached or are about to overshoot a point of culmination and decides to stop his offensive. Lack of adequate *logistical support* is probably one of the most frequent causes of culmination. Poor organization of the forward movement of supplies and lack of transportation assets are only a few examples of such factors.[116] Constant combat and overextension of lines of supply, combined with lack of coordination and shortages of vital supplies such as fuel, water and ammunition, considerably affect arrival at the point of culmination.

The operational commanders might be overly optimistic or pessimistic in their assessment of the operational or strategic situation. Their perception of the enemy operational commander's abilities or intentions might be off the mark by a wide margin. Or they might have unrealistic expectations of their subordinate commanders and the capabilities of friendly forces. Each of these and similar errors might be a significant factor in reaching or overshooting one's own culminating point. The attacker's culmination can occur because the attacking forces move faster than their intelligence can support. A commander with barely adequate forces might then take unnecessary risks or suffer losses and endanger his success.

Poor *intelligence* is often one of the major reasons for the attacker's or the defender's culmination. The attacker and the defender can also reach their respective culminating points because of a lack of adequate intelligence support.[117] Intelligence is the key to identifying and properly evaluating indicators that would lead to a premature culmination of a major operation or campaign. One's intelligence should rely on diverse sources, ranging from technical intelligence to human intelligence.

The premature arrival at and overshooting of one's own culminating point is rarely the result of a single factor, no matter how important or dominant that factor may be. Rather, a *combination of factors* usually causes culmination. The Soviets overshot their point of culmination during their offensive in January–February 1943 in southern Russia because of a combination of logistical difficulties, continued use of greatly exhausted troops, lack of operational reserves, poor intelligence and lack of realistic appreciation of their own capabilities. No operational reserves existed or were created to influence the course and the outcome of the operation. The Soviet commanders also failed to amass whatever combat strength they had, thereby limiting their offensive potential. The commanders of the Soviet armies directed their depleted divisions to continue to fight ineffectively instead of regrouping the remaining tanks, artillery and soldiers into fewer but stronger forces. Higher headquarters constantly pressured subordinates to maintain the momentum and accomplish assigned missions. They allowed units to be bogged down attempting tasks that they were no longer capable of carrying out. The Soviets also wrongly assumed that the Germans were retreating. Otherwise, they might have been able to recognize their forces' arrival at the point of culmination.[118]

Operational/Strategic Deception

Deception is as old as warfare. It is a creative art, not an exact science. The chances of success in a campaign or major operation can be considerably enhanced if a sound, elaborate deception plan is prepared and skillfully executed. Successful deception magnifies the strength or power of the deceiver.[119] The new information technologies and techniques offer more – not fewer – opportunities for successful deception than in the past, for both the stronger side and the weaker side. In theory, the weaker side has more incentive to resort to deception than its stronger enemy[120]. In the past, the attacker has often used deception more frequently and more successfully than the defender. Deception is among the least expensive types of military activity in terms of resources used. It yields a high return for a relatively small investment.[121] Hence, the armed forces should make a determined and sustained effort to incorporate deception systematically into military thinking and practice. However, very often the commanders are unwilling to devote the necessary planning effort and resources to deception. More pernicious is the lack of interest in deception in peacetime. After hostilities break out, it is difficult, if not impossible, to timely and successfully correct the lack of mental preparation and planning techniques for deception. Then deception skills must be learned by trial and error and at great cost in terms of time and resources.

In generic terms, deception can be defined as *an action or series of actions designed to mislead the enemy by distorting, manipulating, or falsifying information available to him, thereby inducing him to do something contrary to his interests*.[122] It is also understood as planned measures for revealing or conveying to the enemy true information, or false information that could be evaluated as true, pertaining to one's strategic plans, strength, dispositions, operations or tactics, with the purpose of causing him to reach false estimates and to act thereon.[123]

Deception is aimed to mislead the enemy about one's real intentions, decisions and courses of action. It can be intended to confuse and disorient the enemy about the time and place of an attack. In the great majority of attacks since the turn of the twentieth century, tactical warning has been opportunely provided. Yet the attacker achieved surprise because deception had been used successfully.[124] Deception can be aimed to create an illusion of one's strength where weakness exists, or to paint a picture of weakness where one's strength exists. It can be intended to induce the enemy to direct his attention and concentrate his forces in the wrong place and thereby violate the principle of concentration of forces in terms of the factor of space.[125]

Friendly center of gravity can be protected by using deception to conceal, or mislead the enemy about, one's true capabilities or type of forces or the location of center of gravity. Deception can mislead an enemy about friendly capabilities by either creating an exaggerated perception of one's numerical strength or by concealing the existence of one's forces. It can be aimed to overload the enemy's collection and analytical capabilities or to deny him critical information. Successful deception invariably introduces "noise" into the collection and analytical work of the enemy's intelligence and weakens the clarity of the signals received. Therefore, it simply makes a lot of sense to use deception if for no other purpose than to create uncertainty into the enemy commander's mind about the true value of the information he receives from his intelligence, an enemy commander once victimized by deception would find it usually difficult to accept any information as reliable.[126]

There are several potential pitfalls in using deception. Among other things, the enemy can simply ignore the signals intended to deceive him, because of either sheer arrogance or stupidity. His intelligence system could be too unsophisticated to take the bait, or the commanders and their staffs might not respect the judgment of their intelligence officers. The Allied efforts to deceive the Japanese in Burma often failed because of the conceit and inflexibility of Japanese commanders and the low esteem in which these commanders held their own intelligence.[127]

Deception efforts differ in terms of objective duration, and resources involved. A distinction is made among deception efforts at the strategic, operational and tactical levels. *Strategic deception* is planned and conducted at the national-strategic level. It is designed to hide one's weaknesses or greatly exaggerate one's military or economic strength in peacetime, to conceal one's preparations or intentions to go to war, or to open hostilities. It can also be used to deceive the enemy about the place and time of opening a new front or initiate a new campaign. It encompasses a wide array of ranging from political, diplomatic, economic and informational measures actions to the threat of use or actual use of military forces.

In planning the cross-Channel invasion in January 1944, the Allies adopted a master strategic deception plan, code-named Bodyguard aimed to deceive Hitler and the Supreme Command of the Wehrmacht (OKW) about the place and time of the planned invasion. Bodyguard, encompassing three subordinate deception plans, was coordinated by the British chiefs of staff. The heart of the Allied deception effort was the plan Fortitude, designed to pose the threat of an Allied landing from the French Atlantic coast to northern Norway. Fortitude's planning and execution were the responsibility of General Dwight D. Eisenhower, Supreme Commander of the Allied Forces. The plan Zeppelin was designed to ensure the success of the Normandy landing by preventing the Germans from moving timely reinforcements from the eastern Mediterranean to northern France. The planners wanted to induce the Germans to believe that the Allies would attack Crete or western Greece and the Dalmatian coast and would conduct further landings on the Greek mainland. To ensure the plan's success, Allied strength in the Mediterranean would be greatly exaggerated.[128] The Soviets agreed to support this effort by mounting a fictitious amphibious landing against the Bulgarian naval base of Varna. To reinforce the plausibility of the Allied deception plan against Norway, the Soviets also planned an actual attack against the German forces deployed on the Kola Peninsula.[129]

Another objective of Bodyguard was designed to convince the Germans that a manpower shortage had forced the British army to resort to cannibalization because several of their large formations lacked administrative and supply units.

Bodyguard was also aimed to convince the Germans that the Allied divisions were fewer than indicated; that the U.S. divisions arriving in the United Kingdom had not yet completed their training; and that the personnel of some British divisions in the Mediterranean were relieved by fresh troops from the United States. The Allies wanted to convince the Germans that there was a shortage of landing craft because of the Pacific operations and that the required number of landing craft could not be made available until summer 1944. The planners also wanted the Germans to believe that the Allied forces in the Mediterranean – particularly the eastern Mediterranean – were larger than they actually were; that French forces were supposedly taking over the responsibility to defend North Africa, thereby freeing Anglo-American forces for operations elsewhere in the spring of 1944; and that certain British divisions and landing craft were being transferred from India to the Middle East, while fresh divisions from the United States were expected to arrive in the Mediterranean.[130]

Theater-strategic deception is a subset of national-strategic deception. It is aimed to deceive the enemy commander about the objectives, place and time of an initial major operation in a new campaign. Deception is usually conducted in a major part of the theater of war. In the spring of 1944, the Allies activated the theater-strategic plan Fortitude, aimed to convince Hitler and the OKW to make faulty strategic dispositions in northwestern Europe before the Normandy invasion by presenting military threats to Norway and the Pas de Calais area in France.[131] The main objective was to pin down German forces as far as possible from the Normandy beaches and to induce the Germans to believe that the invasion would not start until late summer.[132] The plan's scope encompassed the western European and northern theaters of operations.

Fortitude consisted of two subordinate deception plans, Fortitude North and Fortitude South. The first was aimed to convince the Germans to believe that the Allies planned to attack southern Norway on 1 May 1944 and invade northern Norway two weeks later, in conjunction with a Soviet attack. As soon as the Allies were established in southern Norway, an assault would be launched against Denmark.[133] It envisioned simultaneous landings in northern Norway of assault forces notionally held in readiness in Scotland, Iceland and Northern Ireland. Specifically, in the northern assault, two divisions would land at Narvik and advance along the rail line to Gällivare in Sweden, where they would link with Soviet forces advancing overland from Petsamo. A southern assault near Stavanger, Norway, would be carried out with one infantry division, supported by airborne troops. After consolidating their positions, these forces would advance toward Oslo, Bergen, and Trondheim by conducting a series of mutually supporting amphibious landings. Once Norway had been secured and basing rights obtained from Sweden, the Allies would land in Denmark, thereby threatening Germany from the north.[134]

Simultaneously with Fortitude North, the Allies increased their fictitious threat in the Pas de Calais area, where the German Fifteenth Army was deployed. The intent was to tie down that army before and after the Allied landing in Normandy. The deception story for Fortitude South was that the main Allied landing would take place in the Pas de Calais area forty-five days after the Normandy landing.

The latter would be merely a "diversion" to draw German reserves from the adjacent area. Once these reserves were committed, about fifty Allied divisions would launch the main attack on the Pas de Calais.[135]

The deception plan Fortitude South was highly successful. The fictional order of battle contributed materially to the establishment of the Allied bridgehead on the continent. The German Fifteenth Army remained in the Pas de Calais area from 6 June (D-Day) until late July. Field Marshal Gerd von Rundstedt, CINC of the West, did not believe that the Allies would land only south of the Seine. Not until 25 July did the first divisions of the German Fifteenth Army start to move southward in a belated attempt to reinforce the crumbling Normandy front.[136] The Germans were paralyzed by the purported existence of the Fortitude South army, believing that it could invade the Pas de Calais area.[137]

In the strict definition of the term, an *operational deception* pertains to one's actions and measures to deceive the enemy as to the time, place and details of a planned major operation conducted as part of a campaign or an independent major joint/combined operation. Successful operational deception is aimed at protecting the operational commander's intent from the enemy's intelligence-gathering sources and reinforcing the enemy's expectations and preconceptions about one's force. The deception plans for a major operation or campaign are normally multiservice and sometimes require the use of multinational forces.

Sometimes operational deception must be maintained after the initial major operation in a new campaign has been executed but while one's forces are still threatened by enemy forces within striking distance. In that case, the initial deception plan must be modified or completely rewritten. The Allies were faced with such a situation after their landing in Normandy on 6 June 1944. Because of stiff German resistance, the lodgment on the continent was tenuously held for several weeks after the landing. The Allies activated the deception plan Fortitude South II, aimed to persuade Hitler and Rundstedt that there was a continuing Allied threat to the Pas de Calais area. This threat was maintained by a large notional force, composed of the First U.S. Army Group and U.S. XIX Tactical Air Command, on the eastern and southeastern coasts of England. The threat was substantiated by considerable radio activity and about 250 landing craft, tank (LCT) dummies deployed in the harbors and estuaries of eastern and southeastern England.[138] Fortitude South II used forces similar in size to those used in Fortitude South. The big difference was that notional forces were used instead of real ones. The fresh U.S. troops had to be sent as reinforcements to Normandy. Notional forces were composed of two divisions from the 14th U.S. Army and one division from the British 4th Army. These notional assault formations also "carried out" intensive amphibious exercises.[139]

Tactical deception is aimed to deceive the enemy's tactical commander about the time, place, and details of a planned battle or engagement or some other tactical action. Allied planners prepared four tactical deception plans in support of the Normandy landing. In a plan called Operation Titanic, the dropping of dummies and some Special Air Service (SAS) teams was designed to divert German attention from the real airborne drops in the Normandy area on the night of the invasion. Two airborne drops (Titanic-1 and Titanic-2) – north of the Seine and inland of

Cap d'Antifer, and at Dives-sur-Mer (northeast of Caen) – were intended to draw German reserves away from the Allied assault objectives and delay the reserves' westward movement, respectively. The third drop (Titanic-3) was timed with the drop of the British 6th Airborne Division, to divert the German counterattack to the south and west of Caen. The fourth drop (Titanic-4), in the St. Lo area, coincided with the landing of the 101st Airborne Division to the north of Saint-Lô.[140]

Optimally, deception efforts at each level of war should be mutually supporting; the success or failure of an effort at one level considerably affects efforts at other levels. The larger the scale of the deception effort, the more critical it is to the overall success of deception in support of a campaign or major operation. If tactical or sometimes even operational deception plans fail, success may still be achieved if the strategic deception influences the enemy's strategic leadership to conform to one's expectations.

The physical space in which a deception effort can be conducted ranges from an area of operations to two or more theaters of war. Operational deception encompasses the entire theater of operations or a large part of it. In contrast, tactical deception is maintained and sustained over a much smaller area, usually an area of operations, and during a relatively short period. If not coordinated, tactical deception can often disrupt a plan for a major operation or campaign. Operational deception should be maintained over a longer period, with a consequently greater risk of discovery by the enemy. This risk may require a high degree of security to facilitate the deception of the friendly forces involved. There are many commonalities between operational and tactical deception, but also significant differences in timing, scale, and required intelligence support.[141]

Operational Reserve

An *operational reserve* is the principal tool in the hands of operational commanders for exercising and enlarging their freedom to act. The employment of reserves can often spell the difference between victory and defeat. Because of the characteristics of the physical environment, reserves are more commonly used in land warfare than in war at sea and in the air. Yet they can be created and employed for amphibious landings and major defensive air operations.

In generic terms, a *reserve* is a force of a certain size and combat potential aimed at having a decisive impact on the outcome of a given combat action. It is withheld from combat in the initial phase of an action, and its size and composition and the place and time of its employment are directly controlled by the higher commander. Hence, a reserve should not be confused with reinforcements, replacements, or second-echelon forces. Although related, they are not identical in terms of purpose, force size and composition/mix, location, and timing of employment.[142]

Depending on their purpose, tactical, strategic and operational reserves are differentiated. A *tactical reserve* is aimed to have a decisive effect on the outcome of a battle or engagement or some other tactical action. An *operational reserve* is a force of such size and combat potential as to have a decisive impact on the outcome of a campaign or major operation. In land warfare, it is usually composed of the most mobile and combat-ready combined-arms force, prepared and ready for employment at short notice on the order of the operational commander. An operational reserve is usually a one-time-use force.[143] It is intended to ensure and consolidate one's operational success.[144] A *strategic reserve* is intended to have a decisive influence on the outcome of a single or several campaigns in a theater of war. It is controlled by national-strategic authorities or the supreme command. In some instances, it may be controlled by a subordinate theater commander. Strategic reserve is created from available forces or by reducing national commitments or changing strategic priorities in a given theater (of war).

Obviously, if the course of events in combat could be precisely predicted, there would be no need for a commander to create a reserve. In that case, it would not only be foolish but also border on criminality not to employ all available forces initially in the combat. However, the course of any combat action is hard to predict. The enemy can, and most likely will, act in a manner that cannot be fully anticipated. Hence, a prudent commander should have a force held back to react quickly to unforeseen developments and thereby ensure a successful outcome to the actions of friendly forces.

A reserve is mobile and the most effective means at the disposal of the commander.[145] It is the commander's principal means of influencing the outcome of an operation beyond the initial clash of forces.[146] The commanders should have a reserve to minimize the risks to friendly forces. They should employ the maximum force available at the decisive place and time. This means that the commander should find a proper balance between the forces committed to the main effort and those assigned to serve as a reserve. At the tactical level, the commander is usually not concerned with sequels to the action in progress. In contrast, the operational commander must plan for sequels to follow the accomplishment of the operational or major tactical objective.[147]

An operational reserve considerably enlarges the commander's freedom to act. It can have a decisive influence on the outcome of a major operation or even a campaign. Sometimes just the threat of using one's operational reserve can force an enemy commander to react operationally. One's failure to have a reserve because of the need to maximize available strength might lead to a disaster if the enemy does not conform to one's planning assumptions.[148] The main disadvantage of creating a reserve is that it reduces the combat potential of the forces initially committed in combat. This disadvantage is relatively more serious in an offensive campaign or major operation than it is when one's forces are on the defensive. A reserve also requires combat support and combat service support that might be more urgently needed elsewhere. In addition, a reserve requires forces for its protection, which complicates the protection of other forces and installations in a given part of the theater.

Conclusion

No campaign or major operation can be successful if combat forces are not deployed opportunely and in proper sequence into the theater. Mistakes in deployment, and especially in strategic deployment, cannot be corrected easily, if at all, once a campaign or major operation starts. The outcome depends heavily on the outcomes of the initial major operations or major battles because they set the stage for successive major operations or battles.

Time and speed are critical factors in successful deployment; both are largely dependent on the distances and mobility of one's forces. The transportation infrastructure in the theater should be adequate to support the deployment of multi-service and, often, multinational forces. The theater should have numerous and diverse lines of communication linking various transportation modes. The throughput capacity of seaports and airports, rail and road networks, internal waterways and pipelines, especially in the deployment area, significantly affects the success of deployment.

The theoretical foundations behind the center of gravity are relatively new. For the most part it emerged because of renewed emphasis on the study of Clausewitz's *on War.* The subsequent development and application of the term center of gravity was essentially based on a misunderstanding because the English translation of Clausewitz's work equated *Schwerpunkt* with center of gravity. Ironically, the concept of center of gravity is a highly useful tool for planners at all command echelons. This concept can be applied across the spectrum of warfare and all levels of war. The prerequisite, however, is that the center of gravity must be directly related to a given objective; otherwise, the entire concept has little, if any, utility. In the past, the great captains and many less known but successful commanders did not use the term center of gravity but intuitively knew that the quickest and most decisive way to defeat the enemy was to attack the strongest part of his army. This was the way Alexander the Great, Hannibal, Gustav Adolphus, Frederick the Great, Napoleon I, Moltke, Sr, and many other great commanders fought their campaigns. Attacking the enemy center of gravity and protecting the friendly center of gravity are a proven and viable concept despite the somewhat convoluted theoretical underpinnings. This concept needs to be further refined. It is here to stay; no changes in technology are likely to invalidate it.

A full understanding of the theory of the concept of center of gravity is critical if commanders and their staff intend to employ all available sources of power to achieve success in the shortest time and with the least losses for one's forces. It makes sense that, to accomplish the assigned military objective, one must focus the major part) of one's efforts and time against the strongest source of the enemy's power – his center of gravity. That will also ensure observance of the principles of objective, mass, and economy of effort.

An operational maneuver is a critical element of any major operation or campaign. It should not be confused, as it often is, with deployment. New technological advances have not reduced the importance of operational maneuvers. There are some significant differences between an operational maneuver conducted on land, those conducted at sea, and those in the air. Each form of operational maneuver offers distinct advantages and disadvantages. The operational commanders and staff should fully understand all these complexities to make the best decision as to which form of maneuver to apply. Any maneuver, but operational maneuver, is a complex undertaking that requires meticulous planning, preparation, and execution. Any failure to properly plan and carry out an operational maneuver will result in the failure of a major operation or even campaign. Operational maneuver is not a thing of the past. The purpose of such a maneuver will not change in the future. Yet, the size of combat forces conducting such a maneuver will be considerably smaller than in the past. Also, new technological advances will undoubtedly change the types and methods of planning and conducting such maneuvers.

The Clausewitzian concept of a culminating point remains relevant and viable today despite the passage of time. While the theoretical underpinnings of the concept remain essentially the same as Clausewitz postulated, its content has changed significantly. Since the early nineteenth century, the operational level of war and several sublevels of war have emerged. The factors affecting culmination today are also much more diverse and difficult to quantify than they were in the past, making the proper application of the concept more difficult. The factors affecting culmination at the operational and strategic levels are predominantly unquantifiable in character. To successfully anticipate and plan for arrival at the culminating point, it is critical to fully comprehend the theory. Theory can show the way, but it cannot guarantee success. Historical examples can only illustrate certain aspects or problems with theory; they cannot show the path in the future. However, it is difficult to see how the theory can be properly understood without those examples.

Application of the concept of culmination requires the highest degree of skill on the part of commanders and their staffs. This is even more critical at the operational and strategic levels of war, because the consequences of premature culmination or failure to take advantage of the enemy's culmination are far more severe and more durable than at the tactical level. While many factors cause culmination, probably the four most important factors are pursuing simultaneously or nearly simultaneously the accomplishment of multiple operational or major tactical objectives, and inadequate logistics, intelligence and force protection.

Deception is one of the most neglected yet arguably most critical elements for the success of a campaign or major operation. It should be an integral part of any sound operation plan. Deception is a critical part of a campaign or major operation, and operational commanders and their staff should believe in its importance in all their planning and preparations. The target of operational deception is the enemy commander who has the authority and capability to act and react operationally or even strategically. Strategic deception invariably aims to influence decisions made, or to be made, by the highest national or alliance/coalition leadership. The success or failure of strategic and operational deception can have a drastic effect on the outcome of a campaign or major operation.

The new information technologies and techniques increase, rather than reduce, opportunities for deception efforts at all levels of war. They allow both the attacker and the defender far greater choice in deception than they had in the past. Yet technology alone, no matter how sophisticated and how readily available, cannot resolve the basic problem – that is, the wide awareness of and positive attitude toward the critical need to apply deception at all levels of military effort. Nowhere is that more critical to success than in planning, preparing and conducting major operations and campaigns. No success is possible or can be expected without a firm belief in the value of deception as a force multiplier and in the fact that a deception plan must be an integral part of any plan for a major operation or campaign.

The concept of operational reserves is as viable today as it was in the past. Because of the characteristics of the physical environment, reserves are much more important in land warfare than in war at sea or in the air. Current and projected advances in technology can mitigate but not significantly reduce the fog of war and the effects of friction. War will always be full of uncertainties, and no one can realistically expect to obtain information dominance. To meet unforeseen opportunities and overcome crises during a campaign or major operation, the commander should withhold some forces from the initial combat. The operational commanders also must make sound decisions as to the size, force composition/mix and positioning of the reserves they create. To be useful, the operational reserve must have a significant impact on the course of a campaign or major operation. Hence, the size of the operational reserve is not as important as its available combat power.

Notes

1 Headquarters, Department of the Army, *FM 100-17-3: Reception, Staging, Onward Movement, and Integration* (Washington, DC: U.S. Government Printing Office, March 17, 1999), 1–7.
2 E. Hartmann, *Kurzgefasstes Militär-Hand-Wőrterbuch für Armee und Marine* (Leipzig: Verlag von George Lang, 1896), 31.
3 Walter Post, *Unternehmen Barbarossa. Deutsche und sowjetische Angriffspläne 1940/41* (Hamburg/Berlin/Bonn: Verlag E. S. Mittler & Sohn, 1995), 207.
4 Wilhelm Stanger, *Grundzüge der Lehre von der Strategie. Studienbehelf fuer die K.K. Kriegsschule, vol. 1: Theorie mit Kűrzeren Beispielen* (Vienna: Verlag der K.K. Kriegsschule, 1884), 98.
5 Headquarters, Department of the Army, *FM 100-17-3: Reception, Staging, Onward Movement, and Integration* (Washington, DC: U.S. Government Printing Office, March 17, 1999), 1–13.
6 Headquarters, Department of the Army, *FM 100-17-3: Reception, Staging, Onward Movement, and Integration*, 3–9.
7 Headquarters, Department of the Army, *FM 100-16: Army Operational Support* (Washington, DC: U.S. Government Printing Office, 1995), 5–7.
8 Mark Pires, *Strategic Lift: Can the United States Conduct Two Nearly Simultaneous Major Regional Contingencies* (Fort Leavenworth, KS: School of Advanced Military Studies, U.S. Command and General Staff College, 1995), 19.
9 Headquarters, Department of the Army, *FM 100-17-3: Reception, Staging, Onward Movement, and Integration*, 1–4.
10 Pires, *Strategic Lift*, 19.
11 Headquarters, Department of the Army, *FM 100-17-3: Reception, Staging, Onward Movement, and Integration*, I–13.
12 Ihno Krumpelt, "Die Bedeutung des Transportwesen für den Schlachterfolg," *Wehrkunde* 5 (May 1965), 465.
13 P. D. Foxton, *Powering War: Modern Land Force Logistics* (London: Brassey's, 1994), 50.
14 Pires, *Strategic Lift*, 29–31.
15 Headquarters, Department of the Army, *FM 100-17-3: Reception, Staging, Onward Movement, and Integration*, 1–5.

16 Schwerpunkt, *H. Dv 100/900 VS-NfD, Führungsbegriffe (TF/B)* (Bonn: Ministry of Defense, February 1990), Sch-SEA; other related terms include "area of weight of the main effort" [*Schwerpunktraum*], "weight of the main effort in an attack" [*Schwerpunkt des Angriffs*], etc. Huerth, *US-NfD. Militärisches Studienglossary Englisch,* vol. 2/3 (Bonn: Bundessprachenamt, January 1993), 1060; *Handbuch der neuzeitlichen Wehrwisenschaften,* vol. 1: *Wehrpolitik und Kriegführung*, ed. Hermann Franke (Berlin/Leipzig: Verlag von Walter de Grunter, 1936), 649.

17 In the strict definition of the term, center of gravity is defined as "that point of an object around which its weight is evenly distributed or balanced; center of mass; point of equilibrium"; *Webster's New World Dictionary of the American Language*, College ed. (New York, NY: World Publishing, 1960), 237.

18 Thomas M. Kriwanek, *The Operational Center of Gravity* (Ft. Leavenworth, KS: School of Advanced Military Studies, U.S. Army Command and General Staff College, May 1986), 7.

19 Gordon M. Wells, *The Center of Gravity Fad: Consequence of the Absence of an Overarching American Theory of War* (Arlington, VA: AUSA's Institute of Land Warfare, March 2001), 4.

20 Phillip Kevin Giles and Thomas P. Galvin, *Center of Gravity: Determination, Analysis, and Application* (Carlisle Barracks, PA: Center for Strategic Leadership, U.S. Army War College, January 31, 1996), 19.

21 Carl von Clausewitz, *On War,* ed. and trans. Michael Howard and Peter Paret (Princeton, NJ: Princeton University Press, 1984), 596.

22 Giles and Galvin, *Center of Gravity*, 17.

23 Giles and Galvin, *Center of Gravity*, 15.

24 Joseph Schroedel, *The Art and Science of Operational Maneuver* (Fort Leavenworth, KS: School of Advanced Military Studies, U.S. Command and General Staff College, May 18, 1988), 4–5.

25 General Erich von Ludendorff, who was the head of the Deployment Section [*Aufmarschabteilung*] of the German general staff in 1908–1912, claimed that Schlieffen in his memorandum envisaged the employment of 8 infantry divisions and 16 reserve divisions, while in 1905 only 72 infantry divisions were available and there was a shortage of 16 reserve divisions; Terence Zuber, "The Schlieffen Plan Reconsidered," *War in History* 6, no. 3 (July), 1999, 266; Peter Wenning, *Aufmarschplanung als strategische Führungsproblem. Eine vergleichende Untersuchung am Beispiel des Schlieffenplans 1914, der NATO Aufmarschplanung im Rahmen der "Flexible Response" un dem neuen strategischen Konzept der NATO* (Hamburg: Führungsakademie der Bundeswehr, December 30, 1997), 5.

26 R. Ernest Dupuy and Trevor N. Dupuy, *The Encyclopedia of Military History from 3500* BC *to the Present* (New York, NY: Harper & Row, 1986), 93; Holger Herwig, *The First World War: Germany and Austria-Hungary 1914–1918* (London: Arnold, Hodder Headline Group, 1997), 100.

27 Eberhard von Mantey, "Umfassung, Umgehung und Durchbruch. Eine Schlieffen Studie," *Wehr und Wissen* 10 (October 1931), 577.

28 Hermann Franke, ed., *Handbuch der neuzeitlichen Wehrwissenschaften,* vol. 1, *Wehrpolitik und Kriegführung* (Berlin/Leipzig: Walter de Gruyter, 1936), 206.

29 R. Ernest Dupuy, Trevor N. Dupuy, and Paul F. Braim, *Military Heritage of America*, vol. 1 (Dubuque, IA: Kendall/Hunt, 1992), 17.

30 Headquarters, Department of the Army, *FM 100-5: Operations* (Washington, DC: U.S. Government Printing Office, 1994), 7–12.

31 von Bernhardi, *On War of To-Day*, 77–78.

32 Jovo Ninkovic, *Nastanak i Razvoj Ratne Operacije* (Belgrade: Vojno-Izdavački Zavod, 1984), 41.

33 Manfred Rauchensteiner, *Der Tod des Doppeladlers. Österreich-Ungarn und der Erste Weltkrieg* (Graz/Vienna/Cologne: Verlag Styria, 1993), 213–14; Charles A.

Willoughby, *Maneuver in War* (Harrisburg, PA: Military Service Publishing, 1939; reprinted by Headquarters, U.S. Marine Corps, March 17, 1986), 93.
34 Wieker, *Operation, Operative Fűhrung zwischen Taktik und Strategie*, 17.
35 Manfred Rauchensteiner, *Der Tod des Doppeladlers. Ősterreich-Ungarn und der Erste Weltkrieg*, 213–14; Willoughby, *Maneuver in War*, 93.
36 Dupuy and Dupuy, *The Encyclopedia of Military History from 3500* BC *to the Present*, 971.
37 Franke, *Handbuch der neuzeitlichen Wehrwissenschaften*, vol. 1, 15.
38 Historical Evaluation and Research Organization (HERO), *Developing a Methodology to Describe the Relationship of Mobility to Combat Effectiveness, Report Prepared for Research Analysis Organization under Subcontract FY 67-ARD1-1-2* (Fairfax, VA: HERO, December 27, 1966), 37.
39 Historical Evaluation and Research Organization (HERO), *Developing a Methodology to Describe the Relationship of Mobility to Combat Effectiveness*, 37.
40 Militärgeschichtliches Forschungsamt (MGFA), *Development, Planning and Realization of Operational Conceptions in World War I and II* (Herford/Bonn: E. S. Mittler & Sohn, 1989), 9.
41 Franke, *Handbuch der neuzeitlichen Wehrwissenschaften*, vol. 1, 206.
42 von Bernhardi, *On War of To-Day*, 79–80.
43 Franke, *Handbuch der neuzeitlichen Wehrwissenschaften*, vol. 1, 15.
44 Board for Study of Principles of War, *Principles of War*, trans. Dr. Joseph West (Tokyo: Ground Self-Defense Force Staff College, January 1969; reprinted by Combat Studies Institute, U.S. Army Command and General Staff College, Fort Leavenworth, KS, July 1983), 40–41.
45 Dupuy, Dupuy, and Braim, *Military Heritage of America*, 16.
46 Franke, *Handbuch der neuzeitlichen Wehrwissenschaften*, vol. 1, 15.
47 von Mantey, "Umfassung, Umgehung und Durchbruch," 575.
48 Christopher Bellamy, *The Evolution of Modern Land Warfare: Theory and Practice* (London/New York: Routledge, 1990), 67, 71; von Bernhardi, *On War of To-Day*, 68.
49 Board for Study of Principles of War, *Principles of War*, trans. Dr. Joseph West (Tokyo: Ground Self-Defense Force Staff College, January 1969; reprinted by Combat Studies Institute, U.S. Army Command and General Staff College, Fort Leavenworth, KS, July 1983), 40.
50 Bellamy, *The Evolution of Modern Land Warfare: Theory and Practice*, 67, 71; von Bernhardi, *On War of To-Day*, 68.
51 Karl-Heinz Frieser, *Blitzkrieg-Legende. Der Westfeldzug 1940* (Munich: R. Oldenbourg Verlag, 1995), 82, 118.
52 Paul Stuart, *Operational Maneuver: Creator of the Decisive Point* (Fort Leavenworth, KS: School of Advanced Military Studies, U.S. Army Command and General Staff College, May 1987), 17.
53 Headquarters of the U.S. Army, *FM 100-5.1: Operations Terms and Graphics* (Washington, DC: U.S. Government Printing Office, October 21, 1985), 1–60.
54 Bellamy, *The Evolution of Modern Land Warfare: Theory and Practice*, 18.
55 Robert G. Fix, *Operational Encirclement: Quick Decisive Victory or a Bridge Too Far?* (Fort Leavenworth, KS: School of Advanced Military Studies, U.S. Command and General Staff College, Second Term 1993), 6.
56 Dupuy, Dupuy, and Braim, *Military Heritage of America*, 16–17.
57 Alfred Philippi and Ferdinand Heim, *Der Feldzug Gegen Sowjetrussland 1941 bis 1945. Ein operativer Űberblick* (Stuttgart: W. Kohlhammer Verlag, 1962), 78; Seaton, *The Russo-German War 1941–45*, 147.
58 von Mantey, "Umfassung, Umgehung und Durchbruch," 577.
59 Paul T. DeVries, "Maneuver and the Operational Level of War," *Military Review* 2 (February 1983), 25.

60 Militärgeschichtliches Forschungsamt (MGFA), *Development, Planning and Realization of Operational Conceptions in World War I and II*, 25.
61 Militärgeschichtliches Forschungsamt (MGFA), *Development, Planning and Realization of Operational Conceptions in World War I and II*, 25.
62 Militärgeschichtliches Forschungsamt (MGFA), *Development, Planning and Realization of Operational Conceptions in World War I and II*, 17.
63 Wilson A. Heefner, "The Inchon Landing," *Military Review* 2 (March–April 1995), 69–70.
64 Lynn Montross, "The Inchon Landing – Victory over Time and Tide," *Marine Corps Gazette* 7 (July 1951), 27; Malcolm W. Cagle, "Inchon – the Analysis of a Gamble," *Proceedings* 1 (January 1954), 49.
65 Gűnter Roth, "Operatives Denken bei Schlieffen und Manstein," in *Ausgewählte Operationen und ihre militärhistorischen Grundlagen*, ed. Hans-Martin Ottmer and Heiger Ostertag (Herford/Bonn: E. S. Mittler & Sohn, 1993), 292–93.
66 Karl Heinz Frieser, "Der Westfeldzug und die 'Blitzkrieg-Legende'," in *Ausgewählte Operationen und ihre militärhistorischen Grundlagen*, ed. Hans-Martin Ottmer and Heiger Ostertag (Herford/Bonn: E. S. Mittler & Sohn, 1993), 391.
67 Erich Brandenberger, *Der Deutsche Generalstab,* P-031a, ZA/1 1879, Studien der Historical Division Headquarters, United States Army Europe, Foreign Military Studies Branch, Bundesarchiv/Militärarchiv (BA-MA), Freiburg i. Br., 118.
68 Rick Gutwald, *The Effects of Operational Encirclement* (Fort Leavenworth, KS: School of Advanced Military Studies, U.S. Command and General Staff College, 1987), 2.
69 Robert G. Fix, *Operational Encirclement: Quick Decisive Victory or a Bridge Too Far?* (Fort Leavenworth, KS: School of Advanced Military Studies, U.S. Command and General Staff College, Second Term 1993), 7.
70 This number included 20 German and two Romanian divisions and a number of Russian auxiliaries; the Germans had 1,800 guns, 10,000 motor vehicles, and 8,000 horses; Seaton, *The Russo-German War 1941–45*, 321.
71 Bernd Wegner, "Der Krieg Gegen Die Sowjetunion 1942/43," in *Das Deutsche Reich und der Zweite Weltkrieg,* vol. 6: *Der Globale Krieg. Die Ausweitung zum Weltkrieg und der Wechsel der Initiative*, ed. Horst Boog et al. (Stuttgart: Deutsche Verlags-Anstalt, 1990), 1000–1001. Reportedly this force included more than four tanks, three mechanized corps, 14 tank brigades, and several independent tank regiments; these three fronts were supported with some 13,500 guns and mortars, and more than 100 multiple rocket launchers [*Katyushas*]; Seaton, *The Russo-German War 1941–45*, 307.
72 Ziemke and Bauer, *Moscow to Stalingrad*, 468.
73 Wegner, "Der Krieg Gegen Die Sowjetunion 1942/43,"1050–51.
74 Ziemke and Bauer, *Moscow to Stalingrad*, 486.
75 Wegner, "Der Krieg Gegen Die Sowjetunion 1942/43," 1055. Other sources cited Soviet strength at 281,000; Ziemke and Bauer, *Moscow to Stalingrad*, 496.
76 Ziemke and Bauer, *Moscow to Stalingrad*, 501.
77 Cited in S. J. Lewis, "The Battle of Stalingrad," in *Block by Block: The Challenges of Urban Operations*, ed. William G. Robertson and Lawrence A. Yates (Fort Leavenworth, KS: U.S. Army Command and General Staff College Press, 2003), 53.
78 Cited in Gerhard L. Weinberg, "Stalingrad and Berlin: Fighting in Urban Terrain," in *Soldiers in Cities: Military Operations on Urban Terrain*, ed. Michael C. Desch (Carlisle Barracks, PA: Strategic Studies Institute, U.S. Army War College, October 2001), 24–25.
79 Peter Wenning, *Aufmarschplanung als strategische Führungsproblem*, 14.
80 Carl von Clausewitz, *Vom Kriege*, ed. with comments by Dr. Werner Hahlweg, 16th ed. (Bonn: Dümmlers Verlag, 1952), 777.

81 Carl von Clausewitz, *On War*, ed. and trans. Michael Howard and Peter Paret, 8th printing (Princeton, NJ: Princeton University Press, [1976] 1984), 528, 571.
82 Stephen C. Duncan, *Intelligence and Defensive Culminating Point – Piercing the Fog* (Fort Leavenworth, KS: School of Advanced Military Studies, U.S. Army Command and General Staff College, 1995), 4.
83 von Clausewitz, *On War*, 357–58, 566, 571.
84 David Benjamin, *Prerequisite for Victory: The Discovery of the Culminating Point* (Fort Leavenworth, KS: School of Advanced Military Studies, U.S. Army Command and General Staff College, May 13, 1986), 23.
85 Hans-Peter von Kirchbach, *Clausewitz and the Culmination Point of Victory* (Carlisle Barracks, PA: U.S. Army War College, March 15, 1989), 26; Daniel J. Roh, *Operational Pause vs. Offensive Culmination: Lessons in Eisenhower's Broad Front Strategy* (Fort Leavenworth, KS: School of Advanced Military Studies, U.S. Army Command and General Staff College, April 26, 1988), 8–9; von Clausewitz, *On War*, 527; Headquarters, Department of the Army, *FM 100-5: Operations* (Washington, DC: Headquarters, Department of the Army, June 1993), 6–8; Franklin, *Time, Space, and Mass at the Operational Level of War*, 13.
86 Headquarters, Department of the Army, *FM 100-5: Operations* (June 1993), 6–8.
87 von Clausewitz, *On War*, 572; von Kirchbach, *Clausewitz and the Culmination Point of Victory*, 6.
88 James D. Coomler, *The Operational Culminating Point: Can You See It Coming*? (Fort Leavenworth, KS: School of Advanced Military Studies, U.S. Army Command and General Staff College, May 16, 1986), 5.
89 Coomler, *The Operational Culminating Point: Can You See It Coming*? 10–11.
90 Albert Kesselring, *Soldat Bis Zum Letzten Tag* (Bonn: Athenaeum Verlag, 1953), 166.
91 Paul Carell, *Scorched Earth* (New York: Ballantine Books, 1973), 189–90; Frank T. Taddonio, *The Concept of the Culminating Point of Attack* (Fort Leavenworth, KS: School of Advanced Military Studies, U.S. Army Command and General Staff College, May 1986), 6–7; Seaton, *The Russo-German War 1941–45*, 344.
92 Benjamin, *Prerequisite for Victory*, 8.
93 Taddonio, *The Concept of the Culminating Point of Attack*, 11.
94 Militärgeschichtliches Forschungsamt (MGFA), *Development, Planning and Realization of Operational Conceptions in World Wars I and II*, 29–34.
95 Militärgeschichtliches Forschungsamt (MGFA), *Development, Planning and Realization of Operational Conceptions in World Wars I and II*, 29–34; Bruce L. Meisner, *The Culminating Point – A Viable Operational Concept or Some Theoretical Nonsense?* (Fort Leavenworth, KS: School of Advanced Military Studies, U.S. Army Command and General Staff College, March 1986), 11–16.
96 von Clausewitz, *On War*, 357, 383; Barry D. Fulbright, *The Effects of Airpower on the Offensive Culminating Point* (Fort Leavenworth, KS: School of Advanced Military Studies, U.S. Army Command and General Staff College, April 25, 1988), 3; Headquarters, Department of the Army, *FM 100-5: Operations* (June 1993), 6–8.
97 Headquarters, Department of the Army, *FM 100-7: Decisive Force: The Army in Theater Operations* (Washington, DC: Headquarters, Department of the Army, May 1995), 1–3.
98 Headquarters, Department of the Army, *FM 100-5: Operations* (June 1993), 6–8.
99 William W. Hamilton, *Defensive Culmination – When Does the Tactical Commander Counterattack?* (Fort Leavenworth, KS: School of Advanced Military Studies, U.S. Army Command and General Staff College, 1991), 4.
100 Mark J. Redlinger, *Hans Delbrueck and Clausewitz's Culminating Points* (Fort Leavenworth, KS: School of Advanced Military Studies, U.S. Army Command and General Staff College, May 1988), 9.
101 Duncan, *Intelligence and Defensive Culminating Point*, 7–8.

102 Ziemke and Bauer, *Moscow to Stalingrad*, 61; Center for Military History, *The German Campaign in Russia: Planning and Operations (1940–1942)*, 1st ed. (Washington, DC: Center for Military History, United States Army, CMH Pub 104-21, [March 1955] 1988), 80.
103 Alfred Philippi and Ferdinand Heim, *Der Feldzug Gegen Sowjetrussland 1941 bis 1945. Ein operativer Überblick* (Stuttgart: W. Kohlhammer Verlag, 1962), 90–91.
104 Center for Military History, *The German Campaign in Russia: Planning and Operations (1940–1942),* 83.
105 Ziemke and Bauer, *Moscow to Stalingrad*, 61.
106 Alfred Philippi and Ferdinand Heim, *Der Feldzug Gegen Sowjetrussland 1941 bis 1945*, 98–99.
107 von Kirchbach, *Clausewitz and the Culmination Point of Victory*, 9.
108 Redlinger, *Hans Delbrueck and Clausewitz's Culminating Points*, 10–11.
109 von Clausewitz, *On War*, 570.
110 von Clausewitz, *On War*, 570.
111 Headquarters, Department of the Army, *FM 100-5: Operations* (June 1993), 6–8.
112 von Clausewitz, *On War*, 597.
113 U.S. Army Command and General Staff College, *The Principles of Strategy for an Independent Corps or Army in a Theater of Operations* (Fort Leavenworth, KS: Command and General Staff School Press, 1936), 36.
114 Franklin, *Time, Space, and Mass at the Operational Level of War*, 16.
115 Taddonio, *The Concept of the Culminating Point of Attack*, 24; George M. Hall, "Culminating Points," *Military Review* 7 (July 1989), 84; Franklin, *Time, Space, and Mass at the Operational Level of War*, 17.
116 Taddonio, *The Concept of the Culminating Point of Attack*, 6–8.
117 Headquarters, Department of the Army, *FM 100-5: Operations* (June 1993), 6–8.
118 Coomler, *The Operational Culminating Point: Can You See It Coming?* 15–16.
119 Michael Handel, "Intelligence and Deception," *Journal of Strategic Studies* (March 1982), 123.
120 Michael Handel, *War, Strategy and Intelligence* (London: Frank Cass, 1989), 311.
121 Handel, *War, Strategy and Intelligence*, 339.
122 A deception can also be defined as planned measures for revealing or conveying to the enemy true information, or false information that could be evaluated as true, regarding our strategic plans, strength, dispositions, or tactics, with the purpose of causing him to reach false estimates and to act thereon; Strategic Plans Division Branch, OP-607 Support Plans Branch (Series XXII), Box 531, 1.
123 Strategic Plans Division Branch, OP-607 Support Plans Branch (Series XXII), Box 531, Naval Operational Archives, Naval Historical Center, Washington, DC, 1.
124 Headquarters, Department of the Army, *FM 90-2: Battlefield Deception* (Washington, DC: U.S. Government Printing Office, October 3, 1988), 1–1.
125 Handel, "Intelligence and Deception," 126.
126 Handel, *War, Strategy and Intelligence*, 340.
127 Jon Latimer, *Deception in War* (Woodstock/New York, NY: Overlook Press, 2001), 38.
128 Charles Cruickshank, *Deception in World War II* (Oxford: Oxford University Press, 1979), 146–47.
129 "Military Cover and Deception (Strategic)," presentation prepared in the Office of the Chief of Naval Operations, August 5, 1947, Strategic Plans Division Records, OP-607 Support Plans Branch (Series XXII), Box 530, Folder #2; Naval Operational Archives, Naval Historical Center, Washington, DC, 18.
130 Combined Chiefs of Staff, "Plan Bodyguard," 20 January 1944, Strategic Plans Divisions Records, OP-607 Support Plans Branch (Series XXII), Box 531; Naval Operational Archives, Naval Historical Center, Washington, DC, 7–8.

131 Donald J. Sexton, "Phantoms of the North: British Deceptions in Scandinavia, 1941–1944," *Military Affairs* (October 1983), 112.
132 Cover & Deception Report ETO, "Cover & Deception, Synopsis of History," Strategic Plans Divisions Records, OP-607 Support Plans Branch (Series XXII), Box 549; Naval Operational Archives, Naval Historical Center, Washington, DC, 6.
133 Military Cover and Deception (Strategic)," presentation prepared in the Office of the Chief of Naval Operations, August 5, 1947, Strategic Plans Division Records, OP-607 Support Plans Branch (Series XXII), Box 530, Folder #2; Naval Operational Archives, Naval Historical Center, Washington, DC, 20.
134 Sexton, "Phantoms of the North," 112.
135 Military Cover and Deception (Strategic)," presentation prepared in the Office of the Chief of Naval Operations, August 5, 1947, Strategic Plans Division Records, OP-607 Support Plans Branch (Series XXII), Box 530, Folder #2"; Naval Operational Archives, Naval Historical Center, Washington, DC, 21.
136 Military Cover and Deception (Strategic)," presentation prepared in the Office of the Chief of Naval Operations, August 5, 1947, Strategic Plans Division Records, OP-607 Support Plans Branch (Series XXII), Box 530, Folder #2, 24; Cover & Deception Report ETO, "Cover & Deception, Synopsis of History," Strategic Plans Divisions Records, OP-607 Support Plans Branch (Series XXII), Box 549; Naval Operational Archives, Naval Historical Center, Washington, DC, 8.
137 Cover & Deception Report ETO, "Cover & Deception, Synopsis of History," Strategic Plans Divisions Records, OP-607 Support Plans Branch (Series XXII), Box 549; Naval Operational Archives, Naval Historical Center, Washington, DC, 8.
138 SHAEF, "Plan Fortitude South II, 19 July 1944," Strategic Plans Divisions Records, OP-607 Support Plans Branch (Series XXII), Box 531; Naval Operational Archives, Naval Historical Center, Washington, DC, 13.
139 Cover & Deception Report ETO, "Cover & Deception, Synopsis of History," Strategic Plans Divisions Records, OP-607 Support Plans Branch (Series XXII), Box 549; Naval Operational Archives, Naval Historical Center, Washington, DC, 9.
140 "Operational Deception," presentation at the Naval War College, Newport, March 7, 1947, Office of the Chief of Naval Operations; "Deception Evaluation Report Number Two," April 30, 1947, Command File, World War II, Office of the Chief of Naval Operations; Naval Operational Archives, Naval Historical Center, Washington, DC, 17–19.
141 Thomas A. Savoia, *Deception at the Operational Level of War* (Fort Leavenworth, KS: School of Advanced Military Studies, U.S. Army Command and General Staff College, May 1986), 6.
142 *Reinforcements* are, in contrast, intended to replace one's losses and prevent the enemy from gaining more space. They cannot result in obtaining the initiative against the enemy on the attack. Reinforcements are also committed piecemeal. The commander cannot plan at what time or place reinforcements will have to be sent. They are brought in incrementally for the purpose of reinforcing certain parts of the front line. *Replacements* are intended to replace units destroyed or withdrawn from the front line to be reconstituted. Both reinforcements and replacements are not necessarily composed of one's most trained, mobile and equipped forces. *Second-echelon* forces are intended to convert the tactical success achieved by the first-echelon forces into operational success. They are usually deployed in the sector of main effort. The prerequisite for creating second-echelon forces is a rather large numerical superiority.
143 Albert F. Turner, Jr., *The Operational Reserve: What Should It Be Used For?* (Fort Leavenworth, KS: School of Advanced Military Studies, U.S. Army Command and General Staff College, April 6, 1988), 22.

144 Terry W. Bullington, *Considerations for the Organization and Employment of an Operational Reserve* (Fort Leavenworth, KS: School of Advanced Military Studies, U.S. Army Command and General Staff College, May 16, 1986), 3.
145 Friedrich Wentzel, *Kampferfahrungen aus dem Osten,* MS # B-266, ZA/1 617, April 5, 1946, Studien der Historical Division Headquarters, United States Army Europe, Foreign Military Studies Branch, Bundesarchiv/Militärarchiv (BA-MA), 22; Lothar Rendulic, *Entschlussfassung,* MS # D-080a, ZA/1 1429, Studien der Historical Division Headquarters, United States Army Europe, Foreign Military Studies Branch, BA-MA, March 9, 1947, 8.
146 Turner, *The Operational Reserve*, 10.
147 John F. Schmitt, "The Use of the Reserve in Combat," *Marine Corps Gazette* (March 1990), 64.
148 Charles R. Gregory, *Operational Reserves – Renewing the Offensive Spirit* (Fort Leavenworth, KS: School of Advanced Military Studies, U.S. Army Command and General Staff College, April 6, 1988), 24; Turner, *The Operational Reserve*, 11.

6 Operational Command and Control (C2)

Command organization and command and control *(C2)* are related but not identical in their meaning.[1] Yet one cannot exist without the other. Very often, the term C2 is used as pertaining to command structure, and vice versa. C2 and command organizations are closely related and intertwined. Also, C2 is often used to refer to a system, instead of a *process* of planning, preparing and controlling one's forces to accomplish assigned missions in peacetime and in combat. The key prerequisite for an effective C2 process is the establishment and maintenance of a sound command organization.

Command and control is exercised by the commanders at each level of command. The levels of command are established for the employment use of sources of military power to accomplish specific objectives. They exist in times of peace and in wartime. Based on the scale of military objectives, three basic levels of command are differentiated: strategic, operational and tactical. For large countries – such as the United States, China, Russia and India – that deploy forces in two or more theaters, the strategic level of command encompasses two sublevels: national-strategic and theater-strategic. For each theater of operations established in time of war, there would be a corresponding operational level commander. The theater-strategic and operational commanders would be responsible for planning, preparing and executing major operations and campaigns. In some rare cases, operational-tactical commanders (commanders of JTF/CJTF in U.S. terms) would be directed to plan, prepare and execute a major joint operation.

Centralized versus Decentralized C2

There are two principal types of command and control that apply at any level of command: centralized and decentralized. In centralized, or order-oriented, C2, all the authority is concentrated in a single commander and a single headquarters. Centralized C2 is formal and highly structured. It allows the operational commander to make decisions based on more complete information obtained from diverse sources. The operational commanders can better supervise the actions of subordinate commanders, ensuring that their actions are consonant with his intent and directives.[2] Centralized C2 assumes error-free execution of orders.[3] It maximizes effectiveness through the extensive use of information technologies as a means of

DOI: 10.4324/9781003333685-6

eliminating uncertainty. It also ensures unity of effort through unity of command. It greatly facilitates and accelerates the decision-making cycle. It offers considerably more effective use of available forces and resources. By exercising centralized C2 the commander can shift the sector of the main effort quickly.[4]

One of the potential problems in centralized C2 is that a huge volume of information cannot be timely processed. The higher the level of command, the more acute and more difficult the problem of information overload potentially is. Backlogs of reports from subordinate units reduce the speed with which orders or reports are dispatched. Another problem is an inherent cycle of self-generated erroneous information and inappropriate action, which raises the issue of information quality. That quality could be adversely affected by a lack of mutual trust among various echelons of command, a lack of leader continuity or many echelons in an organization.[5]

In centralized C2, subordinate commanders have little, if any, freedom to make their own decisions.[6] Subordinate commanders' ability to perform is likely to suffer because of low motivation.[7] The overemphasis on security and avoiding errors or losses can prevent subordinates from exploiting fleeting opportunities through independent action at a decisive moment. Despite all the technological advances, it is a delusion to believe that an operational commander can effectively control their forces in a timely manner, let alone the enemy or the situation with certainty or precision. Nor can an operational commander possibly have all the facts at hand at the proper time to make a decision.[8]

In a centralized C2 the operational commander issuing the orders dictates not only who, what, when and where but also how. This can lead to unsatisfactory performance on the part of subordinate tactical commanders. In World War II, the Soviets routinely issued binding orders to subordinate commanders in the field. The result was that after accomplishing their assigned missions, the subordinate commanders remained passive and waited on orders from their higher commanders. They usually failed to consolidate and exploit combat success, because they were not educated and trained to act independently without further orders from the higher commander. The Soviets did not perform well at the tactical level because of excessively centralized C2. The Soviet units moved in strict conformance to plans made in advance.[9] Such C2 could work well only against a passive enemy, not against the Germans. The U.S. and British forces in the west in 1944 relied extensively on directive orders.[10] This was one of the several reasons for the slow Allied advance from the Normandy lodgment to the Franco-German border. It took the Allies almost six months to cover that distance, despite their enormous superiority in men and material on the ground, in the air and at sea.

The roots of *decentralized C2* can be found in the changes in the organization and methods of combat of the French army in the late eighteenth century. The French Revolutionary Army created conditions for theater-wide warfare by establishing divisions and army corps. The French generals and their subordinate commanders acted independently within the framework of their mission. Their orders were usually short, while their enemies issued orders several pages long.[11] Napoleon I gave short orders to some marshals, such as Andre Masséna, and more detailed orders to others, such as Emmanuel Grouchy.[12] He moved his army on the battlefield by giving his corps commanders verbal or written orders that assigned a definite task to each corps. How that task would be carried out he almost invariably left to the discretion of the corps commander. Admiral Horatio Nelson practiced what can be considered mission command today. Once Nelson's captains were acquainted with his ideas and intentions, signals became almost unnecessary.[13]

Decentralized C2 was adopted first by the Prussian and German armies in the mid and late nineteenth century because of the combined effects of the advent of the mass armies of the time, the steady expansion of the battlefield, and technological advances. The only links between the high command and the troops in the field were through telegraph. Communications were unable to keep pace with the high intensity and dynamics of combat. The high command, located far behind, was unable to evaluate the situation properly and make timely and sound decisions. This led the Germans to adopt mission command as the best answer to the demands of modern combat.[14]

Field Marshal Helmuth von Moltke, Sr, warned that "it is always dangerous to issue positive orders from a distance. If the higher military authority is not with the army, then it must allow the subordinate commander a free hand." In his view, "war cannot be conducted from the green table. Frequent and rapid decisions can be shaped only on the spot according to estimates of the local conditions."[15] Moltke, Sr, considered the telegraph a means of facilitating C2 and mitigating the impediments forced upon it by the friction of war. He used the telegraph in a limited role for the purpose of obtaining a daily picture of the situation and transmitting concise orders to subordinate operational commanders. He left other decisions entirely in the hands of subordinate operational and tactical commanders.[16]

Decentralized C2 is a rather loose form of decision and action predicated on an understanding of the overall mission requirements rather than on compliance with detailed directions from above. It allows greater flexibility for adapting rapidly to changing battlefield situations, dealing with unforeseen problems, and exploiting fleeting opportunities. Decentralized C2 also facilitates a higher operating tempo. Subordinate tactical commanders do not have to pass reports up the chain of command and wait for instructions to be passed down. Instead, they act on their own initiative. Decentralization is also a lubricant to the ubiquitous uncertainty and friction.[17]

In decentralized C2, the higher commanders assign missions and explain their underlying intent but leave the subordinates as free as possible to choose the manner of their accomplishment. They rely on initiative from subordinates, who must act on their own authority without waiting for instructions. Successful decentralized C2 requires that subordinates have knowledge and full understanding of the commander's intent so that when the situation changes, they can exercise initiative in consonance with the commander's desires.[18]

Decentralized C2 allows subordinate commanders substantial freedom of action and therefore requires them to act with the initiative.[19] It is essentially informal and flexible. Orders and plans are brief and simple. A disadvantage of decentralized C2 is that a subordinate commander can sometimes cause an unwanted escalation of a situation. Lower decision-making thresholds and highly diffuse centers of authority make coordination among command elements more difficult, increasing the risk of loss of control at the most critical time by the operational commander.

One of the aspects of mission command is that the commander accepts errors made in good faith during the execution of the assigned mission.[20] This requires that subordinate commanders be highly educated and trained. They also must have a great deal of confidence in the judgment of the higher commander issuing the intent. Decentralized C2 cannot be successful unless there is good personal leadership from the higher commander and a great deal of trust between him and the subordinate commanders.

In a decentralized C2, dependence on communications is greatly reduced. Reaction time is considerably shortened, because subordinate commanders are allowed to act without constantly asking for advice and consent from their superiors. In the case of a heavy overload or even a total breakdown of communications, lower-command echelons are better prepared to act on their own initiative than in a highly centralized C2. The planning process is usually less cumbersome and more flexible. Speed of action is essential for success and often demands a conscious sacrifice of precision.

Both centralized and decentralized C2 have some advantages and some disadvantages. Neither method is suitable for all situations. Which method should be used would depend primarily on the mission and the situation.[21] Napoleon I applied centralized C2, but his success ultimately depended on decentralized execution and the timely actions of his immediate subordinates. Yet he was often poorly served by his marshals – as, for example, in his ultimately unsuccessful invasion of Russia in 1812.[22]

The less need for the subordinate commanders to exercise initiative, the greater the need for detailed orders and the less need for issuing the commander's intent. Detailed orders are required to reduce the risk of misunderstanding.[23] The greater the need to bring the movements of large forces into harmony, the greater the need to issue detailed orders. In situations that are critical and do not leave room for independent action, decentralized C2 has its limitations. In such cases, it is necessary to command through binding orders. The higher commander takes direct responsibility, guides the employment of forces and means, and orders details of the execution. In that way, unity of actions in the execution and accomplishment of a certain task is ensured.[24]

Centralized C2 is highly effective in planning and execution of the strategic deployment of combat forces and in the destruction/weakening of the enemy and defense and protection of friendly maritime trade. It is also highly effective in one's employment of combat forces for theaterwide defense, sustainment and protection – e.g., air defense, theater ballistic missile defense, operational logistics and operational protection.

Centralized C2 should be applied when the objectives are predominantly non-military. In such cases, the stakes are often too high to allow for mistakes by subordinate tactical commanders. Centralized C2 is often the only viable option in a crisis that might lead to an unwanted escalation or even an outbreak of hostilities. During the Cold War, it was critically important that the highest national authorities on both sides tightly control all the decisions and actions of the top military commanders, including even tactical commanders.[25] Centralized C2 offers more advantages in a situation where combat develops slowly or where there is low-intensity combat, as, for example, in fixed defense.[26]

The commander's knowledge and understanding of the situation have a considerable influence on whether centralized or decentralized C2 is applied. One of the main prerequisites for the application of centralized C2 is that the commander possesses complete, actual, precise, and reliable information on the situation.[27] Advances in communications, such as the introduction of radio, have made it possible for higher headquarters to apply centralized execution. The larger the number of directly subordinate units, the more information that flows to the staff and, hence, the more information the commander must evaluate.[28]

The new information technologies have changed the relationship between factors of space and time. The will of the higher commanders can now be applied directly and quickly. Tactical leaders can be reached via radio, wire, fax or satellite communications. They are linked – but they are tied as well.[29]

In the modern era, a large volume of information in a centralized command organization must be processed, evaluated, and transmitted up and down the chain of command. This flow of information increases the risk of overloading communication channels during times of peak demand. Moreover, operational commanders could easily be overwhelmed by the amount of information they receive. A lengthened system-response time and a corresponding reduction in the quality of information reaching the commander result in a slow and inflexible C2 process in which decisions are often based on outdated information.

Rapid advances in information technologies offer substantial advantages for C2 at all levels. At the same time, there are also considerable disadvantages. The first and the most serious problem is that impressing the operational commanders with the need for "information dominance" would most likely lead them to wait instead of acting. At worst, the belief that there is such a thing as complete knowledge of the situation would breed commanders highly averse to taking high but calculated risks. This is especially critical at the operational and strategic levels, where commanders are forced to make assumptions about not only the current situation but also trends for several weeks or even months ahead. The commanders will rarely have the luxury of waiting for perfect knowledge of the situation. Either they will be forced to act to seize fleeting opportunities, or the enemy will force his hand.

Another major problem with the emphasis on obtaining a complete picture of the situation is that it leads to information overload – at all levels, but especially at the operational and higher levels of command. Data saturation is a continual, real-life problem.[30] The information overload experienced during the wars in Kosovo, Afghanistan and even Iraq has been insignificant compared with what can be expected in a war against a much stronger and resourceful enemy. The situation could be even more serious if networks are used primarily to pass data instead of to process it into information. Then the sheer amount of data collected and transmitted to users will most likely overwhelm the processing capability of lower levels of command. Trying to process more data may saturate commanders and their staffs with a flood of indigestible information.[31]

The challenge today is not in transmitting information but in processing it. If the time one gains through fast transmission is spent receiving, processing and evaluating information, only to discover that the information is untimely, irrelevant or redundant, then that time is wasted, and information technology quickly develops into a drawback.[32]

The handling of information must be learned and practiced. Relevant information should be distinguished from irrelevant information. Information must be collected, assessed and distributed horizontally and vertically in a manner useful to the respective recipient. It is not the mere possession of an information medium. It is having the ability and will to evaluate the information that the medium contains. At no command level can assessment be dispensed with. This means that orders must be adapted to the command level to which they are issued.[33] In a hierarchical command structure, much of the information collected at the lowest level of command is sent back to the next level of command, and so on to the highest level of command.

One of the most serious problems is the tendency toward excessively centralized decision-making and constant interference by operational commanders, and even military or theater-strategic commanders and national-strategic authorities, in the purely tactical decisions and actions of subordinate tactical commanders. Napoleon I's insistence on doing almost everything himself led to a long succession of victories but was also one of the reasons for his eventual downfall. As the war became more extensive and protracted it was not possible even for a person of Napoleon I's genius to manage it effectively without a well-organized general staff. His enemies eventually realized that it was their disunity that was one of the main reasons for their defeats. Napoleon I had two departments: the office of the chief of staff and general staff. Yet their work was not coordinated. The general staff did not take any part in Napoleon I's decisions. Its duty was to unquestioningly obey the emperor's orders.[34] Napoleon I's orders had to be executed whatever the means of command. His habit of undertaking everything with insufficient means, inability to recognize limits, and boundless assurance of success, which in the beginning were the cause of his success, in the end became fatal.[35]

Napoleon I was not well served by his marshals. Many of them had been raised in rank for dynastic, political, and personal reasons. They had been brought up not to command but solely to obey. With very few exceptions, such as Louis-Nicolas Davout, they were followers, not leaders but vassal princes. During his short exile on the island of Elbe (1813–1814), Napoleon I regretted that he had not placed good generals of divisions, with their batons yet to win, in command.[36]

Overly centralized C2 undermines subordinates' morale and motivation. It also breeds leaders who are unwilling or unable to act independently and take responsibility for their own actions. Additionally, when independent action by subordinates is severely or eliminated, there is no good way of promoting junior and middle-level leaders. Leaders who are unaccustomed to exercising initiative are unlikely to do so once they reach senior positions. The Soviet officer corps, with some notable exceptions, was highly bureaucratized and passive. The Soviets emphasized acting strictly according to planned schedules and schemes, as outlined in their plans.

Initiative on the part of subordinates was greatly reduced by the overriding requirements to follow orders and meet various deadlines.[37]

Overly centralized C2 does not necessarily mean that one cannot ultimately win a war against a strong opponent, but there is a heavy price to be paid, as the Soviet example in World War II amply shows. In the initial phase of the German invasion of Soviet Russia in 1941, Stalin and his highest military advisers tried to run the war all by themselves, with catastrophic consequences. The Soviet High Command (Stavka) made all the strategic, operational and even tactical decisions. Subordinate commanders were not allowed to exercise initiative. Stalin personally ordered ill-conceived counteroffensives and prohibited all withdrawals. Yet despite centralizing all the decision-making in their hands, the Stavka had little positive effect on the outcome of the operations and battles. Hundreds of thousands of Soviet troops were encircled and destroyed by the fast-advancing German mechanized forces.[38]

One of the reasons for the early success of the German Wehrmacht in World War II was that Hitler and the Supreme Command of the Wehrmacht (OKW) did not excessively interfere in the authority and responsibilities of the armies and army groups. As in the time of Moltke, Sr, the German army group and army commanders were issued "instructions" (*Weisungen*), not detailed orders. Hitler did not interfere with the authority and responsibilities of subordinate commanders during the campaign against Poland in September 1939.[39] However, all that changed after the invasion of Soviet Russia in June 1941. There were increasing disagreements and, eventually, sharp conflicts between Hitler and the army's general staff (OKH) and the army group commanders on what objectives to accomplish on the eastern front.

In the first major setback of the war, at the front of Moscow in December 1941, Hitler took over command of the German army. Afterward, the OKH became, like the OKW, Hitler's personal staff. Hitler not only centralized all policy and strategic decisions but began to make operational decisions as well. He began increasingly to bypass the army's general staff and army group commanders.[40] Mission command was abandoned, and Hitler began to issue detailed orders all the way to the midlevel command echelons.[41] His destructive influence on the German conduct of the war in Soviet Russia was characterized by his dogmatic thinking and his unwillingness to allow withdrawals from vulnerable and exposed positions to more defensible lines. Hitler robbed the operational commanders of their freedom to act. Frequent changes and dismissals of the army group and army commanders undermined self-confidence and reduced initiative. In short, Hitler degraded German C2 to an excessively centralized machine.[42]

As the war on the eastern front progressed, Hitler constantly interfered in the responsibility of the operational commanders and, in many cases, even the lowest tactical commanders.[43] He issued so-called "Führer's Orders" (*Führerbefehle*) that did not allow for any interpretation and had to be strictly carried out. Very often, Hitler directed army group commanders that a certain city must be held, or corps/division commanders must hold positions behind a certain line. Hitler's experiences as a simple soldier in World War I were the foundation for his decisions and views on events on the eastern front.[44] During the critical battle for Stalingrad, Hitler – intoxicated with the notion of his own infallibility and profiting by almost instant communications with the headquarters of subordinate army commanders – tried to control literally every maneuver from afar.[45] He only rarely allowed withdrawal from some untenable position. This led to very large and unnecessary losses of German troops on the eastern front in 1942–1945.

Another negative result of excessive centralization is that the focus in planning and execution is almost exclusively on targets to be degraded, neutralized or destroyed. Very often, targets are selected first and then the search starts for what objectives are to be accomplished. This violates the very foundation on which the "regressive" planning process rests. In U.S. practice, many commands and agencies, from the Joint Staff in Washington, D.C., to tactical commanders in the field and U.S. agencies, are involved in target development, selection and approval. During Operation Enduring Freedom, some 1,000 personnel worked at the Joint Intelligence Center (JIC) at CENTCOM's headquarters in Tampa alone. The CENTCOM JIC's Target Branch consisted of about 135 targeteers.[46] Political pressure and interference are significant, especially when other nations are involved in the targeting process. The excessive concern to avoid collateral damage requires the extensive and pervasive influence of lawyers at almost all levels of command. The result is a process that is complicated, unwieldy and too slow to meet the exigencies of the situation.

Experience shows that it is invariably bad when the operational commander bypasses his immediate subordinate commanders and issues direct orders to tactical commanders in the field. This situation can occur because of the operational commander's inability or unwillingness to delegate authority. In other cases, the operational commander bypasses his subordinate commanders because he does not trust them, due to their incompetence, personal animosities, or lack of common tactical or operational outlook.[47] Moltke, Sr, wrote that the issuing of orders must observe official channels as far as possible. Bypassing an intermediate authority destroys its effectiveness, because it causes the intermediate authority to appear superfluous. In a case where the direct intervention of the higher commander is necessary to avoid loss of time, bypassed channels must be informed of the relevant orders. Such a deviation from official channels is more justified by the high command than by subordinate command authority.[48]

The advances in information technology allow higher commanders today to observe events in almost real time and at distances of thousands of miles. This has led to a false belief that headquarters, far away from the scene of action, knows and understands a situation better and can make sounder decisions than the tactical commanders. Consequently, not only are tactical commanders required to report directly to the operational commander but also the latter often actually issues orders to the lowest tactical commanders. In the process, intermediate levels of command are not established or are bypassed. This, in turn, relegates the intermediate commander to the position of information administrator, while the higher level involves itself in matters of excessive detail.[49] In short, advances in information technologies, instead of resulting in much greater freedom of action for subordinate commanders, have actually become a highly effective tool in the hands of the strategic leadership and many operational commanders for reducing and even eliminating room for subordinates to exercise the necessary degree of initiative in carrying out their assigned missions. Operational commanders who try to specify everything for their subordinates are bound to get lost in the maze of details and therefore lose operational perspective. They also lose the trust and support of their subordinates and undermine the very basis for their own decisions and actions.[50]

One of the most common problems is the interference of the higher commander with subordinate commanders' authority and responsibility. Today, higher commanders often use new technologies to closely observe and control the actions of subordinate forces, bypassing intermediate commanders. All too often, the operational commanders directly interfere with the authority and responsibilities of their subordinate commanders. Regardless of the reason for such actions, they clearly reflect higher commanders' insufficient trust in subordinate commanders' competence to exercise the initiative in executing the assigned mission.

Interference by a higher commander is highly detrimental to the performance of a command organization. Despite technological advances, C2 still involves the vast and largely unquantifiable realm of human judgment and intellect – at least as much as in Admiral Nelson's days, and perhaps more so. Technology can be a two-edged sword, especially when developments lend themselves to ever-greater centralization of execution and, in extreme cases, to battlefield micromanagement.[51] Moltke, Sr, said that the most unfortunate of all high commanders is the one who is under close supervision, who must give an account of his plans and intentions every hour of every day. This supervision may be exercised through a delegate of the highest authority at his headquarters or a telegraph wire attached to his back. In such a case, all independence, quick decisions and audacious risk, without which no war can be conducted, are sacrificed. An audacious decision can be arrived at by one man only.[52] Moltke, Sr, believed that higher commanders exercise greater influence on a major battle by correctly employing reserves than by such interference with the independence of subordinate commanders in the forward line.[53]

In January 1941, Admiral Ernest J. King, the U.S. CNO/COMINCH, became very much concerned about the tendency of commanders to specify not only what was to be done but also how it was to be done. He cautioned that the U.S. Navy had to change if it was to prevail in the coming war.[54] Admiral Chester W. Nimitz, as the theater commander, and Admiral Raymond A. Spruance, as the (Fifth) fleet commander, realized that their commanders at the scene of action were best suited for making tactical decisions. Nimitz said to leave local commanders alone, because looking over their shoulders would only inhibit them. As long as these commanders had the responsibility, they had to have the freedom to do what they thought best. At the same time, both admirals occasionally made recommendations via radio messages to local commanders if it appeared that the latter were overlooking favorable opportunities.[55]

Theater Commands

Command organization at the theater-strategic and operational levels is usually *area* or *geographically* based. It can encompass the entire theater (of war) or its major part. In general, the principal factors in establishing such a command organization are the nature of the potential strategic threat or strategic objectives to be accomplished, the physical nature of the environment in which friendly forces will operate, the size and capabilities of the potential enemy (or enemies) and the time available. In addition, political, diplomatic, economic, ethnic and, in some cases, psychological factors should be fully taken into account in establishing a theater command and determining its boundaries.

An area-based command organization considerably enhances the integrated employment of service components. It provides for the centralized direction of combat forces operating within a certain geographic area. The operational commander has the necessary authority for control of logistical support and sustainment. Among other advantages, the operational commander and his staff are familiar with geographic, demographic, political, economic and military aspects of the situation in a given theater. The theater structure can be fully developed based on long-range plans and requirements. It can be refined based on lessons learned in peacetime and war. A sound theater organization should allow for optimal employment of joint and combined forces across the spectrum of conflict.

An area-based command organization requires large staffs and technical support systems to function properly. The long distances between headquarters and forward-deployed forces can complicate command and control. Ensuring adequate operations security (OPSEC) is difficult because of the large and complex C4 systems. The new information technologies should result in considerably greater effectiveness of each staff member, thereby creating the real possibility of drastically reducing the size of the frequently bloated theater headquarters. The new technologies should be used to geographically disperse many of the staff's sections and thereby drastically reduce the enemy's ability to decapitate the command organization. In addition, when a theater headquarters is in the territory of a coalition partner, its smaller footprint helps reduce potential political problems.

Any sound command organization requires the existence of an intermediate level of command to ensure proper application of the tenets of mission command. The intermediate level of command ensures the decentralization of efforts and thus allows the necessary delegation of authority by the theater commander. Military or theater-strategic commanders have large responsibilities that encompass both military and nonmilitary aspects of a situation. Whenever possible, they should delegate authority to subordinate area-based or service/functional component commanders. The lack of an intermediate operational level of command was one of the main reasons for the Germans' poor command performance in their invasion of Belgium and France in August–September 1914. The Germans deployed seven armies, with 3.1 million men, in the west; five of those armies, with 1.5 million men, crossed the borders into Belgium and France on 6 August.[56] However, no intermediate level of command existed between the army's Supreme Command (Oberste Heeresleitung – OHL) and the armies in the field. The problem of coordination was somewhat resolved on 25 November 1914, when the OHL divided the Western Army (*Westheer*) into three major groupings controlled by the most senior army commander. However, this solution was found to be unsatisfactory, and in September 1916, the Germans created Army Group *Kronprinz von Bayern* (Crown Prince of Bavaria), composed of three armies. At the urging of General Paul von Hindenburg as the Supreme Commander, East (*Oberbefehlshaber Ost*), the Germans created and unified three army groups on the eastern front in the summer of 1916. By the end of August 1916, Prince Leopold of Bayern had taken command of all the army troops on the eastern front. Afterward, the Germans also created army groups as an intermediate level of command between the OHL and the armies in the field.[57]

For their great offensive in the west in March 1918, the Germans organized their 13 armies into five army groups. Army Group *Crown Prince Ruprecht* and Army Group *Deutsches Kronprinz* (German Crown Prince), with four armies each, were deployed on the front. Three other army groups, with two armies each, covered the area from Bastogne, southern Belgium, to the Swiss border. Although the offensive ultimately failed, the army's Supreme Command had much better control over the operations because of these intermediate headquarters.[58] However, it had a limited ability to control armies through direct instructions, as the means of communication were primitive.

By September 1914, the French front line extended from the English Channel to Switzerland. To streamline his chain of command, General (promoted to marshal in December 1916) Joseph Joffre in late September appointed General Auguste-Yvon-Edmond Dubail to direct the operations of the armies on the French right wing. In early October, he directed General (promoted to Marshal in 1918) Ferdinand Foch to take over responsibilities for the armies on the French left wing. Foch's title was Assistant to Commander-in-Chief. By early October 1914, Foch began to direct the operations of the British/Belgian armies on the left wing. These changes were formally put into effect on 15 January 1915. Afterward, Foch's headquarters referred to itself as the Provisional Group of the North, while Dubail's headquarters referred to itself as the Provisional Group of the East. These two commands operated as de facto army group headquarters. On 13 June 1915 Joffre formally designated them Northern Army Group and Eastern Army Group, respectively. He also created another army group, the Central Army Group, on 23 June 1915.[59]

The Russians were the first who realized the need to establish an intermediate level of command in a theater. One of the leading Russian theoreticians, General Genrykh A. Leer, suggested creating a commander-in-chief responsible for a "front" (army group in Western terms). However, he wanted to limit the commander-in-chief's authority to only informing individual army commanders of his general idea concerning forthcoming operations and the overall situation in the theater of military actions.[60] Leer's ideas laid the foundations for the establishment of the "fronts" in 1900. The Russians afterward divided their Western theater of war into two theaters of military actions (theater of operations, in Western terms), each comprising one Front. These Fronts were directed against Germany and Austria-Hungary, respectively. In World War I, the Russians established two intermediate levels of command on the eastern front: the Northwestern Front, with two armies, and the Southwest Front, with four armies.[61]

The lack of an intermediate theater command makes success more difficult, and sometimes might be even a major reason for lack of success, in a campaign or major operation. The Allies did not establish a joint force command for the Southern Pacific Ocean Area (SOPAC) in time for the Battle of the Coral Sea in May 1942. The nearest operational commander was Admiral Nimitz in Pearl Harbor, more than 4,000 nautical miles away. Despite the operational success in the Battle of the Coral Sea, the Allies' lack of an intermediate command echelon between the tactical commander in the Coral Sea and the theater-strategic commander adversely affected their preparation for and conduct of the Battle of the Coral Sea.

In some cases, an intermediate level of command can be formally established but be ineffective in practice because of services' parochialism or poor personal relationships. This was the case for the Argentines in the Falklands/Malvinas Conflict of 1982. At the beginning of that conflict, the junta established an intermediate operational level of command, Malvinas Theater of Operations (TOM – *Teatro de Operaciones Malvinas*). However, its life was short; it was replaced by a local military command at Port Stanley, East Falklands. On 5 April, the Malvinas Theater of Operations was relocated to the mainland, to Comodoro Rivadavia. Only two days later, this command was replaced by the Southern Atlantic Theater of Operations (TOAS – *Teatro de Operaciones del Atlantico Sur*), led by Vice Admiral Juan Lombardo, with headquarters at Comodoro Rivadavia.[62]

Today, with prior authorization from the national authorities, a theater-strategic commander can establish one or several area-based commands (*subordinate unified commands*, in U.S. terms) encompassing a part of his/her area of responsibility. Such a command would normally be established in an area of continuous crisis and where the probability of an outbreak of hostilities is rather high. The commander of a subordinate area command has authority and responsibility similar to those of the theater commanders within their respective areas of responsibility. In the case of an outbreak of hostilities, such a command, as is U.S. Forces Korea (USFK) or some other subunified command, might be transformed into a theater of operations command. If a regional war breaks out in a part of the theater not covered by a subordinate area command, a new theater of operations might be declared. In the Gulf War of 1990–1991, the United States formally declared a Kuwaiti Theater of Operations (KTO). Likewise, in the war against Iraq in 2003 (Operation Iraqi Freedom) the United States formally declared an Iraqi Theater of Operations (ITO).

Component Commanders

Directly subordinate to the area-based commanders are service and functional component commanders. A *service component command*, as its name implies, is composed of all the forces, organizations and installations of a single service (Army, Navy, Air Force, Marine Corps and Special Forces). It is headed by the respective service component commander. Regardless of the specific command relationship arrangements, each service component commander normally retains certain service-oriented functions. Among other things, service component commanders advise the higher commander on the proper employment of the respective service forces. They select and designate specific units of the parent service component for assignment to other subordinate forces. The service component commanders are responsible for command training and the readiness of the forces under his command. They inform the higher commander in a timely manner of the planned changes in logistic support for the respective service forces. They are responsible for internal administration and discipline and for service logistic functions.

Functional component commands are established by the theater and subordinate theater commanders to employ all the forces operating in the same physical medium. Therefore, such commands are normally, but not necessarily, composed of the forces of two or more services. They may be established in times of peace or war to perform particular missions, which may be of short duration or may extend over a period of time. In U.S. terms, the following functional component commanders may be appointed by a theater commander: joint (combined) force land component commander (JFLCC/CFLFCC), joint (combined) force air component commander (JFACC/CFACC), joint force maritime component commander (JFMCC/CFMCC) and joint force Special Operations commander (JFSOCC/CFSOCC). The functional component commands may be established in peacetime or in wartime to perform particular operational missions that may be of short duration or may extend over a period of time.

The functional component commanders may be assigned a wide range of responsibilities regarding the employment of forces under their control. In planning and executing a campaign, each of the functional component commanders subordinate to a theater commander would be responsible for planning, preparing, and executing respective *major operations*. Among other things, a functional component commander would have the authority for planning, allocation, tasking and coordination among subordinate tactical commanders. Functional component commanders are normally given full tactical and operational control over all assigned and attached forces taking part in a certain operation.

Main Features

A sound operational command organization should have several features that collectively ensure the smooth and uninterrupted functioning of command and control in peacetime and during a war. (1) simplicity; (2) flexibility; (3) unity of command; (4) integration; (5) continuity; (6) homogeneity; (7) balance; and (8) stability. Because of the personal limitations of subordinate commanders or inadequate combat readiness of a force, an operational commander might temporarily violate some requirements for a sound operational organization. But even applying all the requirements of a sound command organization does not necessarily mean that the highest effectiveness of forces is ensured, among other things, poor operational leadership, lack of adequate weapons/equipment, logistical support, or lack of interoperability.

A sound operational command organization should, above all, be simple. *Simplicity* requires that the chain of command be clear and straightforward. Among other things, this means that the responsibilities and authority of the operational commander and subordinate commanders should be clearly delineated. There should be no overlapping responsibilities or authority. No subordinate should be responsible to more than one higher commander at any given time. Each service's capabilities should be integrated to accomplish a common objective but without degrading the combat effectiveness of individual force elements. The lack of a clear-cut command organization at the highest level of the German armed forces in World War I prevented a close and continuous exchange of views between the general staff and the admiralty. The German fleet's task was to carry out an offensive against its principal enemy, the Royal Navy. It was directed to keep the British Expeditionary Force (B.E.F.) away (thereby reducing pressure on the army) but also of obtaining command of the sea. The army's task was to support the navy by quickly seizing control of the French Channel coast.[63]

Command organization is *flexible* when it can expand or contract with changing conditions without serious loss of effectiveness. A highly flexible organization can adapt quickly to a change in the mission and shifting of resources. Its procedures are simple and straightforward. Information is acquired and passed quickly and reliably. A command organization is cohesive if there are strong bonds and feelings of unity. It provides stability and gives rewards for excellent performance.[64] Organizational flexibility is achieved by decentralizing command and control, delegating specific and well-defined functions and responsibilities, and rapidly deploying forces to meet specific situations. Without decentralized C2 it is difficult for a force to be effective when it is faced with a situation demanding quick and timely action. A too-rigid command organization limits the subordinate tactical commanders' freedom of action. In May 1940, the French had such an overly rigid command structure. They emphasized authority and strict obedience to orders at all command echelons. Thus, lower-level commanders had little chance to exercise initiative, and this proved to be a fatal flaw in fast-moving combat. Moreover, the French commanders were inflexible and slow to react to rapid changes in the situation on the battlefield.

Unity of effort is one of the main requirements of successful performance at any level of command. It can be achieved either through unity of command or through cooperation. The highest degree of effectiveness is ensured by having unity of effort through unity of command. This method is usually applied in the command of national or alliance forces. Napoleon I said, "when there is war against but one power there should be but one army, acting on one line, and led by one chief."[65] Unity of command is achieved by having a single commander, and by establishing clear-cur responsibilities, interservice integration, and highest degrere of interoperablity. Unity of command is the key prerequisite for clear and straightforward organization within individual services.[66] It encompasses the command authority to control all available forces for the purpose of attaining success. It is not inherent in command and must be provided if success is to be ensured.

Parochialism of services might present significant difficulties in ensuring unity of command. Sometimes that problem might be resolved through cooperation at the intermediate and lower levels of command. In the German invasion of Norway in April 1940, General Nikolaus von Falkenhorst was only nominally in charge of the campaign. In practice, the Luftwaffe and the Navy were controlled by their respective CINCs. Nevertheless, because of the excellent cooperation between the mid- and lower-level commanders in the field, the Germans were highly successful in the conduct of their campaign.

Unity of command ensures that at every level of command there is only a single source of authority. All efforts are directed toward ensuring that guidance and orders flow from the established authority. Unity of command greatly enhances the quality of planning because all the functions are centralized and thus predictable. It ensures clear delineations of the responsibility for each staff function.[67]

Unity of command is critically important for the success of any military undertaking; without it the commander cannot bring all available forces to bear against the enemy at the decisive place and time. The origins of modern unity of command can be traced back to World War I, when the number of joint and combined operations was considerably greater than in previous wars. Among other things, the advent of aircraft led to the emergence of air–land operations. This made the unity of command a necessity. Hence, instead of a kind of coalition warfare among three services, combined arms action emerged.[68]

Failure to establish unity of command results in a state of divided command. The higher the level of command, the more detrimental to combat performance is the lack of unity of command. The divided command has probably been the cause of more defeats and disasters than any other factor. The defeat of the allied French, Russian and Austrian armies in the Seven Years' War can be attributed to no other cause, for had these armies fought united, they would have been overwhelmingly superior to anything Frederick the Great could have put into the fight against them. The lack of unity of command almost cost the anti-French coalition the victory in the Battle of Leipzig in October 1813. Had the means of communication been equal to the demands Napoleon I's army made on them, the coalition would most likely have been defeated. Even at Waterloo in June 1815, the coalition's victory was assured only because Field Marshal Gebhard Leberecht von Blücher cooperated with the 1st Duke of Wellington (Arthur Wellesley) although there was no unity of command; otherwise, the allies would most likely have lost that battle. In the American Civil War, several armies in the field operated independently because of the lack of a single commander in the Eastern theater. This was one of the reasons that General Irvin McDowell was defeated in the Battle of Bull Run in July 1861. For the next three years, the Union armies operated without unity of command, and the cost in blood and treasure was enormous.[69] Unity of command was finally established in March 1864 when President Abraham Lincoln appointed General Ulysses Grant the General-in-Chief of the Union armies.

One of the factors contributing to the initial Japanese success in the surprise attack on Pearl Harbor was the lack of a U.S. joint force commander in Hawaii. There was little coordination and no integration of Army and Navy forces and efforts in defense of Hawaii. Air reconnaissance was divided among three Navy flag officers and two Army generals.[70] In the Battle of Britain in 1940–1941, the German Luftwaffe suffered from a divided command. The air offensive was mainly conducted by the Second Air Fleet, which operated from the Pas-de-Calais area, the Third Air Fleet, based in northern France, and the Fifth Air Fleet, based in Norway.

Each air fleet drafted its own operation plans and submitted them independently for approval to the CINC of the Luftwaffe, Reichsmarschall Hermann Gőring.[71]

In their campaign against Allied maritime trade in the Atlantic in World War II, the Germans would undoubtedly have been more successful if a theater command had been created to integrate the efforts of the Kriegsmarine and the Luftwaffe. In World War II, the Japanese command organization was usually divided, with a few notable exceptions. In their defense of the Philippines in the fall of 1944, the Japanese never appointed a single commander to control all ground, naval, and air forces in the theater. For the Japanese naval forces involved in the operation, the chain of command ran from the Navy Section of the Imperial General Headquarters (IGHQ) to Admiral Soemu Toyoda, CINC Combined Fleet. In addition, command relationships were poorly defined, except for army and navy land-based aircraft. Toyoda's headquarters was at Hiyoshi, near Tokyo, about 1,700 miles from Manila. The Japanese had trouble controlling their far-flung forces, due to poor communications and the long distances involved.

Throughout World War II in the Pacific, the U.S. and Allied forces had no single commander for the theater. Instead, for parochial political reasons, two de facto theaters of war were established: the Pacific Ocean Area, under Admiral Nimitz in Hawaii, and the Southwest Pacific Area (SWPA), under MacArthur, initially in Brisbane, Australia. On 8 May 1942, Nimitz was appointed CINC of the Pacific Ocean Area (CINCPOA) in addition to his command of the U.S. Pacific Fleet (CINCPac). He was in command of all land, sea and air forces in the Pacific Ocean Area except for the land defense of New Zealand. On 18 April 1942, MacArthur became the Supreme Commander of SWPA. Both Nimitz and MacArthur recruited their staff predominantly from their own services. This solution was a compromise, because the U.S. Navy refused to place the Pacific Fleet under the command of anyone but a naval officer. With two commands in the Pacific, control of the entire theater was vested in the Joint Chiefs of Staff (JCS). No single individual was able to resolve conflicting claims for troops and supplies, assign priorities, shift forces between commands or concentrate the resources of both commands on a single strategic objective. Not surprisingly, the result was an almost continuous competition for resources and a frequent need for the JCS to resolve disputes – even, at times, for the president's intervention.[72] The division of Army and Navy forces in the Pacific was more a means of satisfying interservice rivalries and personal egos than an example of sound military practice.[73] MacArthur believed that perhaps one of the most inexplicable decisions of the entire war was not to unify command in the Pacific. In his view, that decision could not be defended by logic, theory or even common sense. It resulted in divided effort, waste and duplication of resources, and the consequent extension of the war, with added casualties and cost.[74]

The German Wehrmacht established several area-based commands in World War II. Yet none of these commands was truly joint. Formally, the German theater commanders were in control of forces of all three services, yet they only controlled fully the army forces. They did not exercise sufficient authority over the Navy's and the Luftwaffe's units in their areas of responsibility. Nor did the Operations

Section of the OKW exercise sufficient control over the strategic employment of the Kriegsmarine and the Luftwaffe.[75]

Before the Allied invasion of Normandy in June 1944, the German command organization lacked unity of command at the theater level; in addition, it was extremely fragmented. This led to confusion and contradictory orders at critical times before and in the aftermath of the Allied landing in Normandy.[76] Field Marshal Gerd von Rundstedt, the supreme commander of the West (OB West), was responsible for the defense of France and the Low Countries. However, his command was a joint force command in name only. In practice, it was nothing more than an army group headquarters with minimal direct authority over local air and naval commands and even over certain other army commands in the West.[77]

In the early spring of 1943 Rundstedt controlled four armies (1st, 7th, 15th and 19th). The Wehrmacht commander in the Netherlands was subordinate to Rundstedt for preparing coastal defenses in case of an enemy landing.[78] Territorial commanders in France were directly subordinate to Hitler and the OKW and, for logistical matters, to the OKH.[79] Rundstedt had tactical control over other army troops only for preparing defenses against enemy invasion and in case of an actual landing. He also had tactical command over other army troops in case of a threat to law and order in the occupied territories within his areas of responsibility. Reserve divisions and reserve corps deployed in the west were under the control of the Commander, Replacement Army (*Ersatzheer*). The Waffen SS were subordinate to Rundstedt only tactically and for logistics; in terms of administration and discipline, they were subordinate to Reichsführer SS Heinrich Himmler. The Third Air Fleet was subordinate in all administrative and operational matters directly to the Luftwaffe's High Command (Oberkommando der Luftwaffe – OKL). Rundstedt was authorized to request only tactical support from the Luftwaffe; such requests would then be considered by the Third Air Fleet commander. The latter would then decide whether these requests fell within the scope of his overall tasks, and then, if possible, approve them. The Luftwaffe's troops employed for coastal defense were tactically subordinate to Rundstedt and the respective coastal defense sector commanders. The Kriegsmarine's Naval Group Command West (Marinegruppenkommando West) was directly subordinate to the Kriegsmarine's High Command (OKM) for ship movements, minelaying, escort of convoys and reconnaissance. Naval forces were under Rundstedt's tactical control only for coastal defense of ports or naval communications installations. The navy controlled the employment of coastal artillery against enemy ships at sea. The army took control over coastal artillery batteries only after enemy troops stormed the beaches.[80]

In coalition warfare, unity of effort is normally accomplished through *cooperation*. In most cases, political sensibilities, differences in military culture, psychological considerations, and even the personalities of the coalition commanders do not allow the establishment of an organization based on unity of command. For most of World War I, the allies did not have unified command on the Western Front. This problem was only partially solved in the spring of 1918, when the French Marshal Foch was appointed supreme commander. He did not have the authority to issue binding orders to the British Field Marshal, Douglas Haig, and the U.S.

General, John J. Pershing; he had to rely on their cooperation. The problem was further complicated by the lack of a single coalition body on the political level to deal with questions of strategy and policy.[81] The lessons of the lack of an effective operational command were learned in the interwar years. In January 1942, the United States and Britain created the Combined Chiefs of Staff (CCS) to direct the common U.S.-British effort against the Axis powers. Despite a great deal of friction, the CCS proved highly successful. Far less successful, however, was the short-lived ABDA (American, British, Dutch, Australian) Command for defense of the Netherlands East Indies (NEI), created on 3 January 1942.

The German and Italian command organization in the Mediterranean in 1940–1943 was highly fragmented; hence, it was too slow to react to frequent changes in the situation. Two chains of operational command and control were established, one for the Italians and the other for the Germans, with little or no cooperation. On the eastern front in World War II, the Romanian, Hungarian, and Italian armies operated under overall German command but had considerable freedom of action in day-to-day activities.[82]

In the Korean War, 1950–1953, the command structure was a major problem for the effective combat employment of UN forces. On 24 June 1950, General MacArthur established the United Nations Command (UNC) under his command. The UN troops were assigned to the respective U.S. commanders for operational control.[83] On 8 July 1950, President Harry Truman formally appointed MacArthur commander of all military forces assisting the Republic of Korea (ROK). These forces were placed under the unified command of the United States by the UN. MacArthur controlled all UN forces, including the U.S. forces. His title was Commander-in-Chief, Far East (CINCFE). He reported directly to the Joint Chiefs of Staff in Washington, D.C. MacArthur also controlled the Far East Air Forces, and the Army and Naval Forces Far East. He organized the army component command on 7 July 1950, because the original arrangement was not working as envisaged. Afterward, MacArthur directed the army component commander to communicate directly with two other component commanders in requesting necessary support. However, the three component commanders' headquarters were physically separate from one another.[84]

In the Gulf War of 1990–1991, the coalition partners never established a fully integrated command structure. General Norman Schwarzkopf commanded all U.S. forces, while his counterpart, Saudi general Khalid bin Sultan, commanded the Joint Force/Theater of Operations Command, consisting of the Saudi Arabian Armed Forces, the Saudi National Guard and the Arab/Islamic Corps. The French forces operated initially under the operational control of General Khalid and then under CINCCENT. The British forces were under Schwarzkopf's operational control. The only integrated combined structure was the so-called "Coalition Coordination, Communication and Integration Center (C3IC)," established in mid-August 1990 and used mainly to coordinate training, logistics, frequency management and intelligence sharing.

During the war in Afghanistan (Operation Enduring Freedom) in 2001–2022, General Tommy Franks, CINCCENTCOM, did not relocate part of his headquarters to Saudi Arabia but remained at Tampa, Florida. Supposedly, advances in communications and teleconferencing allowed the theater commander to effectively control his forces, regardless of the long distances involved.[85] It was contended that a theater commander today does not need to be physically in the area of operations to exercise effective command and control or to lead and motivate his subordinate forces. Relocating CINCCENTCOM headquarters to Saudi Arabia or some other Gulf Cooperation Council (GCC) countries would also have resulted in a symbolically larger U.S. presence in the region that could potentially lead, in turn, to more political complications and increased security risks.[86]

Some experts asserted that what matters is not where a headquarters is located but whether the commanders can be where they need to be to influence the conduct of the battle.[87] Such and similar views miss a key point in modern command and control at the operational level. The theater commanders should physically remain at their main headquarters, or forward headquarters (but located outside of the boundaries of the newly established theater of operations), so that they can most effectively monitor and control the events in their entire area of responsibility (AOR), not just the part of his theater in which a campaign takes place. The theater commander's ability to use advanced communications and information technologies should not be used as a justification for not establishing an intermediate level of command. Distance and the related factor of time still matter, and the need for a commander to lead and motivate forces remains one of the tenets of successful operational leadership.

In establishing a coalition command structure, differences in military style and traditions and national differences and sensitivities should be taken fully into account. All key political decisions should be resolved before the theater command organization is established. The strongest member of the coalition should treat other members as equal partners in a common cause. Optimally, a single theater commander should be appointed, with authority and responsibility over all the subordinate multinational forces of all services. The operational commander of the strongest member of a coalition should also be the most senior officer in rank. Normally, an operational commander and his staff should not occupy two or more positions simultaneously; in such a case, there is a great danger of placing intolerable burdens on the commander and staff, thereby endangering the accomplishment of their main mission. In a coalition, a subordinate operational commander should not have the right to bypass the chain of command and report directly to the highest national or coalition/alliance authority.

A sound operational command organization should provide for *integration* throughout the chain of command; otherwise, command elements at the operational level can be a source of great weakness in a command's performance. A fragmented operational command structure considerably slows the decision-making cycle. It also often causes confusion. This is an especially great problem when the information reporting and command system cannot keep pace with the fast-changing situation on the battlefield. Integration in a theater organization should be vertical, so that it is fully responsive to the operational commander's will. It should also be integrated laterally, so that the actions of individual force elements can be synchronized.[88] The operational commander should also provide for a coordinating body that ensures smooth and effective coordination among command elements throughout the entire force.

In the initial phase of the Korean War of 1950–1953 because the U.S. Navy and Air Force did not integrate their efforts, UN forces conducted two separate major air offensives. This delayed the start of a comprehensive air-interdiction effort for more than a month. It caused confusion and loss of effectiveness at a time when every single aircraft sortie was critical to the survival of the Eighth Army.[89] The Air Force component commander proposed centralized control over all aircraft in the theater. He requested operational control over naval aircraft operating over Korea, though not for control of aircraft when they were engaged in support of naval forces. Yet the navy strenuously opposed the idea. The result was a compromise, reached on 11 July 1950, by which the air component commander received coordinating authority. This fell far short of operational control over all air elements in the Korean theater.[90]

Continuity suggests that a command organization, once established, should undergo no changes, or only small changes, prior to or during a combat action. The only exception would be when unforeseen and dramatic developments in the situation make such changes necessary for the successful completion of the mission. Whenever possible a change of command organization should be so planned that all the force's elements know the reasons and can thus be prepared to execute it. Continuity of command organization enhances coordination among the various elements of the force. It also facilitates the relationships between a higher commander and his subordinates.

Homogeneity means that all the functions required to accomplish the mission of the organization should be grouped and that individuals should be assigned to these groupings in accordance with their abilities. This is perhaps the most important principle to be considered in developing an organization in general. Unfortunately, it is not consistently applied; worse, it is often completely forgotten. When that happens, the command organization will be inefficient.

Balance between various force elements of the command organization results when those elements are developed strictly in accordance with the relative importance of their contribution to the accomplishment of the overall objectives of the organization. A perfectly balanced organization is seldom found. Sometimes force elements are built up beyond any real need. There are several reasons for a lack of balance in a command organization. A prejudice in favor of a certain force element over others can lead to an imbalance of a given command organization; also, subordinate commanders sometimes acquire responsibility and authority within an organization out of proportion to the services they render. This is usually the result of expropriating the functions of force elements headed by less aggressive, but not necessarily less able, commanders. Uncurbed, individuals can advance their personal interests with complete disregard for the well-being of the organization. Such "empire building" is a waste of time and resources, and the commander should prevent it.

Stability is the ability to fill important positions in an emergency quickly and satisfactorily. Among other things, stability is directly related to structure. It is accomplished by high training among commanders and staff, good morale and discipline, standard instructions and sound administration. A command organization that is stable internally is one that can experience growth and adjust itself promptly to the loss of a commander or other key personnel without a serious decrease in effectiveness.

As they rise through a service, officers should learn to discharge the duties associated with higher ranks. In general, there is a direct correlation between rank and executive effectiveness. However, officers are often promoted beyond the limits of their capacity. Attempts to avoid this by promoting from outside the system are detrimental to morale. A well-developed promotion system is of great value, because it can clearly demonstrate the opportunities for both horizontal and vertical advancement. Vertical promotion or promotion in the direct chain of command is typically the only line of advancement recognized.

C4 System

Command and control cannot be successful without a well-developed, highly effective and survivable theaterwide *command, control, communications, and computers (C4) system*. Intelligence, fires, logistics and protection all depend on responsive C4. It binds all the elements of the theater organization together. A C4 system should provide timely and adequate data and information to plan, direct and control activities at subordinate command echelons, including administration, intelligence, logistics, personnel, planning and operations. The introduction of automated information systems equipment and procedures into the command-and-control process has increased the need for compatibility of equipment and standardized procedures. Automated information systems should be capable of providing rapid, reliable and secure information interchange throughout the chain of command.

A theaterwide C4 system encompasses facilities, equipment, communications, procedures and personnel essential for planning, directing and controlling forces in accordance with assigned missions. It extends to subordinate force elements. The theater C4 system should provide rapid, reliable, and secure information interchange throughout the chain of command. One's combat forces, when engaged in joint operations, should be able to coordinate their combined power to achieve maximum effectiveness. The exchange of information should be continuous, from the recognition of the threat through the assignment of the mission to the force and the commander's operational concept, planning, execution, consolidation, reorganization, continuation and termination of the operation. A theater commander should integrate both military- and nonmilitary-related communications systems and organize organic and component tactical communications systems into interoperable and compatible networks.

Command authority can only be received by, and information conveyed to and from, a commander by way of *communications*. Therefore, communications not only are conduits for information but also bind all the elements of a force together in a cohesive whole. They typically parallel the lines of hierarchy in a theater. Communications are critically important for ensuring a reliable, continuous, and, above all, timely information exchange between the command echelons in a given theater. Thus, an unbroken chain of communications should extend from the national-strategic level to the lowest tactical commanders.

Poor and unreliable communications have often been the main reason that higher headquarters misread or were ignorant of a true operational situation. In the frontier battles in August 1914, the German army's Supreme Command Headquarters in Koblenz had a poor awareness of the operational situation because of the lack of reliable, fast and timely communications with the German left wing and the armies in the center. Communications with the armies on the right wing were conducted via radio – a very unreliable means of communication at that time.[91] In late August, General Helmuth von Moltke, Jr., chief of the German general staff, moved his headquarters to Luxembourg, about 170 miles from General Alexander von Kluck's First Army. Yet this move did not greatly improve communications. There were no telephone communications to army commands because the French and Belgians had destroyed civilian lines. Telegrams often arrived 24 hours late and had to be decoded. Aerial delivery was confused and uncertain. Motorcycle and automobile service was slow and sometimes dangerous.[92] The first telephone communications between Kluck and General Karl von Bűlow, commander of the Second Army, were established on 9 September, some 30 minutes after the First Army commenced its retreat. Another of Moltke, Jr.'s, errors were that during the most critical time at the front, between 5 and 9 September, he did not issue any directive to his two army commanders in the field.[93] The primary reason for these deficiencies was that, in contrast to the very precise and detailed planning of the movements of their armies, the Germans had never prepared a communications plan. Moreover, the deployment area of the German troops did not have a communications network. The army's Supreme Command also declined to establish centralized lines of communications.[94]

Among the reasons for the Allied defeat in France in 1940 were poor intelligence and inadequate communications. The French neither interconnected their main headquarters by teletype nor linked it to armies in the field. The telephone system was unreliable. Telegram message delivery was chronically late. The French high command soon became divorced from reality. Conditions were no better at the tactical level. Deficiencies in communications, interservice coordination, and Allied liaison arrangements resulted in unsynchronized actions that dissipated Allied effectiveness. The resulting fragmented counterattacks posed no serious threat to the German advance.[95] The British, because of their doctrinal (and as it turned out, wrong) belief that the next war would be positional, largely relied on wired communications. This equipment became useless once the campaign entered a mobile phase. The unexpected increase in signal traffic, especially encrypted messages, soon outstripped the system's capacity to handle it. Therefore, intolerable delays in communications were created.[96] In contrast, the Germans had much more modern communications equipment that was fully adaptable to the fluid conditions on the battlefield.

In the Battle of Midway in June 1942, both the U.S. Pacific Fleet and the Japanese Combined Fleet suffered from poor communications. Tactical communications between Nimitz's headquarters and the fleet at sea were never good.[97] However, the Pacific Fleet had the advantage of operating close to its base and having the undersea cable between Hawaii and Midway. In the Leyte Operation, Allied communications were satisfactory overall, but an unprecedented volume of urgent and operational-priority messages caused considerable delays in transmission at critical moments during the operation. The commander of the Seventh Fleet, Vice Admiral Thomas C. Kinkaid, had to issue an order prohibiting administrative radio traffic and requiring such traffic to be sent by airmail. Because of atmospheric conditions and the long distances involved, there were considerable difficulties in the reception of major Third Fleet broadcasts in the west and southwest Pacific. An analysis of dispatches for 17–23 October showed that it took, on average, 30 minutes to 1 hour for delivery of a message to the addressee. If the messages were long and an intermediate station had to be used, urgent messages required about three hours for delivery.[98]

Conclusion

Experience shows that operational commanders must make a choice of which method of C2 is the most suitable for the accomplishment of a given mission. Clearly, it is simply wrong to apply either decentralized or centralized C2 without due regard for the objectives to be accomplished and the military situation. Each method has some significant strengths and great weaknesses. Neither can be applied in all situations. Generally, centralized direction and centralized execution are more suitable in crisis, post-hostilities, and peace operations, where the political leadership or operational commanders simply cannot risk mistakes made by the subordinate commanders that can in turn make it more difficult, and in some cases even fatally undermine, the aims of policy and strategy. In contrast, centralized and decentralized directions should be generally applied in high-intensity conventional war or some other situation in which it is critical that subordinate commanders have the greatest possible room for exercising the initiative. Although mistakes will be made, the result will be a much higher performance from the force. In fighting a war with a highly capable and resourceful enemy, highly centralized C2 has never been an answer if victories are to be won decisively and with the fewest losses for friendly forces. The Soviet example in World War II is not a model to be followed; just the opposite. The Soviets paid exorbitant prices in defeating Nazi Germany. One can strongly argue that success would have been achieved with far fewer losses if the Soviet army had adopted and successfully applied German-style mission command.

Information technologies should not be the sole factor in determining whether the higher commander should rely exclusively on centralized C2. There are other, mostly psychological factors that adversely affect the performance of subordinates in combat. Tactical commanders and their subordinates should not become passive observers and transmitters of detailed orders. If their freedom to act is significantly reduced, or even worse, eliminated, they will not exercise initiative and act independently within the boundaries of the higher commander's intent. A more serious consequence is that junior leaders unaccustomed to acting independently will have great difficulty in doing so and in taking prudent risks once they reach senior command positions. This means, above all, accepting that war is not just a science, but largely an art and a science. Friction and the fog of war are inherent parts of any war. Advanced information technologies can only reduce uncertainty, never eliminate it. Technologies, no matter how capable and advanced, are only means to an end, not ends in themselves. C2 must be focused on those elements of combat potential, and especially on leadership, that will enhance one's ability to fight and win the country's wars decisively and with the fewest losses. Hence, it is critically important that operational commanders and their subordinates be educated and trained to shift quickly from centralized to decentralized C2 and vice versa.

Sound operational command organization should be simple and flexible enough to allow for the full application of the principle of centralized control and decentralized execution. It should provide for unity of effort through unity of command. It should allow the application of the principles of objective and mass when needed. The responsibilities of each command echelon should be clearly established and delineated.

Technological advances in the mobility, range, lethality and information gathering of military forces continue to compress time and space, requiring higher operating tempos and creating a greater demand for information. Operational commanders control multiservice or multinational forces that operate in all mediums: surface, subsurface, land, air and space. Each of these media has distinctive operational characteristics and environmental constraints. In combat, the ever-present threat from the air requires forces on land or at sea to be dispersed over a large part of the theater to reduce their vulnerability to enemy attack. This, especially in a maritime theater, leads to a greatly increased area in which combat actions take place.

The increasing lethality and range of weapons over time have made it necessary for military forces to disperse to survive. Joint forces are not necessarily bigger than in the past, but they have many moving parts and a greater variety of specialized organizations and weapon systems. Thus, they require greater amounts of information. Service and joint forces can move more quickly over greater distances than ever before. They can also engage enemy forces at greater ranges than ever before. The consequence of all this is a fluid, rapidly changing operational situation. The more quickly the situation changes, the greater the need for continuously updated information, and the greater the strain on command and control.

History is replete with examples in which a poor or unsound command organization was the principal reason for a lack of unity of effort, both within a service and among services or multinational forces. These difficulties have sometimes been so severe that they have led to failure at the operational and strategic levels of war. New technologies offer both promise and a potential problem. Higher commanders and political military leaders should not be tempted into believing that technology can or ever will satisfactorily resolve all the problems associated with operational command and control.

Notes

1 The term *command* pertains to the authority a commander legally exercises over subordinates by virtue of rank or assignment; it encompasses the authority and responsibility for effectively using available resources in planning the employment of and

organizing, directing, coordinating, and controlling military forces for the accomplishment of assigned missions. In addition, command authority includes responsibility for the health, welfare, morale, and discipline of assigned personnel; the term *control* refers to creating the conditions on which the commander's directives are implemented; control is a process used to establish limits and provide structure. Its purpose is to deal with the uncertainties inherent in the organization; as a process, it serves primarily as a compensating, correcting device for command; Headquarters, Department of the Army, FM 22–103, *Leadership and Command at Senior Levels* (Washington, DC: U.S. Government Printing Office, 21 June 1987), 41–42.

2 Paul B. Stares, *Command Performance: The Neglected Dimension of European Security* (Washington, DC: The Brookings Institution, 1981), 67.

3 Wolfgang Peischel, "Geistesgeschichtliche Grundlagen operativer Fűhrung im deutschsprachigen Raum," *Ősterreichische Militärische Zeitschrift* 5 (September–October), 2002, 555.

4 Paul K. van Riper and F. G. Hoffman, "Pursuing the Real Revolution in Military Affairs: Exploiting Knowledge-Based Warfare," *National Security Studies Quarterly* (NSSQ) (Summer 1998), 10.

5 Paul D. Hughes, *Mercury's Dilemma: C3I and the Operational Level of War* (Fort Leavenworth, KS: School of Advanced Military Studies, U.S. Army Command and Staff College, 8 May 1988), 28–29.

6 Manfred Bartele, *Beiträge des Generals von Scharnhorst zur Weisungsführung in der preussischen Armee* (Hamburg: Führungsakademie der Bundeswehr, September 1970), 1.

7 Hans Felde and Peter May, "Auftragstaktik oder Befelhstaktik? – Kriterien für die Auswahl eines Führungsprinzips," *Wehrwissenschaftliche Rundschau* 2 (February), 1978, 56.

8 Antulio J. Echevaria II, "Auftragstaktik: In Its Proper Perspective," *Military Review* 10 (October 1986), 51; Riper and Hoffman, "Pursuing the Real Revolution in Military Affairs: Exploiting Knowledge-Based Warfare," 11.

9 Richard D. Hooker, Jr., "The Mythology Surrounding Maneuver Warfare," *Parameters* 23 (Spring 1993), 36.

10 Werner von Raesfeld, "Führung durch Auftrag oder durch bindenden Befehl?" *Wehrkunde* 5 (April), 1960, 166.

11 Heinz Loquai, "Die Auftragstaktik als militärische Führungskonzeption," *Truppenpraxis* 6 (June 1980), 445.

12 Wolfgang Thomas, "Auftragstaktik. Voraussetzungen, Grenzen und Mőglichkeiten," *Wehrkunde* 11 (November 1968), 590.

13 David M. Keithly and Stephen P. Ferris, "Auftragstaktik, or Directive Control, in Joint and Combined Operations," *Parameters* (Autumn 1999), 123.

14 Gottfried Greiner, "Zeitgemässe Führungsprobleme," *Wehrkunde*, March 1961), 123.

15 Daniel J. Hughes, ed., *Moltke on the Art of War: Selected Writings*, trans. Harry Bell and Daniel J. Hughes (Novato, CA: Presidio Press, 1993), 77.

16 Shimon Naveh, *In Pursuit of Military Excellence. The Evolution of Operational Theory* (London: Frank Cass Publishers, 1997), 57–58.

17 van Riper and Hoffman, "Pursuing the Real Revolution in Military Affairs," 11–12.

18 van Riper and Hoffman, "Pursuing the Real Revolution in Military Affairs," 12.

19 Headquarters U.S. Marine Corps, C4I Division, *Command and Control: A U.S. Marine Corps Concept Paper* (February 24, 1994), 63.

20 Peischel, "Geistesgeschichtliche Grundlagen operativer Fűhrung im deutschsprachigen Raum," 555.

21 Bruce Condell and David T. Zabecki, eds., *On the German Art of War: Truppenfuehrung* (Boulder, CO: Lynne Rienner Publishers, 2001), 23.

22 Alan D. Landry, *Time Estimation at the Operational Level of War* (Fort Leavenworth, KS: School of Advanced Military Studies, U.S. Army Command and General Staff College, June 1990), 26.

23 Walter von Lossow, "Auftragstaktik am Beispiel der Operationsführung einer Division des Heeres (Kurzfassung)," in *Möglichkeiten und Grenzen der Auftragstaktik in unserer Zeit*, ed. Hans-Joachim Löser. "Seminar der Clausewitz-Gesellschaft vom 11.-12. Juni 1976 in Rheinbach," *Wehrwissenschaftliche Rundschau* 5 (May 1976), 147.
24 Jochen Bauer, *Führungsinformationssysteme und Führen mit Aufträgen–ein Widerspruch?* (Hamburg: Führungsakademie der Bundeswehr, 11 November 1985), 4.
25 Hans Felde and Peter May, "Auftragstaktik oder Befelhstaktik? – Kriterien für die Auswahl eines Führungsprinzips," 57.
26 Werner Widder, "Auftragstaktik and Innere Fuehrung: Trademarks of German Leadership," *Military Review* 5 (September–October 2002), 7.
27 Felde and May, "Auftragstaktik oder Befelhstaktik? – Kriterien für die Auswahl eines Führungsprinzips," 57.
28 Jobst Rohkamm, "Einfluss der Führungsmittel auf die Auftragstaktik (Kurzfassung)," in *Möglichkeiten und Grenzen der Auftragstaktik in unserer Zeit*, ed. Joachim Löser, 146.
29 Dieter Stockfisch, "Im Spannungsfeld zwischen Technologiefortschritt und Führungsverständniss. Auftragstaktik," *Marineforum* 12 (December 1996), 11.
30 Alan D. Zimm, "Human-Centric Warfare," *Proceedings* (May 1999), 29.
31 Williamson Murray, "Clausewitz Out, Computers In: Military Culture and Technological Hubris," *National Interest* (Summer 1997), 63.
32 Widder, "Auftragstaktik and Innere Fuehrung," 9.
33 Widder, "Auftragstaktik and Innere Fuehrung," 9.
34 Fuller, *The Conduct of War 1789–1961*, 53–54.
35 Fuller, *The Conduct of War 1789–1961*, 54.
36 Fuller, *The Conduct of War 1789–1961*, 55.
37 Richard A. Gabriel, *The Antagonists: A Comparative Combat Assessment of the Soviet and American Soldier* (Westport, CT: Greenwood Press, 1984), 95.
38 Albert Seaton and Joan Seaton, *The Soviet Army 1918 to the Present* (New York/Scarborough: NAL Books, 1986), 118.
39 Günter May, *Zum Probleme der Auftragstaktik in schwierigen Lagen: Der "Haltebefehl" Hitlers, die deutsche Heeresführung und die Entschlussfreiheit des Truppenführers während der Winterkrise des Russlandfeldzuges 1941/42* (Hamburg: Führungsakademie der Bundeswehr, 30 October 1984), 4.
40 May, *Zum Probleme der Auftragstaktik in schwierigen Lagen*, 15–16.
41 Thomas, "Auftragstaktik. Voraussetzungen, Grenzen und Möglichkeiten," 593.
42 Alfred Philippi and Ferdinand Heim, *Der Feldzug Gegen Sowjetrussland 1941 bis 1945. Ein operatives Überblick* (Stuttgart: W. Kohlhammer Verlag, 1962), 171–72.
43 Burkhart Müller-Hillebrand, *Deutsche und Sowjetrussische Militärische Führung im Zweiten Weltkriege*, ZA 1/56 P-145, Studien der Historical Division Headquarters, United States Army Europe, Foreign Military Studies Branch, Bundesarchiv/Militärarchiv (BA-MA), Freiburg i. Br., 1954, 131–32.
44 May, *Zum Probleme der Auftragstaktik in schwierigen Lagen*, 18, 16.
45 Antony Beevor, *Stalingrad* (London: Penguin Books, 1999), 68.
46 John M. Donnelly, "A Top Target Planner Gives Inside Look at 'Enduring Freedom," *Defense Week*, July 15, 2002, 1.
47 Löser, ed., "Möglichkeiten und Grenzen der Auftragstaktik in unserer Zeit," 142.
48 Hughes, Moltke *on the Art of War*, 183–84.
49 Widder, "Auftragstaktik and Innere Fuehrung," 10.
50 Widder, "Auftragstaktik and Innere Fuehrung," 10.
51 Keithly and Ferris, "Auftragstaktik, or Directive Control, in Joint and Combined Operations," 124.
52 Hughes, Moltke *on the Art of War*, 77.
53 Hughes, Moltke *on the Art of War*, 157.
54 Thomas E. Buell, *Master of Sea Power: A Biography of Admiral Ernest J. King* (Boston/Toronto: Little, Brown and Company, 1980), appendix 1, 521–23; and appendix

2, 524–25; Gerard D. Roncolato's letter to the editor, re "Taking Maneuver Warfare to Sea," *Proceedings* 3 (March 1996), 22.

55 Thomas E. Buell, *The Quiet Warrior: A Biography of Admiral Raymond A. Spruance* (Annapolis, MD: Naval Institute Press, 2nd printing, 1988), 173.

56 B. H. Liddell Hart, *The Real War 1914–1918* (Boston/New York/Toronto/London: 1st published 1930, rev. ed. 1964), 54; Correlli Barnett, *The Swordbearers: Supreme Command in the First World War* (Bloomington, IN/London: Indiana University Press, 1963), 26–27.

57 Volker Wieker, *Operation, Operative Führung zwischen Taktik und Strategie – Bedeutung und Entwicklung des Begriffes seit Moltke d. Ae. – Bearbeitung aus der Sicht der Landstreitkräfteführung* (Hamburg: Führungsakademie der Bundeswehr, 1987), 16.

58 Karl-Volker Neugebauer, *Grundzüge der deutschen Militärgeschichte*, vol. 1: *Historisches Überblick* (Freiburg, i.Br: Rombach Verlag, 1993), 247.

59 Robert A. Doughty, *Pyrrhic Victory: French Strategy and Operations in the Great War* (Cambridge, MA: Belknap Press of Harvard University Press, 2005), 111.

60 Harrison, *The Russian Way of War: Operational Art, 1904–1940*, 25.

61 Jovo Ninkovič, *Nastanak i Razvoj Ratne Operacije* (Belgrade: Vojno-Izdavački Zavod, 1984), 35.

62 Raymond. A. Burden, *Falklands: The Air War* (London/New York, NY: Arms and Armour Press, 1986), 15.

63 Bruno von Uthmann, Ulrich Kessler, and Gerhard Matzky, *Wie hat sich das Fehlen eines gemeinsameen Oberbefehls über die verschiedenen Wehrmacht-teile bei den Mittelmächten und bei der Entente im Weltkrieg ausgewirkt?* Lecture at the Wehrmachtakademie, May 1936, RW 13/v 36, Bundesarchiv-Militärarchiv (BA-MA), Freiburg, i.Br., 59.

64 Headquarters, Department of the Army, FM 22–103, *Leadership and Command at Senior Levels*, 49.

65 Walter Krueger, *Command*, Record Group 4, Box 28, Folder 1165, Naval War College Historical Collection (Newport, RI: Naval War College, 12 September 1925), 19.

66 von Uthmann et al., *Wie hat sich das Fehlen eines gemeinsameen Oberbefehls über die verschiedenen Wehrmacht-teile bei den Mittelmächten und bei der Entente im Weltkrieg ausgewirkt?* 19–20.

67 Headquarters, Department of the Army, FM 22–103, *Leadership and Command at Senior Levels*, 52.

68 von Uthmann et al., *Wie hat sich das Fehlen eines gemeinsameen Oberbefehls über die verschiedenen Wehrmacht-teile bei den Mittelmächten und bei der Entente im Weltkrieg ausgewirkt?* 42–43.

69 Krueger, *Command*, 20.

70 Samuel E. Morison, *History of United States Naval Operations in World War II*, vol. 3: *The Rising Sun in the Pacific, 1931–April 1942* (Boston, MA: Little, Brown, [1948] 1959), 134.

71 *Command and Control* (Washington, DC: Headquarters, United States Marine Corps, 1994), 36.

72 *Command and Control* (1994), 15.

73 Jason R. Barlow, "Interservice Rivalry in the Pacific," *Joint Force Quarterly*, Spring 1994, 76.

74 Barlow, "Interservice Rivalry in the Pacific," 76.

75 Walter Warlimont, The Unification Problem: Some Lessons from the German Experience, vol. 6, Part 3: "Command Structures, Continued," in *World War II German Military Studies*, ed. Donald S. Detwiler, Charles B. Burdick, and Jürgen Rohwer, 24 vols. (New York/London: Garland Publishing, 1979), 37.

76 Blumenson and Stokesbury, *Masters of the Art of Command*, 337.

77 Jeffrey J. Clarke and Robert Ross Smith, *Riviera to the Rhine*: *The European Theater of Operations, United States Army in World War II* (Washington, DC: Center of Military History, United States Army, Washington, 1993), 56.

78 Bodo Zimmermann, *Bericht über die Operationen des OB West von den Vorbereitungen gegen die Invasion bis zum Rückzug auf den Westwall*, ZA/1 658 B-308, vol. 4 (Freiburg im Breisgau: Studien der Historical Division Headquarters, United States Army Europe, Foreign Military Branch, BA-MA, 1947), 5–6.
79 Zimmermann, *Bericht über die Operationen des OB West von den Vorbereitungen gegen die Invasion bis zum Rückzug auf den Westwall*, 6.
80 Zimmermann, *Bericht über die Operationen des OB West von den Vorbereitungen gegen die Invasion bis zum Rückzug auf den Westwall*, 8–9.
81 Blumenson and Stokesbury, *Masters of the Art of Command*, 306.
82 Blumenson and Stokesbury, *Masters of the Art of Command*, 65.
83 Thomas A. Cardwell III, *Command Structure for Theater Warfare: The Quest for Unity of Command* (Maxwell AFB, Alabama: Air University Press, 1984), 13.
84 Cardwell III, *Command Structure for Theater Warfare*, 15.
85 Thomas E. Ricks, "A War That's Commanded at a Distance," *Washington Post*, December 27, 2001, A16.
86 Richard Hart Sinnreich, "The Long Arm of Command," *Washington Post*, December 31, 2001, A17.
87 Sinnreich, "The Long Arm of Command," A17.
88 Stares, *Command Performance*, 61.
89 Stares, *Command Performance*, 63.
90 Cardwell III, *Command Structure for Theater Warfare*, 15.
91 Wolf Weth, "Führungsgrundlagen und ihre Bedeutung. Betrachtung über einige historische Ereignisse und die mőgliche Nutzanwendung der dabei gewonnen Erkentnisse," *Wehrkunde* 2 (February 1967), 157.
92 Robert B. Asprey, *The German High Command at War: Hindenburg and Ludendorff Conduct of World War I* (New York: Quill William Morrow, 1991), 100–101; Peter Wenning, *Aufmarschplanung als strategische Führungsproblem. Eine vergleichende Untersuchung am Beispiel des Schlieffenplans 1914, der NATO Aufmarschplanung im Rahmen der "Flexible Response" un dem neuen strategischen Konzept der NATO* (Hamburg: Führungsakademie der Bundeswehr, 30 December 1997), 14.
93 Holger Herwig, *The First World War: Germany and Austria-Hungary 1914–1918* (London: Arnold, Hodder Headline Group, 1997), 101, 103.
94 Stefan Kaufmann, *Kommunikationstechnik und Kriegführung 1815–1945. Stufen telemedialer Rüstung* (Munich: Fink, 1995), 150–51.
95 Stares, *Command Performance*, 28, 31.
96 Stares, *Command Performance*, 29.
97 Stares, *Command Performance*, 41.
98 "Allied Communications Arrangement," CINCPOA, COMSOWESPAC, COMTHIRDFLT, and COMSEVENTHFLT, Record Group 23, Special Projects Department, Naval Historical Collection Division, U.S. Naval War College, Newport, RI; Milan Vego, *The Battle For Leyte, 1944. Allied and Japanese Plans, Preparations, and Execution* (Annapolis, MD: Naval Institute Press, 2006), 337.

7 Operational Support

For maximum effectiveness in the employment of one's combat forces in a major operation or campaign, several supporting structures and activities, arbitrarily called "operational support," should exist. Optimally, they should be fully established already in peacetime. The main components of operational support are operational intelligence, information warfare (IW), fires, logistics and protection, respectively. The list of what constitutes operational support should not be considered something unchangeable. It is quite possible that some existing structures and activities may acquire radically improved capabilities in the near or distant future and be elevated to the operational level. Each component of operational support encompasses several diverse related elements. These should be fully integrated to ensure the highest degree of effectiveness. The operational commander is responsible for properly sequencing and synchronizing not only joint forces but also all components of operational support.

Operational Intelligence

Operational commanders and their staff would have great difficulty planning, preparing and executing major operations or campaigns or other actions without accurate, reliable, timely and relevant intelligence. It is essential that the operational commander integrates intelligence into the other components of operational support. Intelligence is only one – and not necessarily the most important – factor in making a sound decision and then preparing and executing operation plans.

In one of its many definitions, intelligence is described as the product resulting from the collection, processing, integration, analysis, evaluation and interpretation of available information on potential opponents or enemies. It can be defined as knowledge about the enemy obtained through observation, investigation, analysis or understanding. Intelligence helps the commander make sound decisions by providing accurate, timely, and relevant information. It focuses on improving one's knowledge and understanding of the enemy and of one's vulnerabilities to enemy intelligence. It plays a critical role in conducting intelligence preparation of the battlespace/theater and battle damage assessment.

At each level of war, a specific intelligence picture is required for the respective commander to make sound decisions. The higher the level of command the broader

DOI: 10.4324/9781003333685-7

intelligence picture and further ahead the trends in the situation must be projected. The higher the level of war, the more intangible and nonmilitary aspects of the situation must be processed, analyzed and presented to the users.

In terms of scale, strategic, operational and tactical intelligence is differentiated. Intelligence required by the highest political and military leadership to make sound decisions differs considerably from the intelligence used by the operational or tactical commanders. *Tactical intelligence*, as the name implies, supports the planning, preparation and execution of tactical actions. In conventional combat tactical intelligence focuses on a much smaller physical space (combat zone/sector), shorter time (several hours to 96 hours) and deals predominantly with the tangible factors of the situation. Tactical commanders and their staff are predominantly concerned with the tangible aspects of the situation. Tactical information is mostly transitory in nature. *Strategic intelligence* supports the national leadership or theater-strategic commander. It provides input for the formulation of national policy, national security and military strategy or theater strategy.

Operational intelligence supports the planning, preparation and execution of a campaign or major operation. It is a composite of tactical and strategic intelligence.[1] One of the principal tasks of intelligence officers is to depict the operational situation by taking a vast amount of information on the tactical situation and meshing it with information obtained by the national-strategic or theater-strategic level of command. While the tactical commanders can relatively quickly adjust their options to unforeseen changes on the battlefield, the operational commander cannot, without major loss of time and great risk, change the employment of friendly forces once a major operation or campaign is underway. Operational intelligence should anticipate events several weeks or even months ahead. In contrast, tactical commanders are concerned with events that take place over a day, or several days at the most (there are, of course, exceptions to this).

The scope of operational intelligence in terms of space, time and force is considerably larger than that of tactical intelligence. Among other things, operational intelligence is concerned with both tactical and operational characteristics of the operating area. It visualizes possible developments in the situation several weeks or months ahead. It should analyze the capabilities of not only the enemy forces currently in combat but also those that might be fielded or introduced into the theater several weeks or even months in the future.[2] In contrast to tactical intelligence, operational intelligence should be focused on *operationally significant* elements of the factors of space, time and force in a given theater of operations.

Operational commanders need to understand both the true capabilities and vulnerabilities of friendly and enemy's forces. Operational intelligence should provide the operational commander with timely, accurate and relevant information about the enemy forces' order of battle (OOB) and capabilities, and the enemy's critical strengths and weaknesses. It should deduce *patterns* and *probable trends* in the military and non-military situation for several weeks or even months in the future. Another task of operational intelligence is to determine enemy critical vulnerabilities that one can attack, directly or indirectly, to neutralize, destroy or degrade the enemy's operational or strategic center of gravity.

Operational intelligence is concerned with many hard-to-quantify aspects of the situation, such as the quality of the operational commanders, soundness of doctrine, degree of cooperation among friendly forces, state of morale, discipline and combat readiness of the enemy's major force elements. The operational commanders should have information on the personal traits of the opposing commanders, their leadership skills, experience and relationships with subordinate and higher commanders. They should know how their enemy counterparts visualize the theater, what options are open to him, and which of these he is most likely to adopt. As opposed to tactical intelligence, which is concerned primarily with immediate and near-term enemy capabilities, operational intelligence focuses on the enemy's intentions over time.

Because war is a contest of will between two sides, the enemy's actions influence and shape one's actions or reactions. Operational intelligence should – in contrast to tactical intelligence – view and analyze the situation in a broader context and for longer periods and consider how those actions affect the enemy's future decisions. Political, economic, social, ethnic, cultural religious, legal, environmental and other factors could considerably affect the enemy's operational decisions. The operational commanders must know and understand the essential elements of non-military aspects of the situation to make their decisions. This information is most often provided only through strategic intelligence.

The success of operational intelligence is contingent on already having an all-encompassing and fully deployed intelligence organization in peacetime. It is too late to collect and evaluate data on a theater just prior to the start of hostilities. Doctrine for the use of intelligence should be written and should be fully understood by commanders and their staff. Interagency coordination agreements should be worked out and procedures established before the start of hostilities.[3]

Operational intelligence cannot be successful without an *operational perspective* on the part of intelligence officers, commanders and their staff. Intelligence officers would acquire operational perspective through education and training well before their forces become involved in hostilities.[4] Optimally, all intelligence efforts should be highly centralized; otherwise, it is very difficult to use available intelligence sources most effectively. One of the perennial problems of the United States in the Vietnam War was that little, if any, central direction was given to the intelligence effort.[5]

One of the most challenging tasks of intelligence at any level is to acquire and properly evaluate *intangible* aspects of the enemy's situation, such as morale and discipline, will to fight, quality of operational leadership and soundness of service or joint doctrine. Modern technology cannot satisfactorily find the answers in evaluating such and similar aspects of the operational or strategic situation. The soundness of the enemy's doctrine can be assessed only after long and painstaking efforts in collecting information on the enemy's battle experiments, exercises, and maneuvers in peacetime and properly evaluating the enemy's lessons learned.

Very often too much emphasis is given to technical means of information collection and to intelligence concerning the tangible factors of a situation, while human intelligence (HUMINT) is given short shrift. Yet it is precisely HUMINT that is the key to obtaining good operational and strategic intelligence. Such hard-to-quantify aspects of the situation as the enemy's commanders skills, personality traits, possible intentions, and actions or reactions cannot be usually obtained through technical collectors. If HUMINT is neglected for too long, it is not easy or cheap to restore; rebuilding a network of agents and informers takes a long time. HUMINT-gathering efforts should be broadly based. Moreover, they should be tailored to regional political traditions and society, cultural nuances, language, and local mindset. This is the only way to obtain a proper understanding of the area and anticipate likely enemy actions and reactions.

An overemphasis on the offensive aspect of warfare is invariably detrimental to achieving the necessary level of proficiency in intelligence gathering and evaluation. The IJN had little interest in aerial reconnaissance in World War II; that task was considered too "defensive." The result was that the Imperial Japanese Navy (IJN) suffered throughout the war from a shortage of reconnaissance aircraft and trained observers. It had only two reconnaissance squadrons, both land based. All photographic reconnaissance was conducted by these two squadrons, although their pilots did not have any special training in photography. The U.S. fast carrier forces repeatedly delivered successful strikes before and during the Leyte Operation because of the Japanese failure to maintain a regular and thorough air search and reconnaissance.[6]

Operational intelligence should be believable to operational commanders and their staff. This means overcoming the preconceived ideas of the operational commanders, and their staff. Sometimes the biggest problem is the operational commanders who want to be their own intelligence officers. They might feel more comfortable with their own experiences and be prone to ignore changes in the situation that might make that experience irrelevant. Alternatively, the operational commanders might have strong cultural, political, or other biases that would cause them to reject accurate intelligence presented by the intelligence officer. They might also choose to ignore valuable intelligence or to accept such intelligence too late. Other causes of failure have been the politicization of intelligence agencies, ambiguous assessments (due to bureaucratic reluctance to accept risks) and users' resistance to accept intelligence warnings. The leadership should always be ready to accept legitimate differences in the interpretation of any information and intelligence received.[7]

The tendency to collect more and more information without regard to what is required to make a sound decision is certain to overtax any staff. The situation is even worse if a vast amount of information is collected and then posted before being processed. These tendencies must be resisted by the operational commanders and their planners. The focus should always be to reduce the complexities of the situation to its essentials. This means that operational intelligence should depict an *operational* not tactical picture of the situation.

All three levels of intelligence are intertwined and mutually dependent. Good tactical intelligence can significantly enhance chances at the operational level. Poor or inadequate tactical intelligence adversely affects planning for a major operation or campaign. In addition, deficiencies in information collection at the tactical level affect operational intelligence and possibly strategic intelligence as well.

Because the strategic framework is much broader and more critical for ultimate success, a failure of strategic intelligence invariably has highly adverse consequences at the operational and tactical levels. Errors in anticipating events and assessing enemy actions and possible reactions at the strategic level cannot usually be overcome by intelligence at the tactical and operational levels.

Intelligence is not a science but rather an art or a craft.[8] Sound intelligence should be definitive when the available information supports such a conclusion; otherwise, judgments should be qualified. It is incumbent on the part of the users to know and understand the nature of the intelligence so that it can be optimally used for making decisions and preparing plans. In any large organization, finished intelligence is almost invariably a product of competing views and often represents a consensus on a given issue. In the final analysis, the users of intelligence should not have any doubt that the intelligence product is a result of a team effort.[9]

One of the major problems with intelligence at the operational and strategic levels is interference by policy makers who, by either overt or subtle pressure, want an intelligence product that supports a strategic decision that has been made or is to be made. Politicization of intelligence also occurs when an intelligence product is unacceptable because it runs counter to a certain policy or is too critical of higher officials or foreign leaders. Experience shows that the overpoliticization of intelligence can seriously undermine a political or military rationale for war. Overselling a threat before a war leads to overreacting during the hostilities. It results in major credibility problems in the aftermath of the war. This then can interfere with nation-building and limit domestic and international support in a future war.[10]

Poor operational intelligence can be the result of relying on the assumed enemy's intentions instead of on his real capabilities. This has probably been the cause of major military failures more often than any other intelligence deficiency. Unfortunately, one's limited resources frequently drive planning toward what the enemy is most likely to do. In late January 1940 the British intelligence service made a classic bad evaluation of the enemy's intentions regarding a possible German landing in Norway. The British correctly estimated that the Germans had about 400 transport aircraft in the northeastern part of Germany and could assemble, on short notice, about 1,000 machines capable of transporting about 10,000 men. However, British intelligence firmly believed that the Germans would not intervene in Scandinavia as long as their iron ore import was not endangered, or the Allies did not establish a naval base on the Norwegian coast. MI 2 Section of British intelligence stated:

> There is no reason to expect them to be more successful than we have been in the past. There is no reason to believe that the Germans will be able to land and maintain forces more rapidly than we can so long as our communications through Trondheim are not destroyed. The Germans, of course, have a much greater military on which to draw.[11]

Another cause of operational intelligence failure that has often led to operational setbacks is overconfidence, including belief in the superiority of one's own culture over another.[12] Prior to May 1940 the British and French intelligence services had collected a significant amount of information about the German planned advance through the Ardennes. Because this information did not conform to their expectations, they simply ignored it. As the German offensive unfolded, French military intelligence had difficulty deducing not only the extent of the German gains but also the whereabouts of their own forces. This was mainly due to a shortage of reconnaissance aircraft, which were being progressively destroyed by the Germans.[13]

Allied intelligence possessed information on the impending German attack in the Ardennes in December 1944 but failed to correctly interpret signs of it. The prevailing view was that the fighting spirit of the Germans was so low that no offensive could be expected.[14] The Allies expected that Ultra would tell them why Hitler was shifting fighter aircraft westward and building up a large reserve with SS panzer divisions at its core. However, they failed to penetrate Hitler's masterful deception, in which the Germans paraded the assembly of the Sixth Panzer Army in the north while secretly preparing to attack in the Ardennes.[15]

Operational IW

The concept of information warfare (IW)[16] is not new; it has existed in various forms and applications since the beginning of warfare. However, in recent years, capabilities for conducting IW have grown significantly because of ever-greater advances in weapons, sensors, and communications. In general, IW is understood as the integrated use of operations security, military deception, psychological operations, electronic warfare and physical destruction, all supported by intelligence, to influence, degrade, deny information to, or destroy the enemy's command and control capabilities while protecting one's command and control against such actions. It is applicable across the entire spectrum of conflict.[17] To be fully effective, IW must be fully synchronized in support of the basic plan for a tactical action or campaign/major operation.

Depending on the objectives to be accomplished, distinctions should be made among tactical, operational, and strategic IW, respectively. *Tactical IW* is planned in support of tactical actions. *Operational IW* is planned and conducted by operational commanders in support of a campaign or major operation. It is aimed to disrupt the enemy's decision cycle and, if possible, decapitate his command structure from the main body of his combat forces in the field. Operational IW should prevent the enemy operational commander from deducing operational commander's real intentions. It should enhance the operational commander's flexibility in making a sound decision. *Strategic IW* is aimed to support national strategic objectives in peacetime or in war. It is planned and executed by the highest national military command echelon.

Regardless of scale or purpose, IW traditionally encompasses five key components: psychological operations, military deception, operations security, electronic warfare and physical destruction. Physical destruction can be part of lethal operational fires, while when used offensively other components can be part of operational fires. Operations security, defensive elements of electronic warfare, psychological operations and deception are at the same time integral parts of operational protection. Each component of IW can be used to supplement the effect generated by other elements; often, however, the use of one component conflicts with the use of other components.

Psychological operations (PSYOPs) are an essential component of both offensive and defensive IW. They are aimed to discourage or dissuade the enemy. They are designed to enhance the actual and perceived mission, intent and combat potential of the joint force for appropriate foreign audiences. PSYOPs are aimed at degrading the morale of the enemy troops and populace. They attempt to influence the enemy commander's decision-making process over time. Sometimes the distinction between PSYOPs and public affairs is blurred. Although these two activities use similar techniques and procedures, their objectives and target audiences are diametrically opposite. One of the main objectives of public affairs is to shape public opinion through the dissemination of information understandable to the public on the decisions and actions of friendly forces. Public affairs are conducted to inform and educate friendly forces and friendly nations and neutrals based on the facts of the situation. Their aim is to enhance the free flow and exchange of information. Public affairs should not be used for spreading misinformation, propaganda, and deception. This undermines their credibility and opens gaps that can be successfully exploited by the enemy.[18]

PSYOPs can multiply the effects of deception, reinforce apparent perceptions, plant seeds of doubt about the enemy leadership to their followers, proliferate discreet messages to enemy intelligence collectors and enhance and combine live-fire demonstrations with surrender appeals – all to magnify one's superiority and encourage the enemy to give up the fight. They can undermine enemy command and control by lowering morale, instilling fear and sowing confusion. PSYOPs can also provide insights into the enemy's reactions. They can be used to direct other IW activities, such as deception, into areas where they are most likely to succeed. They are coordinated at the strategic level to ensure the proper sequencing and timing of all elements of operations. PSYOPs require extensive intelligence information. A variety of actions are used, such as political and diplomatic communiqués, leaflet drops, loudspeaker broadcasts and various other means of transmitting information. In addition, a strong psychological impact can be derived from various military activities in peacetime, such as moving large forces or holding large exercises or maneuvers.[19]

Military deception (MILDEC) is conducted in peacetime and in war. Its main purpose is to surprise or mislead the enemy and induce him to act in a way that can be exploited to one's advantage. Another purpose is to give the enemy false ideas about the activities and intentions of one's forces. A prerequisite for success is effective countersurveillance aimed at preventing the enemy from observing indications of friendly forces' activity. Successful deception multiplies force effectiveness. It reduces expenditures of time, effort and resources and can have far-reaching psychological effects.

Offensive deception involves actively disseminating false information to mislead the enemy about one's intentions. Defensive deception (or counter deception), in contrast, tries to provide false information to deceive or mislead the enemy, who has the initiative. Offensive and defensive military deception as part of operational IW is related but not identical to deception planned as part of the operational idea (scheme). They provide broader support for the success of the deception planned for a campaign or major operation. They are also often conducted in peacetime to enhance the protection of one's forces and various installations deployed in a given theater.

The defensive component of operational IW encompasses *operations security* (OPSEC). At the operational level, operations security is aimed at denying the enemy's operational commander information concerning one's capabilities, intentions and vulnerabilities. Its purpose is to identify which actions enemy intelligence collection can observe, and which indicators of one's actions can be interpreted or pieced together to ascertain one's intent. OPSEC is an excellent means of achieving surprise.

In peacetime, OPSEC's main purpose is to protect classified information, ensure the physical security of facilities and personnel, keep the commander's decisions secret from the enemy and prevent the enemy from conducting deception and effectively using his counterintelligence. In a crisis, OPSEC's tasks encompass evaluating indicators and vulnerabilities, conducting camouflage, delaying information necessary for the enemy to make decisions, and degrading the effectiveness of the enemy's decisions. In combat, OPSEC is focused on assessing indicators and vulnerabilities, denying the enemy knowledge of critical nodes and one's intentions, ensuring the dispersal of enemy intelligence, applying information security and conducting counterintelligence.[20]

OPSEC encompasses diverse active and passive measures aimed at denying the enemy knowledge of the disposition, capabilities and intentions of both friendly and enemy forces. It focuses on identifying and protecting those parts of the plan that could provide early indications of impending action or reveal friendly vulnerabilities. OPSEC's assessment should be conducted from the enemy's viewpoint. The key to identifying critical information is to determine what can be deduced from a particular event or piece of information – hence the importance of protecting essential elements of friendly intelligence.

OPSEC should identify critical elements of intelligence that the enemy could use to disrupt or even defeat one's plan for a campaign or major operation. These efforts should be coordinated at the highest level. It is the essential IW element for hiding one's critical information from the enemy. PSYOPs and military deception can augment the effort by influencing the enemy in such a way that he will not look for the vital truth one's forces seek to hide. If the enemy's focus is manipulated away from the truth, he may overlook it. By using jamming, communications security and associated security disciplines, electronic warfare has a vital role in concealing information. Physical destruction can ensure denial of information by removing the enemy's physical means for obtaining information.[21]

Planners should first identify critical information, activities or indicators that, when collected and analyzed by enemy intelligence, may reveal one's plans. They should develop a thorough understanding of the enemy's intelligence capabilities, including his means of collecting information and his ability to analyze and disseminate it. The analysts should determine whether the enemy could collect the elements of critical information. Then the operational commander should decide what actions to take to conceal his critical information or, conversely, accept the risk of doing nothing at all. Afterward, the planners recommend the appropriate countermeasures.[22] News media portrayals of military activities before combat can possibly prevent hostilities or build public or international support for the employment of one's forces. The presence of the media can also be a valuable source of information for the enemy.

Electronic warfare (EW) is aimed at degrading the performance of the enemy's electronics and weapons systems, providing warning of enemy action, and locating and identifying enemy sensors, thereby reducing their effectiveness against friendly forces. Electronic warfare plays a significant role in the process of information degradation or destruction because information transmitted by electromagnetic means is extremely vulnerable to electromagnetic or high-energy weapons.

EW in support of operational IW consists of three main components: electronic attack, electronic protection, and electronic warfare support. *Electronic attack* (EA) is aimed to deny the enemy's operational commander the use of the electronic spectrum he needs for effective C2 of his forces. It uses jamming, electronic deception and directed energy (DE) to degrade, exploit or destroy the enemy's electromagnetic spectrum (EMS). It tries to deny the enemy operational commander the use of the EMS to effectively command and control his forces. *Electronic protection* (EP), in contrast, is designed to ensure that the operational commander can use the EMS for command and control. It protects friendly forces against both enemy uses of the electromagnetic spectrum and undesirable effects from one's own use. *Electronic support* (ES) involves actions and measures to search, intercept, identify, and locate the enemy's sources of intentional or unintentional energy and detect immediate threats. It can provide the real-time information required to locate and identify the enemy's C3 nodes and supporting systems during electronic attack missions.

Physical destruction involves the employment of one's weapons systems and their associated sensors to seek, identify and disable or destroy the enemy's sensors, communications and C4 nodes and thereby either delay or deny information to the enemy. The use of "hard-kill" weapons against designated targets as part of an integrated IW effort includes direct or indirect fires. In many cases, special forces teams could possibly be used to destroy or neutralize specific elements of the enemy's C4. This does not necessarily mean that the targets of attack must be physically destroyed; fire demonstrations and weapons effects and selective degradation of certain parts of the enemy's command and control are examples of alternatives to complete destruction. Destruction can also be inflicted on elements of the enemy force that support C4 nodes.[23] It means that the enemy's C4 node or some other elements of his command and control cannot be performed, either permanently or temporarily.

The employment of special forces is potentially one of the most effective ways to destroy or neutralize the enemy's command and control. The long-term advantage of destroying the enemy's C4 nodes cannot be assumed. The timing of the physical destruction is critical to ensuring that subsequent action to exploit short-term effects is successful. In peacetime, the aim is to identify targets, determine the feasibility of an attack, and plan for redundancy and damage control. In conflict, the task is to degrade, neutralize or destroy the enemy's C4 nodes while protecting friendly C4 nodes from enemy attack. The long-term effects of these attacks are generally limited.[24] Physical destruction is often the operational commander's only available option. In cyber-attacks, it is sometimes more advantageous to leave many vital parts of the enemy's C4 functioning than to destroy or neutralize them.

Offensive IW is designed to disrupt the enemy's decision-making and planning process by denying information to the enemy commander so that he cannot obtain an accurate picture of the situation. This would greatly affect his estimate of the situation. The objective is to disrupt the enemy commander's decision cycle and thereby slow the enemy's tempo of operations. Offensive IW is also aimed to disrupt, and optimally to sever, the links between the enemy's operational commander and his subordinates. Electronic warfare is a crucial tool for degrading the enemy's key points for command and control, while physical destruction severs those links more permanently. Offensive IW is of little use unless it supports the operational commander's plan. Properly planned and executed, it should force the enemy to be reactive. It should induce the enemy's operational commander to use his forces at the wrong time and place.

The aim of offensive IW is to degrade the enemy's informational systems and force the enemy to react. Offensive IW requires sound balancing and integration of all its components. The planners must consider the effect that degradation of the enemy command and control may have on ongoing efforts to influence the enemy operational commander's actions. Sometimes degrading the enemy C4 links too severely adversely affects these efforts.[25]

To attack the enemy C4 systems, planners must have comprehensive knowledge and understanding of how the nodes within that system fit together to make the system work. Within a C4 system, one or more nodes may be of such critical importance that their destruction immediately degrades the functioning of the entire system. To provide a viable target, the C4 node should be not only critical but also vulnerable. It should be susceptible to degradation or have an exploitable weakness.[26]

Military deception, as part of offensive IW, can have potentially positive effects on the conduct of OPSEC. It can influence the enemy not to collect certain information or to underestimate the capabilities of one's operations security. It can contribute to the success of PSYOPs by providing information in harmony with their main theme. It can enhance the effectiveness of one's EW by giving the enemy false perceptions of friendly electronic attack/electronic support capabilities or leading him to defend the wrong elements of his own command and control structure. Military deception should be carefully planned and executed, because its requirements often conflict with the requirements of other elements of offensive IW. Another potential problem can be a lack of coordination between plans for operational deception as part of an operational idea for a campaign/major operation, and deception efforts as part of IW. If the plans are not coordinated, both deception efforts could end in failure.

OPSEC's requirements support military deception by concealing observable parts of one's forces. They often may limit information that can be used to enhance the success of the deception story. OPSEC supports PSYOPs by concealing competing information. It can also have an adverse effect by restricting information that could be used to develop themes for the conduct of psychological operations. It supports physical destruction by concealing one's forces assigned for offensive IW and denying the enemy information about the true nature of one's capability to physically destroy his C4 systems. It also supports electronic warfare by denying the enemy information on one's EA and ES capabilities.

EW plays a significant role in offensive IW. Traditional methods of attacking enemy command and control, such as jamming, and new methods, such as cyber-attacks, are used in conducting electronic attacks today. EW enhances operations security effectiveness by neutralizing the enemy's reconnaissance and surveillance. Electronic attacks and electronic support can be used extensively as part of military deception.

EW supports PSYOPs by making it difficult or impossible for the enemy to observe and report conflicting information. PSYOPs could be even more complicated, because an electronic attack might deny the use of certain of the enemy's frequencies and thereby restrict one's ability to target or reach the audience for the PSYOP themes. Electronic support can also provide information on the enemy's C4 nodes and other targets for physical destruction.

Physical destruction in offensive IW is aimed at destroying enemy command posts and C4. The enemy's information capabilities can be neutralized or destroyed via the physical destruction of nodes and links. This method is most effective when the enemy operational commander relies on certain C4 functions or when the focus is on resource-intensive and hard-to-reconstitute targets. Physical destruction supports OPSEC by degrading or neutralizing the enemy's reconnaissance and surveillance of friendly forces. Physical destruction could separate the enemy's operational commander from his forces in the field at critical times and thereby enhance the chances of deception. Similarly, physical destruction enhances the chances of PSYOPs by degrading or neutralizing the enemy's ability to evaluate the situation or by isolating the targeted enemy's audience from conflicting information.

Defensive IW is aimed to maintain effective C2 or to neutralize the enemy's efforts to deny information to or degrade or interrupt friendly C2. Failure to adequately protect one's command and control means loss of freedom to act, losing the initiative, and perhaps, in the end, dooming a military enterprise. Defensive IW is also a part of operational protection, because it is used to protect friendly information. Despite the "defensive" in the term, this component of IW contains both offensive and defensive components. Offensive actions usually protect four elements of offensive IW, while defensive actions and measures are aimed at reducing one's C4's vulnerability to enemy attack.

The main purpose of protecting C2 is to obtain decision advantage by ensuring the unhindered collection, processing and evaluation of information; obtaining an accurate picture of the operational or strategic situation; and maintaining continuous and reliable communications between operational commanders and subordinate commanders. This can be achieved by drastically reducing the enemy's ability to attack C4 nodes. Defensive IW should also be able to reduce command and control vulnerabilities by applying various protective measures. Another aim is to reduce interference with C4 systems throughout the electromagnetic spectrum by properly coordinating and deconflicting one's work.

OPSEC is an essential element for protecting C2. It should function continuously. It must protect the location of C4 elements, and techniques and procedures for using informational systems. In offensive IW, the aim is to counter the enemy's propaganda by discrediting the source of media attacks and maintaining favorable public opinion. EW in defensive IW is used to degrade and destroy enemy information systems. Signal intelligence (SIGINT) data, in the form of signal security monitoring to identify an enemy's potential source of information about friendly C4 systems, can be used when an enemy attack on one's C4 nodes is impending. Deception is used to neutralize or exploit the enemy's counter-C2 by misleading the enemy's operational commander about one's capabilities or limitations, thereby causing him to misallocate his forces and resources.[27] Deception in defensive IW is designed to give the enemy commander a false picture and thereby degrade his ability to make appropriate command decisions. One of the most effective methods of protecting one's C4 system is to use cover and concealment. Physical destruction in defensive IW is aimed to destroy or neutralize targets that pose a threat to one's C4 nodes. Targeting enemy counter-C2 directly enhances the survivability of one's command and control.

Operational Fires

Operational fires are often one of the main prerequisites for the success of the initial phase of a campaign or major operation. To be fully effective they must be sequenced and synchronized with other elements of operational support, particularly intelligence and logistics. Operational fires were used for the first time in World War II, when great increases in the range, speed and payload of land- and carrier-based aircraft allowed joint commanders to hit targets deep in the enemy's operational or even strategic depth. They have been used in several regional wars since 1945. However, their theoretical underpinnings and practical utility have not always been well understood.

Depending on the effects to be achieved, strategic, tactical and operational fires can be differentiated. The most important difference between fires is their ultimate purpose. *Tactical fires* are designed to support maneuver forces in direct contact with the enemy forces.[28] They are designed to have a decisive impact on the outcome of an engagement or battle and are planned and conducted by tactical commanders. Tactical fires usually take place within a given area of operations, combat zone, or sector,[29] concurrently with or shortly before a battle or engagement. They are aimed to hit targets in the *tactical depth* of the enemy's or one's defenses.

Strategic fires are intended to achieve a major effect on the course and outcome of a campaign or, in some cases, even a war.[30] They are planned by theater-strategic commanders and their staffs. Strategic fires usually take place outside the boundaries of a given theater of operations where a major operation or campaign would start or is underway. One example of strategic fires was the Allied Oil Plan, conducted as a complement to the landing in Normandy (Operation Neptune) and aimed to shut down the German war economy.[31] German industrial capacity was heavily damaged, although the Allies suffered severe losses in the air.[32]

Operational fires are described as the application of lethal fires aimed to generate a decisive impact on the course and outcome of a campaign or major operation. They should not be confused with strategic bombing aimed to weaken the enemy's military-economic potential in general. The success of an operational maneuver is not necessarily dependent on operational fires. Yet they can facilitate one's operational maneuver.[33] They are conducted in the operational and/or strategic depths of the enemy's defenses.[34]

Operational fires are conducted in the operational (and sometimes strategic) depths of the enemy's defenses.[35] They are planned to accomplish an operational objective. Sometimes the objective is to force the enemy commander to react operationally – forcing him to commit his operational reserves earlier than planned or deceiving him as to the one's sector of the main effort. The intensity of fires might even lead the enemy to mistakenly react on the strategic level, as the Japanese did in October 1944. The intensity of the attacks and strikes by Task Force 38 (TF-38) of the U.S. Third Fleet against Japanese airfields and other air installations, ports, shipping and warships on Formosa and in the Ryukyus prompted the Japanese on 10 October to issue alert orders for the defense of Formosa (Sho-2 Plan) and the Philippines (Sho-1 Plan). These orders led to a decision to transfer aircraft from four carriers and two half-converted battleship carriers to land bases on Formosa.[36]

Operational fires can be differentiated, according to their delivery mode, as ground-, air-, space- and sea-based fires, respectively. Identical weapons platforms can be used for operational and tactical fires. The longer the range of the weapons platform and the more destructive it is, the more suitable it is for conducting operational fires. Air-based fires not only offer a longer range than other types of fire but also are highly lethal and precise. Operational fires are conducted beyond a declared (or undeclared) area of operations in which an initial major operation or new campaign is conducted. Tactical fires, in contrast, are conducted within the declared (or undeclared) boundaries of a given area of operations.

In terms of timing, operational fires are normally conducted sometime before the beginning of a major operation or the opening of a new campaign. Once a major operation/campaign is underway, operational fires are often combined with tactical fires. Tactical fires, in contrast, start shortly before the start of a major operation and reach their full intensity as the major operation unfolds. In some cases, there is no clear distinction between operational and tactical fires in terms of timing and forces employed, because the same forces can conduct operational and tactical fires in the same period. In terms of duration, however, operational fires are normally conducted for a much longer period than tactical fires.

The methods of combat force employment in conducting operational and tactical fires may differ considerably. Operational fires can be planned (or seen in retrospect) as a major naval or air operation. This is usually the case in opening a new front or a new campaign, as the landing at Leyte in October 1944 illustrates. Constituting both a preliminary major naval operation (in the then newly opened Philippines campaign) and operational fires was TF-38's series of strikes from late August to mid-October against the Japanese airfields, air facilities, ports and shipping in the Ryukyus, on Formosa and on Luzon prior to the landing at Leyte. From 27 October to 27 November, TF-38 delivered operational fires in support of the Leyte Operation, attacking Japanese airfields and shipping on the island of Luzon. Likewise, on 12 October, the U.S. Fifth Air Force started operational fires by striking targets on the western Visayas and Mindanao in the Philippines. These actions, in combination with strikes against Japanese air and naval facilities in northeastern Borneo and the Sulu Archipelago, amounted to a major preliminary air operation.[37]

Operational fires can be conducted to accomplish a single or several purposes. Optimally, the planners should focus on a certain purpose to be achieved by conducting operational fires within a given time frame. In general, operational fires can be planned and conducted to accomplish one or more of the following purposes:

- Isolate/shape an area of operations.
- Destroy enemy operational reserve.
- Delay arrival of the enemy operational reserve.
- Destroy the enemy's critical capabilities.
- Facilitate movement of one's amphibious forces.
- Isolate an amphibious objective area.
- Deceive the enemy about the place/time of a new major amphibious landing.
- Protect the newly acquired naval bases/facilities.
- Prevent the enemy's retreat or withdrawal.
- Other.

Operational fires are often used to *isolate/shape the area of operations* by preventing the arrival or slowing down the movement of enemy troops, air, or naval reinforcements or operational reserves in the area where one's major operation/campaign would be conducted or is already in progress.[38] In the preliminaries for the Allied landing on Sicily on 10 July 1943 (Operation Husky) in the last two weeks of June, Allied heavy and medium bombers conducted massive bombing of various targets along Italy's western coast and on Sicily to prevent the movement of the Axis troop reinforcements and supplies to Sicily. Between 18 and 30 June the Northwest African Air Force (NAAF) flew 317 heavy and 566 medium bomber sorties, with the help of 107 sorties by the Ninth Bomber Command, attacking key supply points, terminal ports and marshaling yards along Italy's western coast and smaller ports in Sicily.[39] Most of the Allied attention was focused on the port of Messina, which then served as the main terminal for Axis troop reinforcements and supplies from Italy's mainland. These attacks were complemented by attacks on Reggio di Calabria and San Giovanni (across the strait of Messina). In addition, between 12 June and 2 July Allied bombers attacked a wide range of targets between Naples and Leghorn on Italy's western coast and on Sardinia.[40]

Allied operational fires conducted before the Normandy landing on 6 June 1944 were largely successful in *delaying the movement of German operational reserves* toward the landing beaches. The destruction of the rail system west of Paris and of the bridges across the Seine and Loire rivers forced the Germans to make long marches, with many detours, under continuous aerial attack. The Germans virtually abandoned daylight road movement, and their night movements were restricted by the short (six-hour) summer nights. This greatly inhibited their ability to concentrate enough combat power to destroy the Allied bridgehead in Normandy.[41]

Operational fires can be intended to *deceive the enemy as to the place and/or time* of initiating new major operations or campaigns. Because of the long operational reach of land-based bombers, carrier groups, and submarines, it is possible to attack many targets within a given theater of operations, thereby making it difficult for the defender to correctly deduce the time and place of a major operation or campaign. In World War II, as part of the deception effort, Allied air forces were used on many occasions to attack a variety of targets over a large part of a theater. One of the most successful applications of operational fires in a deception plan preceded the invasion of Normandy (Plan Fortitude South). For several weeks before the landing, Allied air forces conducted intensive strikes on rail networks in the area east and north of the Seine River. The destruction of the bridges across the Seine west of Paris convinced the Germans that the Allies wanted to cut off supplies toward the Pas de Calais area, where they expected the landing to take place. About 40 percent of the Allied bombing missions on D-3 and D-2 were conducted in the Pas de Calais area. The main targets of these attacks were the German coastal defenses and road and rail communications in the area.[42]

Another purpose of operational fires can be to *isolate amphibious objective area* by destroying or neutralizing enemy forces capable of interfering with one's amphibious landing. In most cases, the greatest threat to the success of one's amphibious landing, especially in its initial phase, is posed by the enemy's land-based aircraft. In some cases, operational fires can *continue after a major amphibious landing took place.* Then the objective would be to prevent the enemy from moving additional troops and logistical supplies into the amphibious objective area.

Operational fires can be planned to *prevent the enemy's retreat or withdrawal.* This usually occurs after the major ground operation starts, as in Operation Desert Storm. Because a major part of the Iraqi ground forces was deployed in the KTO, the coalition task was not so much to prevent the arrival of reinforcements as to prevent the Iraqis from withdrawing their forces once the coalition ground operation began.[43] The general retreat of the Iraqis from Kuwait started on the evening of 25 February (two days after the ground offensive started). Coalition aircraft interdicted retreating columns of Iraqi troops moving on the roads in trucks, cars, or any means available. Iraqi forces deployed in the northern part of Kuwait had the best chances of escape. After air attacks had halted the Iraqi forward elements, coalition aircraft attacked the remaining several miles of vehicles. For some two miles, part of the road (called "the Highway of Death") was crowded with about 1,400 burning vehicles.[44] A second choke point for the retreating Iraqi forces was a multiline causeway over the lakes and marshlands at Hawr al-Hammar, northwest of Basra. By the evening of 26 February, damaged vehicles blocked traffic over the causeway. Air strikes continued the next day, and eventually an estimated 550–600 vehicles were abandoned by the Iraqis.[45] However, the coalition forces failed to prevent the escape of the bulk of the crack Iraqi divisions from southern Iraq and Kuwait. The remnants of three infantry divisions and one armored division escaped, in addition to the entire first-rate Hammurabi and Medinah armored divisions and parts of five infantry divisions of the Republican Guards, with 845 tanks (a quarter of all the tanks in the theater) and about 1,410 armored vehicles (half of all such vehicles in the theater). Yet about 70 percent of the Republican Guards escaped destruction by coalition forces.[46] At least 365 T-72 modern tanks, or half the tanks with the Republican Guards, escaped (the IRG started the war with 786 tanks).[47]

Operational Logistics

Logistical support and sustainment are perhaps two of the most critical factors for the success of a campaign or major operation. Failure to establish sound logistical organization in the theater, a lack of readily available supplies of all kinds, and an inability to provide protection to both the elements of logistical organization and the lines of communication will lead to major setbacks and often to defeats. Logistics can be described as both a science and an art. It deals with planning and carrying out the movement and maintenance of forces.[48] Among other things, logistics encompasses design and development, acquisition, storage, movement, distribution, maintenance, evacuation, and disposition of materiel; movement, evacuation and hospitalization of personnel; acquisition or construction, maintenance, operation and disposition of facilities; and acquisition or provision of services.[49]

Tactical, operational, and strategic logistical support and sustainment can be differentiated by the scale of military action. *Tactical logistics* is concerned with planning for and providing support and sustainment of battles, engagements, strikes and other tactical actions. *Operational logistics* encompasses logistical organization in a formally declared or undeclared theater of operations. It links strategic logistics to tactical logistics. Its main purpose is to ensure that one's actions are continuous through all phases of a campaign or major operation. Operational logistics should balance current consumption with the need to build up logistics support for subsequent operations. It should provide for lengthening the lines of communication and staging logistics support forward to maintain the desired operational tempo.

The adequacy of operational logistics is measured in terms of its ability to perform its sustainment function and distribute supplies inherent to those functions.[50] Sustainability is the measure of the ability to maintain logistics support for all users throughout the theater for the duration of the operations. Therefore, long-term support is the logisticians' great challenge. Operational sustainment supports combat forces throughout all phases of a campaign or major operation. Depending on the location of the theater of operations, this may require simultaneous employment of land, air, and sea lines of communications. Then the primary limiting factors on the operational commander's freedom of action are the availability of transportation and the ability to protect lines of supplies from the enemy's action.

Strategic logistics deals with planning, preparing, and providing resources in support of the country's armed forces. Specifically, it provides logistical support to industrial base mobilization, installation base support, consumer logistics, intertheater strategic concentration, and acquisitions and construction of facilities, services and host-nation support.[51] Strategic logistics is the bridge between the national economy and the combat forces. Strategic logistics plans for and sustain military forces. It also supplies the national means to support the forces and their operations. The national capability to provide adequate logistical support for the armed forces is the most critical element in generating and maintaining the combat potential necessary to prosecute a war.

Logistics should be fully integrated into the plan of a campaign or major operation. At the same time, too much focus on logistics can unduly restrict options to the operational commander. A proper balance should be found between the requirements for adequate logistical support and sustainment and purely operational requirements. On many occasions, the German army in World War II did not sufficiently integrate operations and logistics. This cost the Germans dearly when many of their major operations were stopped prematurely. Despite the provisioning of supplies in the homeland, supplies did not reach the German troops at the front in time. The main reasons were the inadequate transportation network, destruction of railroads, inadequate security in the rear and unfavorable weather conditions.[52]

In contrast to the Germans, the Allies gave perhaps too much emphasis to logistical requirements in their operational planning in World War II. In their invasion of Normandy (Operation Neptune), logistics dominated the determination of the objectives, the choice of landing sites, the size of the assault forces and the plans for building up initial forces and the subsequent push inland.[53] This overemphasis on logistics and the corresponding need to secure geographic objectives, as opposed to destroying enemy forces, was one of the main reasons that it took longer than it should have to defeat the German forces.

Logistics considerably affects the *operational reach* of friendly combat forces – the distance over which one's power can be massed and employed decisively. This goes beyond merely conducting long-range reconnaissance or strikes. Operational reach is affected by the effective range and endurance of combat forces, and the length and degree of protection of lines of supplies. It is influenced by the intervening physical environment separating the two sides in hostilities. Operational reach could be extended by establishing new bases and deploying combat forces into forward areas, increasing the effective range of weapons and improving the transportation system. Operational logistics determines characteristics of operational reach, including the size of combat forces, depth of attack and speed of advance. It has a finite distance beyond which logistics cannot support combat forces; at that point, the offensive becomes logistically overextended.[54]

For the Allies, the war in the Pacific in 1941–1945 revolved around capturing Japanese island positions from which to threaten the remaining enemy positions. The U.S. forces greatly extended their operational reach by capturing the enemy-controlled Gilbert Islands in November 1943. By then, the U.S. operational logistics in the Pacific Ocean Area (POA) extended all the way from the Solomons in the southern Pacific to the Gilberts in the central Pacific and, via Pearl Harbor, Hawaii, to the U.S. West Coast. The main logistics center was in the San Francisco area, with warehouses in Alameda and Oakland. The largest storage depot was built at Clearfield, Utah. The distance between Pearl Harbor and the Gilberts is about 2,000 miles – a 4,000-mile round trip for supply ships. Advanced bases were established to store supplies for the fleet.

The main elements of logistics are bases of operations and theater transportation networks. The availability of diverse bases theaterwide is a critical factor in the application of one's combat potential. The operational commander is responsible for establishing a theater base. Base availability directly affects the number of sorties and the rate of resupply. The geometric arrangement and successive positioning of bases directly influence the operational commander's ability to defend and conduct offensive actions against the enemy.

The *main base of operations* is usually located in the home territory. Yet sometimes it can be located on the territory of a friendly country. It encompasses several ports and airfields or airports, supply depots, hospitals and repair workshops. It serves as an entry and exit point for all personnel and materiel to be transported by ships, aircraft or railways and trucks forward to the combat area.

A theater of operations should be logistically self-sufficient, with a *theater sustainment base* adequate for contemplated operations. Once a sustaining base is established, its expansion, defense and protection should be solely the operational commander's responsibility. The operational commander should ensure a smooth and secure flow of logistical requirements, regardless of changing priorities and conditions, throughout the course of a major operation or campaign.

A theater sustainment base on land is usually established between the main base of operations and the combat logistics area and, if possible, beyond the effective range of the enemy's long-range fires. The *intermediate base of operations* is often located near a railhead, to save time when loading and unloading personnel and supplies. It serves as a distribution point for the resupply of one's forces in the combat zone. Those stocks, tied directly to the support of combat units on the front, are assembled in the *combat logistical area*.[55]

Geographic or other conditions in the theater may preclude the initial establishment of an in-theater sustainment base. Hence, the initial major operations may have to be sustained from bases external to the theater or from offshore. This was true in all the campaigns and major operations conducted on the opposing shores in the Pacific and the European theater in World War II. In the Normandy landing, the base buildup was intended to bring in sufficient logistical resources to the lodgment. The landing, establishment of a theater logistics base and envelopment of the Ruhr were all heavily influenced by logistics. The effects of logistics were evident in watercraft availability and turnaround time, beach sustainability, port availability and shipping.[56]

Logistical bases outside the theater may impose severe restrictions on the operational commander's freedom of action and the conduct of a campaign or major operation. This is especially true in an undeveloped theater. Logistical limitations were one of the main reasons the Allies were unable to resume their offensive in the southern Pacific for five months after the end of the struggle for control of Guadalcanal in February 1943. Between January and June 1943, the Allies mounted no major actions in the area.[57]

The *transportation network* is another critically important element of the logistical infrastructure. The more straightforward and unified the transportation, the greater its effectiveness. Determining factors include size and quality of means of transport; number, effectiveness, and security of transport routes; quality of services; capability of transport organization; and weather and climatological conditions.[58]

A well-developed theater usually possesses large, efficient and diverse systems of rail and road communications, inland waterways, seagoing/riverine ships and air transport. Not only should the basic modes of transport be in place but also within each transport mode a variety of means should be available to augment or supplement each other. Roads, canals, rail lines and air routes should be kept open to allow movement. Failing to keep routes open reduces the sustainment of the forces.[59]

No campaign or major operation can be successful without adequate, reliable, and timely *operational sustainment* of forces in combat. Logistical sustainment is the extension of logistical support from the start of combat actions until the assigned military objective has been accomplished. It may also be a key restraint by limiting the commander's options for conducting operational maneuvers. An early identification of critical logistical constraints associated with planned major operations is indispensable for achieving strategic or operational success. Logistically, the mobile force should be as self-contained as possible. Choices of logistics base sites and lines of communication are crucial and should be made by the operational commander in relation to the decisive points. Lack of sustainment or improperly sited bases and lines of communication can cause operational pauses that interrupt the continuity of maneuver.[60]

Depending on the location of the theater of operations, this may require simultaneous employment of land, air, and sea lines of communications. The primary limiting factors on the operational commander's freedom of action are the availability of transportation means and the commander's ability to protect these lines of supply from enemy action. In the fall of 1944, the Allied advance stopped on the German border because logistics was unable to keep up. Despite the greatly effective air transport, the ground transportation system could not bring sufficient supplies to the front in time.[61]

After 1943, the U.S. Navy had an excellent system of logistical support and sustainment for its striking forces in the Pacific. By 1944, the U.S. Navy's logistical sustainment capabilities allowed fast carrier groups operated at sea for many weeks without needing to return to their bases. Operational sustainment for the Third or Fifth Fleet was the responsibility of the At-Sea Logistics Service Group – composed in October 1944 of thirty-four fleet oilers, eleven escort carriers, nineteen destroyers, twenty-six destroyer escorts and several seagoing tugs. This force was divided into ten to twelve task units to fuel the fleet in echelons. A group of nine to twelve tankers was deployed near the Third Fleet's TF-38, refueling one carrier group after another. About every three or four days a task unit of three oilers with escorts joined this refueling group to relieve those that were almost empty. These tankers then transferred the remainder of their oil to other tankers that were partly full and returned to Ulithi to take new fuel loads from the commercial tankers. Each of these groups was accompanied by an escort carrier with replacement aircraft and pilots from Eniwetok (Marshalls), Guam (Marianas) and Manus (Admiralties). Combat air patrols (CAP) and antisubmarine (A/S) patrols for the oiler groups were conducted by a group of eight escort carriers with their escorts. Rescuing damaged ships was the responsibility of two fleet tugs accompanying each refueling group.[62] The TF-38 – composed of four carrier groups with nine heavy and eight light carriers, six battleships, three heavy and six light cruisers, three AA light cruisers and fifty-eight destroyers – were regularly replenished at sea by TF-30 logistics ships.[63] Fast carrier forces were at sea for thirteen of sixteen weeks; this was possible only because of excellent logistical organization.[64]

Operational Protection

Protection of combat forces, civilian population and military/civilian and infrastructure from a wide range of threats in peacetime and in war is one of the most important responsibilities of a military commander. In a war, threats to friendly combat forces are more diverse than in peacetime. The enemy can use both conventional and nonconventional weapons against combat forces, the civilian population, and important military-economic installations/facilities. The range of possible threats, especially in a "mature" (developed) theater, is rather large. At the same time, one's ability to provide full defense and protection is usually limited. Therefore, commanders must make hard choices in determining which forces and which part of the theater infrastructure need more protection than others. The problem becomes more complicated when one's forces are deployed in a theater far away from home bases.

A distinction is made between tactical, operational and strategic protection. *Tactical protection* pertains to the defense and protection of combat forces deployed or located in a combat zone or area of operations. Broadly defined, the term *operational protection* refers to a series of actions and measures aimed at preserving the effectiveness and survivability of military and nonmilitary sources of power located within the boundaries of a given theater of operations. *Strategic protection* pertains to the protection of military and nonmilitary sources of power in a given theater of war (theater-strategic level) or for the country as a whole (national-strategic level).

Operational protection in a land theater in time of crisis or war encompasses the following main components:

- Threat indications and warning (I&W).
- Airspace control & air defense.
- Force protection.

- Rear area defense.
- Protection of information systems.
- Protection against weapons of mass destruction (WMD).

In addition to the above-described elements of operational protection, a maritime theater or land theater bordering the sea/ocean, operational protection would include a number of elements related to ensuring control of the coastal area and adjacent waters necessary for basing/deployment area control of one's maritime forces. Defense of the coast is a broad term referring to direct or indirect defense and protection of one's coastal area, offshore islands/archipelagoes and adjacent waters from the enemy's attack across the sea and from the air.

No component of operational protection should stand alone. In practice, there are no clear boundaries between them, and they can overlap considerably. Air defense, force protection, OPSEC and physical security are integral parts of all elements of operational protection. Likewise, the protection of information systems encompasses all information systems in a given theater of operations.

Threat indications and warning is aimed at providing timely warning so that a theater commander can take timely action to preempt, counter or neutralize enemy's attack. The main tasks of I&W are to observe and evaluate enemy's capabilities and intentions, preparations for mobilization, changes in combat readiness and deployment patterns. Another major task is to deduce patterns in the development of non-military aspects of the strategic situation. I&W information should be shared throughout all the levels of command.

Operational airspace control is established to ensure close coordination among airspace control, air traffic control, and air defense elements. It is defined as "capabilities and procedures used to increase operational effectiveness by promoting safe, efficient, and flexible use of airspace."[65] It includes coordinating, integrating and regulating aerospace.[66] In U.S. terms, airspace coordination measures (ACM) are intended to facilitate the effective use of airspace and provide safety to friendly forces. In U.S. military, the principal coordination measures are fire support coordination measures (FSCM) for facilitating rapid engagements of targets; maneuver control measures (MCM) aimed to define lines of responsibility in support of friendly maneuver (they are normally established by ground force commander); air reference measures (ARM) aimed to help C2 coordination and define a point over the ground or volume of airspace for reference; air defense measures (ADM) are aimed to facilitate detection, identification and tracking to engage enemy aerial targets and missiles; maritime defense measures (MDM) are used to facilitate maritime actions; and air traffic control measures (ATCM) are aimed to expedite the safe movement of aircraft in conjunction with civil and military Air Charter Association (ACA) and International Civil Aviation Organization (ICAO).[67]

To be truly effective, airspace control must be integrated throughout the chain of command. Its aim is to ensure the combat effectiveness of ground, air and maritime forces during preparations for major operations/campaigns. Another purpose is to eliminate (or at least reduce) the chance of fratricide during the employment of friendly combat forces. Combat airspace control's requirements should be identical to those of air defense for identification of friend or foe. Accordingly, combat airspace control, air defense and supporting C4 systems must have a high degree of interoperability and must be mutually supporting.

Operational air defense encompasses the defense and protection of combat forces, civilian population, and military/civilian infrastructure from enemy air attacks (including attacks from or through space). This is accomplished through the destruction or neutralization of enemy's air-attack capabilities, both in the air and on the ground, and protection of critical military/civilian facilities/installations in a given theater of operations. Air defense is also aimed at defending and protecting ground forces from enemy attacks during their movement to concentration areas and operational maneuvers.

Air defense consists of active and passive measures. Active air defense is aimed to destroy, nullify, or reduce the effectiveness of hostile air and missile attacks. It should be multi-layered and established in great depth.[68] Passive measures of air defense are aimed to provide timely detection and warning of enemy air attacks and thereby provide maximum reaction time for friendly forces to seek shelter or take appropriate action. *General warning* indicates that attacks are imminent or have occurred. *Specific warnings* are issued to signify that specific units of areas are in danger of attack.

Passive measures of air defense include dispersal of friendly forces, camouflage, concealment and deception (CCD), hardening, redundancy and mobility.[69] *Dispersal* is aimed to complicate the enemy's ability to locate, target and attack friendly forces. *Hardening* in construction of some key elements of military/civilian infrastructure is aimed to enhance protection against protect against physical attack, electromagnetic pulse and transient radiation. *Redundancy* aims to duplicate critical capabilities and enable the essential systems to continue operating even after critical nodes have been destroyed or damaged. *Mobility* aims to reduce vulnerability and increases survivability by complicating enemy surveillance, reconnaissance and targeting.

In terms of the factor of space, point and area air defense are differentiated.[70] *Point defense* is aimed at protecting limited area and the key elements of military/civilian infrastructure such as airfields, naval bases, major cities, industrial plants, etc. Area defense encompasses a major part of even the entire theater of operations. The first theater air defense system was created in Britain in the late 1930s. The British government assigned more resources to the development of radar in 1935. The first coastal radar station chain became operational in the spring of 1938. The early version of coastal radars was able to detect aircraft flying from 1,000 to 25,000 feet. The problem of detecting enemy aircraft at lower altitudes was partially resolved by adding portable radar stations capable of detecting aircraft flying at an altitude of 500 feet.[71]

The U.S. joint doctrine includes among threats from the air not only manned and unmanned aircraft, aerodynamic and ballistic missiles but also ground-based rockets, artillery and mortars.[72] Air defense consists of a combination of counterair and Integrated Air and Missile Defense (IAMD).[73] Offensive and defensive counter air are differentiated. Offensive counter air (OCA) aims to destroy or neutralize enemy aircraft, missiles, launch platforms, and their supporting structures and systems both before and after launch; aim to prevent or disrupt the launch.[74] The main methods used in OCA are fighter escort, and fighter sweep. Defensive counter air (DCA) and air missile defense (AMD) includes all defensive measures within the theater aimed to neutralize or destroy enemy forces trying to penetrate or attack through friendly airspace.[75] AMD is aimed to destroy nullify or reduce the effectiveness of hostile air and ballistic missile threats against friendly forces.[76]

Ballistic missile defense (BMD) (formerly theater missile defense – TMD) is an integral part of area air defense. Its primary purpose is to protect population centers and military/economic infrastructure against the enemy's theater ballistic missiles. BMD is one of the key elements of regional stability.[77] It cannot be successful without the participation of two or more services. Like defense against enemy air attacks, BMD consists of a series of defensive and offensive lethal actions and passive measures. Defensive lethal actions are aimed at destroying the enemy's ballistic missiles in various phases of flight. Offensive actions are intended to destroy the enemy's research centers, ballistic missile production facilities, and launchers, with associated sensors and supporting structures. Passive measures are designed to minimize the effects of an enemy ballistic missile attack and reduce the vulnerability of friendly forces, theater infrastructure, and civilian populations to such an attack. Ballistic missiles can use depressed trajectories or release maneuvering warheads or can carry penetration aids that attempt to deceive, obscure or jam defender tracking sensors.[78]

Operational force protection is aimed at protecting large forces and formations from enemy attacks from the air (or through space), on the ground or across the sea. It provides protection to friendly personnel against various hazards on the battlefield, maximizing the survivability of personnel and their weapons and equipment; and preserving the health of personnel. In the U.S. joint doctrine, force protection is defined as "preservation of the effectiveness and survivability of mission-related military and nonmilitary personnel, equipment, facilities, information, and infrastructure deployed or located within or outside the boundaries of a given operating area."[79] Force protection is multidimensional and multilayered. It is aimed at protecting military members and civilian employees, their families, contract employees and visitors on an installation.[80] Spectrum of threats include conventional and irregular forces, paramilitary forces, subversion, sabotage, intelligence activities and evasion and escape networks.[81] Sound force protection is based on being comprehensive, multilayered, integrated, redundant, enduring and continuous activity.[82] Force protection encompasses buildup of fortified areas, as well as the use of defensive EW, OPSEC, physical security, defense against WMD, and defense against terrorist acts. Physical security – is designed to safeguard personnel and prevent unauthorized access to equipment, installations, material and documents and safeguard them against espionage, sabotage, damage, theft and terrorism.[83]

The range and types of threats depend on the predominant physical characteristics of a theater and whether the theater is developed or undeveloped. In an undeveloped theater, the focus is predominantly on force protection while in a developed theater, the theater commander has the additional responsibility of protecting the most important elements of the theater infrastructure and population. However, the national political leadership may sometimes decide force protection is the most important task, regardless of whether a theater is developed or undeveloped.

In a war, the operational commander is responsible for establishing and maintaining security in the rear so that combat forces can be secure while preparing and conducting campaigns/major operations. The *rear area defense* is aimed at creating a secure environment for the stationing and movements of combat forces and their logistical support and sustainment. Another objective is to ensure the smooth and uninterrupted functioning of the economy and other activities for the needs of one's forces and the civilian population. In U.S. joint doctrine rear area security is aimed at preventing or minimizing disruption of support, enemy interference with C2, protecting personnel and facilities, protecting lines of communication, area damage control and combating terrorism.[84]

The theater commander is normally responsible for the defense of all air, army, and naval bases/ports within his area of responsibility. *Base cluster* is a collection of bases, geographically grouped for mutual protection and command control. In U.S. terms, a base cluster consists of two to seven bases.[85] Base clusters are designated when their number exceeds the commander's normal span of control. Bases are located in close geographic proximity.[86] Base security forces would provide local security to a base. The main defensive measures include establishing a defense perimeter, vehicle and personnel entry points, an attack warning system, developing a fire plan, personnel survivability shelters, security patrols and defenses of air bases from air attacks.[87]

Defense is organized against hostile forces trying to penetrate the base and defense against standoff attack.[88] Aircraft are most vulnerable during take-off, landing, approach, departure and taxiing, and when parked on an airfield.[89] The individual tactical commanders are responsible for the defense of their respective base or a base cluster. The base boundary is a line that delineates the surface area of a base for the purpose of facilitating coordination and deconfliction of operations between adjacent units, formations or areas.[90] The base boundary should be negotiated between the base commander and the area commander or host-nation authority.[91] In a joint operating area, an air base is protected in layers and in depth. Base security forces are responsible for defense from the base boundary inward. Mobile forces could provide defense of air base boundary outward.[92] There is a need to coordinate with area commanders to ensure that base boundaries are adjusted to provide adequate protection from rocket, artillery, and mortar attacks.[93] Because resources are finite, not all bases can be defended equally well. Therefore, the operational commander needs to determine which basing area must be protected better than another.

Logistics installations are high-value targets and should be adequately protected. This is accomplished by assigning adequate forces to the theaterwide protection of key elements of logistical support and sustainment – specifically, main, and intermediate bases of operation, supply depots and distribution points, transportation nodes and lines of communication. Passive measures of protection include decentralization of logistical infrastructure and dispersal of logistical installations and facilities; cover, concealment and deception (CCD); and physical protection of personnel and equipment.

Defense of lines of communication is the responsibility of an operational commander. The primary threats are mines, ambushes and improvised conventional weapons and munitions.[94] Active security measures include patrols, snipers, fighting positions, checkpoints and route sweeps.[95] Passive measures include convoy formations and march controls, security- related driving techniques, route variance, variants of convoys departure times and OPSEC countermeasures.[96] Security of lines of communication is enhanced by conducting reconnaissance and surveillance by using manned aircraft or UAVs, military or police physical reconnaissance and traffic control patrols and waterway patrols.[97]

In today's environment, terrorist acts are a constant threat to one's forces, facilities, and the civilian populace. The problem is further complicated because terrorist acts can be directed against military installations, personnel, and their dependents located far beyond the boundaries of a given theater. The theater commander should presume that civil authorities and host governments will take a major part of the responsibility for antiterrorist actions and measures. However, the commander is usually responsible for coordinating those measures aimed specifically at protecting military facilities, equipment, personnel and family members. Service component commanders are responsible for publishing the theater commander's rules of engagement (ROE) and ensuring that they are followed to provide flexible self-defense and deterrent posture.

Military forces and infrastructure are increasingly vulnerable to hostile actions against their information systems (INFOSYS). The operational commander is responsible for employing *defensive information warfare* against asymmetrical attacks on information systems, infrastructure and other critical areas vulnerable to nontraditional means of attack or disruption. It is aimed at providing tactical warning, attack assessment, emergency response and infrastructure restoration. Threats to the theaterwide information infrastructure are potentially worldwide in origin. The globalization of networked communications has resulted in vulnerabilities due to increased access to one's information infrastructure from points around the world. Types of attacks against elements of the information infrastructure can range from unauthorized access to physical destruction. Unauthorized access can be aimed either at gaining information or at inserting false or misleading information. Malicious software might be inserted to cause a computer to operate differently than its user intended. Various viruses can be inserted to alter or corrupt data or to make it misleading or useless. Some viruses are designed to bypass protective programs.

Information attacks, even when they consist of nothing more than isolated incidents, create a perception of vulnerability. The fact of such attacks can have an impact far greater than the actual damage inflicted. The attacks do not need to destroy the computer system completely; all that is needed is to disrupt the work of a given system and thereby create a perception of failure and a loss of confidence in the system's reliability. Redundancies in information systems are only a partial solution to this growing problem.

Protection against WMD (or CBRN – Chemical, Biological, Radiological, Nuclear in U.S. terms) poses unique problems for the operational commanders and is difficult to organize on a theaterwide basis. Passive and active measures of avoidance are one of the main principles in defense against WMD. Passive measures include dispersion, hardening and cover and concealment. Warning, detection of contamination and reporting are the main active measures of avoidance. Protection of personnel includes special clothing, protective masks, collective protection shelters and medical treatments. After the use of WMD, various decontamination methods restore combat power and prevent additional casualties.

In military terms, the sea coastal or littoral area consists of: (1) the seaward area (or forward coastal area), (2) the coastline and (3) the rear area. The *Seaward area* extends from the coastline to a certain distance outward. The *coastline* encompasses the beaches and very narrow land strip. Beyond the coastal strip is the *rear area.*

Defense of the coast should be fully established already in peacetime. In wartime, the main objectives are to secure control of basing/deployment area; uninterrupted flow of coastal sea traffic, prevent large-scale enemy seaborne landings, and major attacks on important coastal facilities/installations and military-economic centers in depth of one's territory. Normally, defense of the coast is the primary responsibility of the army and coastal defense forces. Yet experience also shows that defense of the coast cannot be successful without full and effective involvement of the navy and air force. This is especially critical for a much weaker side at sea because the outcome of a war depends on its ability to fully integrate all participating forces and prevent the enemy's large-scale landing.

Defense of the coast can be generically divided into: (1) seaward defense zone; (2) coastal defense zone; and (3) rear defense zone. These zones overlap and are complementary to each other. The *seaward defense zone* (or coastal forward zone) extends from one's coastline to a certain distance outward. Its defense should be primarily the navy's responsibility. Forward boundaries of the seaward zone depend primarily on the distances of offshore islands/archipelagoes from the mainland, the effective range of one's coastal surveillance radars and land-based aircraft. The depth of a seaward defense zone is usually greater than the ability of a weaker side to effectively ensure full protection and safety of naval forces and aircraft and coastal shipping operating within a given zone.

Seaward defense is aimed to prevent surprise attacks and ensure the safety of one's naval forces, commercial shipping, naval/air bases/ports, and coastal facilities/installations. These actions should be conducted by *local sea defense forces.* In combat, the principal responsibilities of the local sea defense forces are to conduct: (1) anti-submarine defense; (2) anti-combat craft defense; (3) air defense; (4) defense & protection of coastal sea traffic; (5) defensive mining; (6) anti-mine defense or mine countermeasures (MCM); (7) defense against commando raids; and (8) defense against terrorist acts. The main prerequisite for effective employment of local sea defense forces is reliable and continuous *coastal reconnaissance* conducted both in peacetime and in times of war. Coastal reconnaissance is aimed at detecting, locating, identifying and monitoring potentially hostile or enemy surface ships, aircraft, submarines, and minelaying within a given seaward defense zone. It is conducted by coastal surveillance posts, land-based aircraft, small surface combatants, submarines, unmanned vehicles (UAVs/USVs/UUVs) plus commercial vessels and fishing boats.

Conclusion

Good intelligence is one of the most important prerequisites for making a sound decision and then preparing a sound plan and skillfully implementing it. The significance of good intelligence should be neither overestimated nor underrated. It is relatively greater at the operational and strategic levels because of the much larger objectives to be accomplished and the correspondingly greater factors of space, time, and force. However, tactical intelligence is often crucial for achieving tactical victories. Intelligence in a theater must be fully integrated; otherwise, deficiencies at one level can seriously undermine the effectiveness of intelligence at other levels of command.

Significant differences exist among tactical, strategic, and operational intelligence because of the very different scale of the objectives to be accomplished. These differences should be fully understood by both intelligence analysts and commanders and their staff. Good strategic intelligence is of little use if tactical intelligence is poor or inadequate. The opposite is equally true.

Because of much larger factors of space, time and force, operational intelligence necessarily must deal with a much larger scope and diversity of uncertainties in many aspects of the situation. It is more focused than tactical intelligence on such hard-to-quantify elements as the personalities and peculiarities of the enemy's operational commanders and their subordinate commanders, their skills and experience, the enemy's decision-making cycle and its exploitable vulnerabilities, service/joint doctrine, morale and discipline, and quality of training. Yet it is information on precisely these elements of the enemy's situation that is hardest to obtain and properly evaluate and analyze. The new information technologies are no great help in this matter, despite what their most enthusiastic proponents assert. While these technologies have greatly improved and will continue to improve the operational commander's knowledge and understanding of the physical aspects of the theater, not much progress is being made in improving his knowledge of the intangible elements. There is no substitute for the human element, in either intelligence collection or analysis, to resolve this critical and integral part of war.

Over the past two or three decades, and especially in recent years, the significant increase in the capabilities of reconnaissance and surveillance sensors, computers and precision munitions has brought about qualitatively new capabilities for attacking enemy command and control and protecting one's own. IW should be conducted both in peacetime and in times of crisis and war, and across the entire spectrum of conflict. While increasingly more important in peacetime and in war,

IW should be considered something so new that it is used outside the traditional methods of warfare. It is the lethal force – not the nonmilitary sources of power – of which IW is a part that determines the outcome of war.

IW is becoming an increasingly critical part of all levels of war. The higher the level, the more important and potentially more decisive IW is. At the operational level, IW is conducted with forces of two or more services, and often the services of friendly armed forces as well. The operational commander is the only one who can properly sequence and synchronize the use of IW in support of a campaign or major operation. IW can also be quite effective in operations short of war and in times of crisis. Among IW's greatest advantages are its diversity and flexibility, compared with the use of lethal power. Any component of IW can be used individually or in combination, depending on the assigned objectives and the situation. Each component can supplement the effects of other elements. They can also conflict with each other. Therefore, the employment of several components of IW in support of a campaign or major operation requires a judicious balancing of often-contradictory requirements. IW planning and its incorporation into plans for a campaign or major operation are extremely complex and require the highest degree of professional knowledge and skills on the part of operational commanders and their staff.

Operational fires are an important means by which operational commanders enhance the chances for a successful outcome of a campaign or major operation, especially in the initial phase. Yet, by themselves, operational fires cannot ensure success in a major operation or campaign. They differ considerably from other fires in their main purpose, timing, and duration, the area in which they are conducted, and the command echelon that plans and conducts them. They are intended to have an operational or even strategic impact on the enemy.

The increased range, accuracy, and lethality of modern weapons offer a corresponding increase in their operational employment. As the range of weapons used to support tactical maneuvers increases, weapons are bound to play a more important role in the delivery of operational fires. The biggest changes in terms of capability can be expected in ground-delivered lethal fires. Hence, the diversity of weapons systems capable of delivering operational fires will continue to increase, allowing the operational commander more options in planning and executing operational fires.

Logistical support and sustainment are two of the most important components of operational support. Its ultimate purpose is to extend operational reach or to prevent the enemy from extending operational reach for his own combat forces. Many major operations and even campaigns have failed because of inadequate or poor logistical support and sustainment. Operational commanders are solely responsible for the smooth and effective functioning of logistics in the theater. They should have full authority over logistics for their subordinate forces; otherwise, it is difficult to synchronize various logistical elements with operations during the planning, preparation and execution phase of a campaign or major operation.

The theater's logistics organization should be simple, flexible and effective. It should also be fully integrated with other theaterwide functions. Once hostilities start, the operational commander should ensure responsiveness, continuity, simplicity, flexibility, economy of effort and security. Logistics plans should be coordinated and fully integrated with the plans for all phases of the campaign or major operation. However, operational commanders should avoid the danger of allowing logistical considerations to predominate or dictate the design of a campaign or major operation. While logistics is vital for success, it can also be an obstacle. Optimally, the logistical requirements for a campaign or major operation should be properly calculated. An excessive logistical buildup should be avoided. However, some surplus in ammunition, POL, spare parts, and food is far preferable to having shortages in these most critical categories. War is more an art than a science, and one can never accurately predict how a campaign or major operation will unfold.

Operational protection is inherently a multi-service responsibility. It should be organized, planned and conducted by a single operational commander. Ideally, all sources of power, both military and nonmilitary, should be fully defended and protected. Yet this is not possible in practice because one's forces and resources are not indefinite. Therefore, the operational commander must decide which elements of operational protection should have the highest priority versus those having lesser priority. In a high-intensity conventional war, the protection of combat forces in a given theater of operations would have the higher priority than the protection of civilian population and infrastructure. A growing problem for an operational commander is how to effectively protect the information systems. This highly complicated issue defies easy, fast, or cheap solutions. The problem is further complicated when some critical elements of the information infrastructure are physically located not in a given theater but hundreds or even thousands of miles away.

Notes

1 Lanning M. Porter, *Preconceptions, Predilections, and Experience: Problems for Operational Level Intelligence and Decisionmaking* (Fort Leavenworth, KS: School of Advanced Military Studies, U.S. Army Command and General Staff College, 12 May 1986), 5.
2 Porter, *Preconceptions, Predilections, and Experience*, 29.
3 Porter, *Preconceptions, Predilections, and Experience*, 28.
4 Porter, *Preconceptions, Predilections, and Experience*, 8–9.
5 Stares, *Command Performance*, 22–23.
6 United States Strategic Bombing Survey (Pacific), *Japanese Military and Naval Intelligence Division: Japanese Intelligence Section G-2*, Dates of Survey: 1 November 1945 through 1 February 1946 (Washington, DC: U.S. Government Printing Office, 1946), 24–25, 46.
7 United States Strategic Bombing Survey (Pacific), *Japanese Military and Naval Intelligence Division*, 5; Ronald L. Burgess, *Operational Intelligence: Is It a Panacea?* (Fort Leavenworth, KS: School of Advanced Military Studies, U.S. Army Command and General Staff College, 21 April 1987), 26.
8 Efraim Halevy, "In Defence of the Intelligence Services," *The Economist,* July 31, 2004, 21.
9 Halevy, "In Defence of the Intelligence Services," 21.
10 Anthony H. Cordesman, *Intelligence Failure in the Iraqi War* (Washington, DC: Center for Strategic and International Studies, 16 July 2003), 29.
11 Olivier Desarzens, *Nachrichtendienstliche Aspekte der "Weserübung" 1940* (Osnabrück: Biblio Verlag, 1988), 79, 89, 91.
12 Burgess, *Operational Intelligence*, 28.

13 Stares, *Command Performance*, 23.
14 Fielding L. Greaves, "Intelligence, Key to Victory," *Military Review* 9 (September 1965), 35.
15 Charles B. MacDonald, *A Time for Trumpets* (New York: Bantam Books, 1985), 78.
16 Formerly "Command, Control, and Communications Countermeasures (C3CM)."
17 Joint Pub 3–13.1, *Joint Doctrine for Command and Control Warfare (C2W)* (Washington, DC: U.S. Government Printing Office, 7 February 1996), I–4.
18 Derik W. Crotts, "Operational Implications of Public Affairs – Factors, Functions, and Challenges of the Information Battlefield," *IOsphere* (Winter 2006), 13–14.
19 Joint Pub 3–13.1, *Joint Doctrine for Command and Control Warfare (C2W)*, II–3 and II–4.
20 Elizabeth A. Hurst, *Shaping the Battlefield with Command and Control Warfare* (Fort Leavenworth, KS: U.S. Army Command and General Staff College, 7 June 1996), 26–28.
21 Hurst, *Shaping the Battlefield with Command and Control Warfare*, 22–23.
22 Hurst, *Shaping the Battlefield with Command and Control Warfare*, 27.
23 Joint Pub 3–13.1, *Joint Doctrine for Command and Control Warfare*, II–7.
24 Hurst, *Shaping the Battlefield with Command and Control Warfare*, 41.
25 Hurst, *Shaping the Battlefield with Command and Control Warfare*, 22–23.
26 Hurst, *Shaping the Battlefield with Command and Control Warfare*, 20–21.
27 Joint Pub 3–13.1, *Joint Doctrine for Command and Control Warfare*, II–5.
28 Leonard G. Tokar, *U.S. Doctrine for Command and Control of Operational Fires* (Fort Leavenworth, KS: School of Advanced Military Studies, U.S. Army Command and General Staff College, April 1996), 8.
29 David M. Annen, *Joint Interdiction: The Gray Area* (Carlisle Barracks, PA: U.S. Army War College, 1996), 3.
30 Tokar, *U.S. Doctrine for Command and Control of Operational Fires*, 8–9.
31 William J. Rice, *Operational Fires: What's in a Name?* (Fort Leavenworth, KS: School of Advanced Military Studies, U.S. Army Command and General Staff College, June 1990), 18–20.
32 The Allied air forces had lost 2,000 aircraft and 12,000 airmen in the two months preceding D-Day. Russell F. Weigley, *Eisenhower's Lieutenants* (Bloomington, IN: Indiana University Press, 1981), 94.
33 Thomas A. Kolditz, *Exploring the Conditions for Decisive Operational Fires* (Fort Leavenworth, KS: School of Advanced Military Studies, U.S. Command and General Staff College, June 1993), 3; Blueprint of the Battlefield, TRADOC Pamphlet 11–9, 13.
34 Ralph G. Reece, *Operational Fires* (Maxwell AFB, AL: Air War College, 1989), 7, 10.
35 Reece, *Operational Fires*, 7, 10.
36 Samuel Eliot Morison, *History of United States Naval Operations in World War II,* vol. 12: *Leyte: June 1944–January 1945* (Boston, MA: Little, Brown, reprint 1984), 91.
37 "5th USAAF Letter of Instruction, No. 6," 29 September 1944; Record Group 23: Battle Evaluation Group (BEG), Box 56, Folder L; Fifth, Thirteenth Air Force, OP-INST/LOI. Naval War College Archives.
38 Annen, *Joint Interdiction*, 6.
39 Wesley F. Craven and James L. Cate, *The Army Air Forces in World War II,* vol. 2: *Europe: Torch to Pointblank* (Chicago, IL: University of Chicago Press, 1949), 435.
40 Craven and Cate, *The Army Air Forces in World War II*, 437.
41 Robert W. Madden, *A Thousand Points of Light: Integrating Operational Fires into Campaign Design* (Fort Leavenworth, KS: School of Advanced Military Studies, U.S. Army Command and General Staff College, May 1990), 15–16.
42 Jerry D. Garrett, *Strategic Airpower as Operational Fires: Integrating Long-Range Bombers into Campaign Design* (Fort Leavenworth, KS: School of Advanced Military Studies, U.S. Army Command and General Staff College, May 1991), 18.

43 Thomas A. Keaney and Eliot A. Cohen, eds., *Gulf War Air Power Survey: Summary Report* (Washington, DC: U.S. Government Printing Office, 1993), 92.
44 Keaney and Cohen, *Gulf War Air Power Survey: Summary Report*, 113.
45 Keaney and Cohen, *Gulf War Air Power Survey: Summary Report*, 115.
46 Hartmut Zehrer, ed., *Der Golfkonflikt: Dokumentation, Analyse und Bewertung aus militärischer Sicht* (Herford/Bonn: Verlag E. S. Mittler & Sohn, 1992), 111.
47 Michael R. Gordon and Bernard E. Trainor, *The Generals' War: The Inside Story of the Conflict in the Gulf* (Boston: Little, Brown, 1995), 429.
48 The term *logistics* appeared for the first time in 1800, but with a different meaning than today; Hans Kettgen, "Gedanken über Logistik in Theorie und Praxis (I)," *Wehrkunde* 8 (August 1964), 3.
49 Henry E. Eccles, *Logistics in the National Defense*, with introduction and comments by Captain John E. Jackson (Newport, RI: Naval War College Press, [1959] 1997), 42.
50 Ted O. Kostich, *Operational Sustainment in an Immature Theater: Considerations for Planning and Sustaining a Campaign in a Mid- to High-intensity Conflict* (Fort Leavenworth, KS: School of Advanced Military Studies, U.S. Army Command and General Staff College, 1990), 4.
51 Headquarters, Department of the Army, FM 100–16, *Army Operational Support* (Washington, DC: U.S. Government Printing Office, 1995), 3–6.
52 Krumpelt, "Die Bedeutung des Transportwesen für den Schlachterfolg," 466.
53 James A. Huston, *The Sinews of War: Army Logistics 1775–1953* (Washington, DC: U.S. Government Printing Office, 1966), 523.
54 Kevin R. Wheelock, "Review Criteria for the Logistic Plan," *Joint Force Quarterly* (Spring 1997), 131–32.
55 P. D. Foxton, *Powering War: Modern Land Force Logistics* (London: Brassey's, 1994), 7.
56 Robert T. Dail, *Does the U.S. Army Really Understand Operational War? A Logistics Perspective* (Fort Leavenworth, KS: School of Advanced Military Studies, U.S. Army Command and General Staff College, May 2, 1988), 21–22.
57 Center for Military History, *Logistics in World War II, Final Report of the Army Service Forces* (Washington, DC: Center of Military History, 1st printing CMH Pub 70–29, [1948] 1993), 50.
58 Ihno Krumpelt, "Die Bedeutung des Transportwesen für den Schlachterfolg." *Wehrkunde* 5 (May), 1965, 466.
59 Howard V. Nichols, *Operational Level Logistics: An Examination of U.S. Army Logistical Doctrine for the Operational Level of War* (Fort Leavenworth, KS: School of Advanced Military Studies, U.S. Army Command and General Staff College, May 16, 1986), 3.
60 Schroedel, *The Art and Science of Operational Maneuver*, 14.
61 Krumpelt, "Die Bedeutung des Transportwesen für den Schlachterfolg," 466.
62 Morison, *History of United States Naval Operations in World War II*, vol. 12, 76.
63 Worrall Reed Carter, *Beans, Bullets, and Black Oil* (Washington, DC: Department of the Navy, 1953), 249.
64 Morison, *History of United States Naval Operations in World War II*, vol. 12, 74–75.
65 Air Force Doctrine Publication 3–52: *Airspace Control*, 31 December 2021, 4.
66 Joint Publication 3–52: *Joint Airspace Control*, 13 November 2014, vii.
67 Air Force Doctrine Publication 3–52: *Airspace Control*, 31 December 2021, 10–11.
68 Joint Pub 3–01, *Joint Doctrine for Countering Air and Missile Threats* (Washington, DC: U.S. Government Printing Office, 19 October 1999), xi.
69 Air Force Doctrine Publication 3–01: *Counterair Operations*, U.S. Air Force, 15 June 2023, 6–7.
70 Joint Pub 3–01, *Joint Doctrine for Countering Air and Missile Threats* (Washington, DC: U.S. Government Printing Office, 19 October 1999), vi.
71 Justin Roger Lynch, "The Chain Home Early Warning Radar System. A Case Study in Defense Innovation," *Joint Force Quarterly,* No 95, 4th Quarter 2019, 101–02.

72 Joint Publication 3–01, *Countering Air and Missile Threats,* 21 April 2017, ix.
73 Joint Publication 3–01, *Countering Air and Missile Threats,* 21 April 2017, ix.
74 Joint Publication 3–01, *Countering Air and Missile Threats,* 21 April 2017, I–5.
75 Joint Publication 3–01, *Countering Air and Missile Threats,* 21 April 2017, I–6.
76 Air Force Doctrine Publication 3–01: *Counterair Operations*, U.S. Air Force, 15 June 2023, 5; Joint Publication 3–01: *Countering Air and Missile Threats,* 21 April 2017, I–6.
77 Willard G. Fallon, "Combating the Ballistic Missile Threat," *Proceedings* 7 (July 1994), 33; Dennis McDowall, "Theater Missile Defense," *Joint Force Quarterly* (Winter 1993–94), 82.
78 Joseph T. Buontempo and Joseph E. Ringer, "Airbase Defense Falls Between the Cracks," *Joint Force Quarterly,* no. 97 (2nd Quarter 2020), 118.
79 Cited in Richard E. Berkebile, "Thoughts on Force Protection," *Joint Force Quarterly* (2nd Quarter 2016), 140–41.
80 Cited in Air Force Doctrine Publication 3–10: *Force Protection*, 1 June 2023, 2.
81 Air Force Doctrine Publication 3–10: *Force Protection*, 1 June 2023, 9.
82 ADP 3–37: *Protection,* Headquarters, Department of the Army, 31 July 2019, 1–3.
83 ADP 3–37: *Protection,* Headquarters, Department of the Army, 31 July 2019, 2–7.
84 Joint Publication 3–10, *Joint Doctrine for Rear Area Operations*, 28 May 1996, ix.
85 Joint Publication 3–10, *Joint Security Operations in Theater*, 25 July 2019, Validated on 6 August 2021, IV–2.
86 Joint Publication 3–10, *Joint Doctrine for Rear Area Operations,* 28 May 1996, II–12.
87 Joint Publication 3–10, *Joint Security Operations in Theater*, 25 July 2019, Validated on 6 August 2021, IV–4.
88 Joint Publication 3–10, *Joint Security Operations in Theater*, 25 July 2019, Validated on 6 August 2021, IV-7.
89 Joint Publication 3–10, *Joint Security Operations in Theater*, 25 July 2019, Validated on 6 August 2021, IV–15; Joseph T. Buontempo and Joseph E. Ringer, "Airbase Defense Falls Between the Cracks, *Joint Force Quarterly,* No. 97, 2nd Quarter 2020, 115.
90 Air Force Doctrine Publication 3–10, *Force Protection*, 1 June 2023, 16.
91 Air Force Doctrine Publication 3–10, *Force Protection*, 1 June 2023, 16.
92 Joseph T. Buontempo and Joseph E. Ringer, "Airbase Defense Falls Between the Cracks," *Joint Force Quarterly*, no. 97 (2nd Quarter 2020), 115.
93 Buontempo and Ringer, "Airbase Defense Falls Between the Cracks, 115.
94 Joint Publication 3–10, *Joint Security Operations in Theater*, 25 July 2019, Validated on 6 August 2021, IV–21.
95 Joint Publication 3–10, *Joint Security Operations in Theater*, 25 July 2019, Validated on 6 August 2021, IV–22.
96 Joint Publication 3–10, *Joint Security Operations in Theater*, 25 July 2019, Validated on 6 August 2021, IV–24.
97 Joint Publication 3–10, *Joint Security Operations in Theater*, 25 July 2019, Validated on 6 August 2021, IV–24.

8 Operational Design

The term "operational design" originated in writings of the German military theoreticians in the late nineteenth century, where "basic thought" (*Grundgedanke*) or "idea" was used to describe Napoleon I's and Helmuth von Moltke Sr's focus on the destruction of the enemy army. In Napoleon I's crossing of the Alps in 1800, his basic thought was to cut off the enemy's line of retreat; in 1815 it was to keep separate the British and Prussian armies. In 1866, the basic idea of Moltke Sr was the concentration of the forces from the very extended end point of railways and then forward movement into Bohemia. In 1870, his basic idea was to lure the French army toward the north. The Germans made a distinction between a "scheme" or "sketch" (*Entwurf*) – referring to the assembly of forces in peacetime – and "operation scheme" (*Operationsentwurf*), wartime concentration because of unexpected and rapid changes in a situation, as happened in March, April, and May 1866.[1]

Any plan for combat employment of one's forces should be developed within a certain broader framework, arbitrarily called "design." The scope of design is determined by the scale of a given military objective.

Tactical and operational design, respectively, can be differentiated. Tactical design should be used as a framework for planning a tactical action, while operational design should be used for planning a campaign or major operation. Operational design is not a campaign plan or a plan for a major operation. Rather, it is a mental process by the commander and the staff in considering elements for starting decision-making and planning process.[2] This process should normally start after the operational commanders receive the mission from the higher commander's authority. There is no consensus about the principal elements of operational design. Generically, the main elements of a sound operational design are as follows:

- Ultimate/intermediate objectives.
- Balancing operational factors versus the strategic/operational objective.
- Identification of the enemy's/friendly operational/strategic center of gravity.
- Initial geostrategic positions/lines of operation.
- Operational/strategic direction.
- Operational idea.

DOI: 10.4324/9781003333685-8

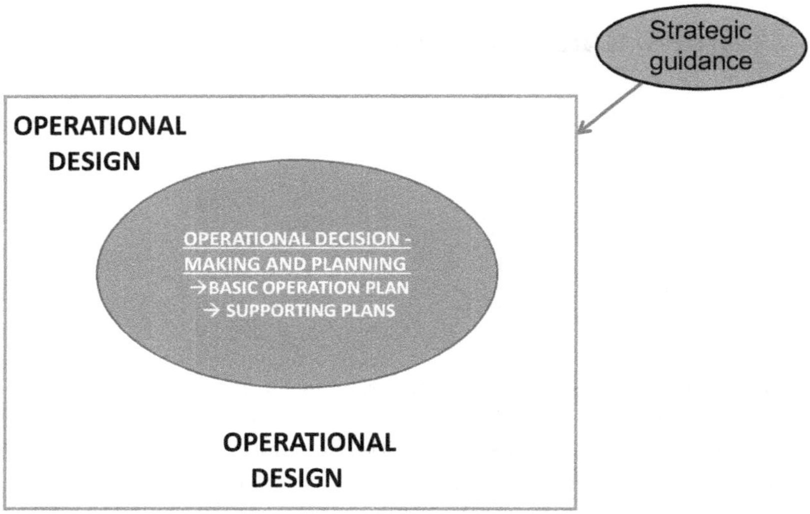

Figure 8.1 Operational Design

Because of the much larger scope of the objective to be accomplished and the corresponding differences in the factors of space, time, and force, the main elements of design for a campaign differ qualitatively from those for a major operation. Each of these elements consists of several sub-elements. How many of them should be considered would depend on the personality and experience of the commanders and their knowledge and understanding of operational art and the time available. Obviously, the mission and the situation should dictate which elements of the operational design should be part of the basic plan for a campaign or major operation.

For a campaign, the ultimate objective is invariably military- or theater-strategic, while for a major operation is operational in scale.

In their final plan to invade Soviet Russia in June 1941, the Germans did not properly determine the ultimate strategic and initial operational objectives. This was one of the major reasons for the ultimate failure of the entire campaign. The initial German operational objectives focused on destroying the mass of the Soviet forces deployed in western Russia. The planners envisaged employing strong, fast forces to quickly penetrate the depth of the Soviet front north and south of the Pripyat Marshes and then destroy the thus-divided enemy groups. Army Group North was directed to destroy Soviet forces in the Baltic and capture Leningrad and Kronstadt. Support of the panzer forces of the Army Group Center would be determined by the army's general staff. Army Group Center's initial operational objective was to destroy enemy forces in White Russia and then reach the Smolensk area and create favorable conditions for cooperation with Army Group North in the Baltic. The OKH would decide whether to attack Moscow without diverting forces

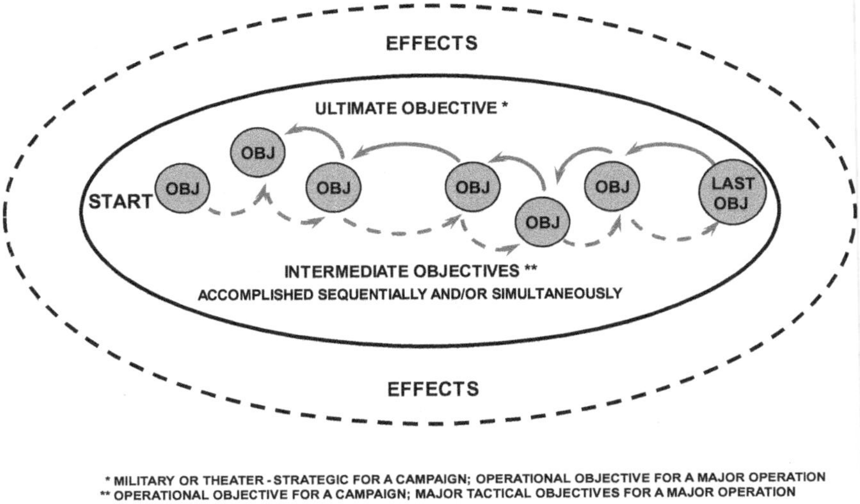

Figure 8.2 Ultimate versus Intermediate Objectives

northward. Army Group South would advance toward Kyiv, destroy Soviet forces in western Ukraine and Galicia, and capture the Dnieper's crossings south of Kyiv.[3]

The operational commanders and their staff should devote some time to determining the possible *effects* that the accomplishment of the ultimate objective will have on the enemy, the friendly side, and neutrals. This analysis should also take into account the possibility that one's objective is only partially accomplished, or the entire operation ends in failure. In predicting possible effects, reliance on various quantifiable methods of analysis should be generally avoided. The operational commander should not adhere rigidly to the initially determined ultimate objective if it cannot be accomplished with the forces available or becoming available. Then a strategic decision must be made either to adjust the objectives according to what is realistically achievable by military action or to apply a different combination of instruments of national power.[4] The more "mature" (or developed) the theater, the more complex, and hence the more difficult it is to accomplish all aspects of a given military- or theater-strategic objective.

Normally, the operational commander should try to accomplish the ultimate objective of the campaign through the destruction of enemy combat forces. If a campaign is primarily aimed to capture some geographic area because of the political, economic or logistical significance then the entire effort would result in wasting time and effort. In planning for the Normandy invasion, the Allies focused almost exclusively on the lodgment on the continent and several major ports from which further offensive operations could be conducted. Not until follow-on operations did the German forces become the Allied objective.[5]

For a campaign, *intermediate objectives* are normally operational in scale. The accomplishment of each of these objectives would normally require the planning and execution of a major operation. For a major operation, intermediate objectives are major and/or minor tactical. The number of intermediate objectives depends on the size/composition of friendly and enemy combat forces, distances, characteristics of the operating area, and logistical capabilities. In land warfare, the planners should consider the factor of distance in selecting intermediate operational objectives. The longer the distance between the two successive operational objectives on land, the longer the lines of operation and lines of supply. The larger the number of these objectives, the shorter the lines of operation and the lines of supply. The larger the number of intermediate objectives, the slower the operational tempo. On the other hand, too few operational objectives would result in taking on too much of the enemy's strength and, unless the sequencing schedule is changed, would risk overreaching the point of culmination before the assigned objective is accomplished. The planners must find a way to avoid both extremes. Another problem is that they might determine only the initial operational objectives, and then select successive operational objectives depending on the operational situation. This enhances the flexibility of the plan. Yet this can cause some serious problems in the execution of the campaign as the German experience in their invasion of Soviet Russia in the summer of 1941 exemplifies.

The need for an intermediate base of operations, logistics or purely political or economic considerations should not dominate the determination of the intermediate objectives. The Allies selected the ports of Anzio-Nettuno (31 miles southwest of Rome) as the amphibious landing place in January 1944 primarily because of political and prestige considerations and British Prime Minister Winston S. Churchill's personal interference in the theater planning process. It proved to be a poor decision for a number of reasons. Churchill firmly believed that landing at Anzio would speed the advance of the U.S. Fifth Army toward Rome and that the capture of Rome was desirable at almost any cost. The initial problem was further compounded when the Anzio landing was transformed during planning from a subsidiary attack on the left flank of the Fifth Army to a main attack into the German rear.[6]

Failure to contemplate the ultimate objective and then plan for a series of intermediate objectives invariably results in fragmentation of one's efforts, time and resources. Such campaigns last much longer, as the example of the Allied campaign in Italy (1943–1945) conclusively shows. The Allied operational planning in the Mediterranean in 1943–1944 was highly improvised. The Allies never developed a campaign plan for the next phase of the war in the Mediterranean theater after they destroyed the Axis forces in North Africa in May 1943. They focused instead on the next operational objectives, unrealistically expecting their accomplishment to lead to the end of the war in the Mediterranean. Moreover, they repeatedly underestimated the German will to fight. The Allies, strangely, failed to grasp that it was Germany's power, not Italy's, which had to be broken to end the war in the Mediterranean on Allied terms. The Allies invaded Sicily in July 1943 hoping to eliminate Italy as a belligerent. When that did not happen, they chose to invade southern Italy in September 1943. The Germans offered stiff resistance on Italy's mainland until May 1945, when the German forces surrendered on all fronts. The main reason for the Allied difficulties in the Mediterranean was the lack of a well-prepared and well-executed campaign plan. Instead, the Allies planned one or two major operations at a time, and each successive phase was dictated by political and military developments as they arose.

Intermediate operational objectives should not be pursued simultaneously or in a short sequence. The exception is when the attacker possesses overwhelming superiority, as the Japanese did in their almost-simultaneous conquests of Malaya, Singapore, the Philippines and the Netherlands East Indies (NEI) in December 1941–April 1942.

In selecting intermediate operational objectives, operational commanders and their staffs should avoid being either too rigid or too predictable Allied operational objectives in the Mediterranean theater in 1943–1944 were invariably selected to be within the effective range of Allied fighter aircraft. Thus, the Germans were largely successful in predicting where the Allies would land next.

In designing a major operation, after determining the ultimate objective, the next step is to select several intermediate – normally major tactical – objectives. In the UN landing at Inchon (Operation Chromite) on 15 September 1950, the initial operational objective was capturing port of Inchon plus Kimpo airfield (some 16 miles from the beachhead). Afterward, the UN forces would cross the Han River and then capture the South Korean capital of Seoul – successive operational objective.[7]

The most complicated element of the operational design is *balancing* (or *harmonizing*) *operational factors against a given operational/strategic objectives*. The factors of space and time can be usually calculated with a high degree of precision. This is not true of the factor of force. The true strengths and weaknesses of a given force encompass many intangible elements that are hard if not impossible to quantify. Over the years, the factor of time has been gradually compressed, while that of space has steadily expanded.[8] The process of balancing should start with quantifiable factors – that is, space and time.[9] The factor of time is perhaps the most critical of all three operational factors. The time lost can never be recovered, but space lost can be regained. Losses suffered in personnel and materiel can be replaced or regenerated.

Each operational factor contains numerous elements. Obviously, it would simply be impossible to balance all these elements with the respective military-/theater-strategic or operational objective. Hence, the operational commanders should select for their assessment only those elements of the factors space, time and force, having the largest effect on the accomplishment of a given objective. The factor of time should have central role in reconciling the advantages and disadvantages of the factors of space and force. Operational commanders should fully consider effect of information on each of the traditional operational factors.[10] The result of balancing process would require a major change or modification, or scaling down of a given objective. In the worst case, the initially determined military-/theater-strategic or operational objective must be abandoned and new one determined.

The concept of a center of gravity is, along with the determination of the ultimate and intermediate objectives, the most critical part of any military planning process. Determination of the enemy's center of gravity is a critical element in establishing clarity of purpose, focusing effort, and generating synergistic effects in the employment of friendly combat forces. A plan for a campaign or major operation should be clearly focused on destroying or neutralizing the enemy's center of gravity while adequately protecting the friendly center of gravity; otherwise, the ultimate objectives would require more time and resources than available, and even the entire expedition may fail.

After determining the ultimate strategic or operational objective, the operational commanders should determine a corresponding enemy and friendly center of gravity for the ultimate objective of a campaign/major operation plus initial intermediate objectives. Centers of gravity for successive intermediate objectives should be identified during a campaign/major operation.

To identify all intermediate objectives prior to the start of a campaign/major operation would result in a waste of time and effort.

An analytical construct is needed to identify operational/strategic centers of gravity; otherwise, the risk is too high that many critically important elements in the military situation may be partially omitted or even ignored. Obviously, the analytical process by itself cannot ensure that a center of gravity will be properly determined. The knowledge and understanding and, even more importantly, the judgment and wisdom of the commanders and their staffs are the key factors in determining the proper center of gravity and selecting the most effective methods for attacking or protecting it.

The process of identifying an enemy center of gravity should start with an analysis of one's *operational* or *strategic objective.* The scale and complexity of the

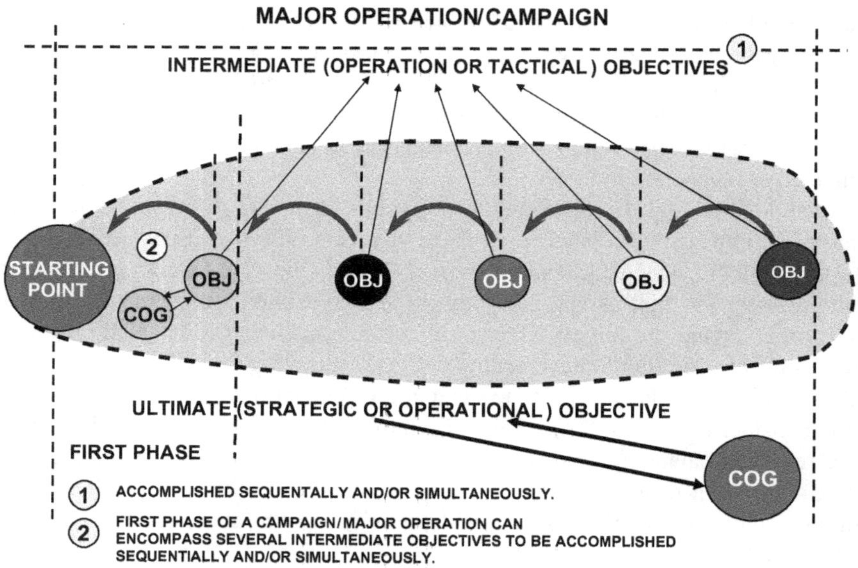

Figure 8.3 Process of Identifying Centers of Gravity

objective determine the *scale of the situation.* Any situation encompasses many physical or tangible and "abstract" or intangible elements. Tangible elements are relatively easy to quantify. Yet, the operational commander's knowledge of tangible elements is rarely perfect. The enemy can hide or use deception to deny the number and/or quality of tangible elements. Intangible elements are inherently hard or even impossible to quantify with any degree of precision.

The operational commanders and their staff are naturally interested in obtaining as much information as possible on all aspects of the situation. The larger the scale of the military objective, the more diverse its components and hence the more complex and difficult it is to properly evaluate a given military situation. And the more information is available, there is definitively a limit to what can be transmitted. The larger the volume of information received, the more difficult is for the commander and the staff to timely understand it and then make a sound decision.

Regardless of how much information one possesses, the key is to focus on evaluating those aspects of the situation – arbitrarily called *critical factors* – considered *essential* for the accomplishment of a *given military objective.* Tangible critical factors vary from geographical in nature (positions, mountain passes, valleys, plains, straits, bays/gulfs, lines of operation/communication, etc.) to the military forces (armed forces as a whole, individual services, theater/numbered armies and naval/air fleets, divisions, brigades, wings, squadrons, etc.) to the nonmilitary (state organization, diplomacy, economy, finances, agriculture, mineral resources, technology, culture, etc.). Intangible critical factors include such elements as alliance/coalition cohesion, quality of leadership, soundness of doctrine, morale and discipline, unit cohesion, public support and will fight. Note that the analysis of critical factors should be limited to only those directly or indirectly related to the military objective to be accomplished; otherwise, the entire effort would lack focus and would result in a waste of time and resources.

The analysis of critical factors should be narrowed down by dividing them into two major categories: critical strengths and critical weaknesses. Very often what is critical or less critical is difficult to determine with any degree of precision. It is all essentially a matter of the operational commander's good judgment and experience. The operational commanders and their staff should evaluate only those critical strengths and weaknesses that have affected or potentially might affect the accomplishment of the enemy's or friendly objective; to do otherwise would lead to a waste of time and resources.

Critical strengths are understood as friendly or enemy capabilities considered *essential* for accomplishing a given or assumed *military objective.* In military terms, critical strengths are primarily physical and/or moral strengths of a given combat force. For a military force, critical strength is massed in available combat potential. This is now changing, due to the steadily increasing importance of information. Hence, information should be considered an integral part of the source of one's combat potential.

Critical weaknesses are those sources of military power considered essential for the accomplishment of a given objective but are at the same time grossly inadequate to perform their intended function or task. At the national or alliance/coalition level, critical weaknesses could include a defeatist attitude on the part of the leadership, low morale of the population, weak public support for war or some other military action, and fragility of the coalition or serious disagreement on some fundamental issue among the alliance members.

For a military force, examples of critical weaknesses are inadequate mobility or firepower, incompetent leadership, rigid doctrine, low morale and discipline, weak unit cohesion, inadequate air defense or force protection, and insufficient protection of information systems. A critical strength can become a critical weakness due to enemy actions during a war. A high degree of combat readiness or morale can be reduced so that it becomes a critical weakness. Likewise, the unit cohesion of one's combat forces can be high at the beginning of a major operation or campaign but become a critical weakness due to higher-than-expected losses and poor leadership.

In the next step, the planners need to identify for each critical strength and critical weakness, some elements – called *critical vulnerabilities* – that allow open attack, control, leverage or exploitation by the enemy. In most cases, the enemy's or friendly critical vulnerabilities are related to critical weaknesses. Yet sometimes a critical strength can become a critical vulnerability if structurally or organizationally it lacks sufficient protection or support. This situation might exist in the case of some critical elements of logistical support and sustainment, C4 nodes, or information. Critical vulnerabilities play a major role in selecting methods of defeating the enemy's or protecting friendly center of gravity.

Critical factors are subject to considerable changes resulting from the actions of enemy and friendly forces. The enemy's critical strengths and weaknesses cannot be properly determined if one does not fully know and understand all the essential aspects of the enemy's society and military. A major problem for proper understanding of the enemy is cultural bias and racial stereotyping. The key is to avoid mirror-imaging – an extremely difficult thing to do – in the process of analysis. Thus, intelligence has a critical role in monitoring these changes on the enemy side and in providing planners with timely information. On the other hand, a lack of realism or overconfidence or extreme self-esteem could prevent a sound evaluation of the true critical strengths and critical weaknesses of friendly forces.

Critical strengths and weaknesses can differ considerably depending on the scale of the military objective and the corresponding levels of war. The higher the level of war, the more critical for ultimate success the critical factors are. What constitutes a critical strength at the tactical level can become a critical weakness if the required resources are inadequate to ensure the favorable outcome of a campaign or major operation. Logistical support and sustainment can be a critical strength at the tactical level, but it can become a critical weakness at the operational level because of one's inability to properly protect elements of logistical infrastructure, or because of insufficient logistical resources to support and sustain much larger forces engaged in a campaign or major operation. Alternatively, morale and the will to fight at the tactical level might be high or very high, but the highest politico-military leadership's will to fight or the strength of public support might be weak or sorely lacking.

Based on the analysis of the military situation, a *list of enemy and friendly critical strengths and critical weaknesses* should be compiled. Only those critical strengths that can prevent the accomplishment of friendly objective and the enemy's objective should be listed. Those elements of a given critical strength that serve as integrators, protectors and supporters/sustainers should not be considered the potential strategic center of gravity. Likewise, the military critical strengths related to the factor of space, such as theater geometry, and the elements of the outer core of the enemy and friendly combat potential/power that cannot endanger the respective centers of gravity (e.g., intelligence, IW, logistics and C2 posts) should not be considered potential centers of gravity. Yet such critical strengths should be retained for determining methods of indirectly attacking the enemy's or protecting the friendly center of gravity.

In the final step, the commander must decide which of the critical strengths is the *strongest* in terms of its combat potential to prevent the accomplishment of friendly or the enemy's objective. That critical strength should be then considered as the enemy's and friendly *center of gravity*, respectively.

An alternative and potentially much simpler method of determining the friendly center of gravity is to focus all efforts on identifying enemy's center of gravity. Afterward, the operational commander and the staff should divide friendly forces into the main and supporting forces. The best led, trained, and equipped forces – called main forces – friendly operational/strategic center of gravity – should be assigned to defeat the enemy's center of gravity. This method is much quicker than the traditional method used for identifying the center of gravity. The pitfall is that if a wrong enemy's operational/strategic center of gravity is selected, then the same error would be made in selecting a friendly center of gravity.

Before a final decision is made about the enemy's and then the friendly operational/strategic center of gravity, the operational commander should conduct *mental tests for validity*. The first question to ask is whether destroying or neutralizing the enemy's center of gravity would prevent friendly forces from accomplishing their respective operational/strategic objective. The second question is whether friendly forces are adequate to destroy or neutralize the enemy's operational/strategic center of gravity, given existing military and nonmilitary limitations. The answers to both questions should be affirmative; otherwise, previously identified critical factors should be reassessed and a different operational/strategic center of gravity might emerge.[11] The third question to ask, is whether the selected enemy operational/strategic center of gravity will considerably hinder or prevent consolidation of one's operational or strategic success. The same questions, but from the enemy's operational commander's view, should be asked regarding the selected friendly operational/strategic center of gravity.

The *initial geostrategic position* used by one's forces has a considerable effect on the initial course of a campaign. It determines the direction/axis of advance, integration of the logistical support and sustainment, and operational reach of friendly forces. It determines whether component major operations in a campaign will be conducted from the interior or exterior lines of operations. The operational commander and the staff should evaluate the advantages and disadvantages of the geostrategic position occupied by friendly forces and the enemy's positions. Depending on whether a campaign is offensive or defensive, the ratio of forces on both sides, and the theaterwide supporting structure, the operational commander must decide how to maximize the advantages of the position friendly forces occupy and minimize or drastically reduce the value of the position the enemy forces occupy. The position of friendly forces should be improved in geostrategic terms after the completion of each campaign phase and certainly after the successful completion of the campaign.

In the operational design for a major operation the focus should be on the selection of the *lines of operation* along which one's forces would operate. The initial position of friendly forces determines whether the planned major operation will be conducted using interior or exterior lines of operations. Each of these lines offers some advantages and disadvantages. They need to be properly evaluated so that the operational commander and his staff can try to maximize the respective advantages and minimize disadvantages.

A major offensive operation conducted from the interior line of operations normally aims to concentrate combat forces and then defeat the enemy forces in succession. The great speed of movement and short but devastating strikes can ensure success. When friendly forces are inferior, the best option is to strike the enemy force, as early as possible, in the process of concentration. Optimally, friendly forces should operate along relatively short, multiple and interior lines. Such lines generally favor smaller but highly mobile attacking (or defending) forces. Therefore, it is often highly advantageous either to occupy a central position before the start of hostilities or to obtain it during combat. A wide movement on land is best conducted by operating along interior lines of operation.

Using exterior lines offers opposite advantages and disadvantages from those gained by operating along interior lines. Exterior lines require the employment of relatively larger forces and facilitate an enveloping maneuver. They demand greater speed of movement and agility from the operational commander. The commander must act with initiative.[12] Prior to the landing, most of the amphibious landing operations are conducted by operating along exterior lines of operation. The operational commander's ability to protect lines of communication and resupply friendly forces depends on selection of initial bases of deployment and lines of operation. Extended lines of communication can reduce the tempo, which then can result in failure to achieve leverage.[13] Favorable lines of operation and communication facilitate the protection of a friendly center of gravity.

An important element of a design for a campaign or major operation is a *direction* – a swath of physical space of a certain length, width and depth extending from one's base of operations to the ultimate physical objective via selected intermediate points. According to their importance, tactical, operational, and strategic directions or axes are differentiated. They vary in importance depending on the characteristics of the physical environment: land, sea or air. Each direction should be evaluated in detail in terms of the factors of space and force versus the ultimate objective. They should not be regarded as just a piece of land or sea/ocean area on the chart. This is especially the case in the selection and evaluation of the axes on land, where not only physical but also human elements should be fully considered.

Depending on the initial geostrategic position, a campaign can be conducted along a single axis called a strategic axis, several operational axes, or a combination of a single strategic and several operational axes. A *strategic axis* extends from one's base of operations to the last intermediate (operational) objective in a campaign. The great advantage of using a single strategic axis of advance is that it facilitates the synchronization of combat forces and operational support. A disadvantage is that the enemy can easily discern the path of advance toward the ultimate strategic objective and thus carry out timely counteractions.

Each of the Allied campaigns in the Pacific in World War II was conducted along a single strategic axis. In the Solomons campaign of 1942–1943, the strategic axis extended from the base of operations in Fiji and New Caledonia through the Solomons Archipelago – New Britain – New Ireland. It consisted of four consecutive operational axes: Fiji – New Caledonia – Guadalcanal – New Georgia – Bougainville – New Britain/New Ireland area. In the Central Pacific campaign, November 1943–September 1944, the Allied line of advance extended from the main base of operations in Hawaii to the Western Carolines via five intermediate operational objectives: the Gilbert Islands, the Marshalls, the Marianas, the Central Carolines and the Western Carolines. The operational axes roughly coincided with the general line of advance and extended from one to another intermediate operational objective. The other strategic axis of advance was in the southwest Pacific, along the north coast of Papua New Guinea and into the islands between northwestern Papua New Guinea and Mindanao, the southernmost large island of the Philippine archipelago.[14]

Sometimes a campaign is conducted along several *operational axes* of advance, and without clear distinctions among them in terms of forces and combat support. This makes it more difficult for the enemy to determine where the next blow will come from. Normally, the exterior geostrategic position offers multiple operational thrusts that are mutually enhanced. The German campaign plan for the invasion of Poland in September 1939 and Yugoslavia in April 1941 envisaged a series of deep thrusts launched from an arc-like base of operations surrounding the respective country. To invade Poland, the Germans advanced along several broadly dispersed thrusts, each aimed at destroying the cohesion of the Polish defenses.[15] In the invasion of Yugoslavia, the Axis forces advanced along several operational axes toward the capital of Belgrade. The most important operational thrust, from Bulgaria toward Skopje, was aimed to cut off any possible linking of the Yugoslav forces with the Greek and British.

Generally, the planners should select a single strategic and several operational axes in case one's forces occupy a central geostrategic position; otherwise, the campaign may result in a slow operational tempo. The situation is even more unfavorable if friendly forces advance into the depth of the enemy's territory beyond their mutual supporting distance, as the Germans did in their invasion of Soviet Russia in June 1941. The German plan contemplated three army groups advancing along three divergent axes and beyond mutual supporting distances.[16]

The Allied campaign in France in the late summer and fall of 1944 was conducted on a broad front all the way to the Franco-German border, with the British

21st Army Group and the U.S. 12th Army Group each having its own operational axis of advance. General Bernard L. Montgomery was not satisfied with the slow pace of advance and in August 1944 proposed a bold plan to advance to the Rhine either north or south of the Ruhr – the industrial heart of Germany, the area between the cities of Frankfurt and Karlsruhe. In either case, a strategic axis of advance would be created where all the Allied forces would be under a single command and where a major part of logistical support/sustainment would be assigned. The practical consequence of General Dwight Eisenhower's acceptance of Montgomery's plan would have been that the advance of the U.S. armies deployed on the right flank of the 21st Army Group would have slowed down or even halted. Eisenhower was also not as sure as Montgomery that the Germans were near collapse and that the concentrated push toward the Ruhr would end the war quickly. He believed the better solution to the logistical bottlenecks was to expand the capacity of the newly liberated ports on the Channel coast as fast as possible and to advance on a broad front toward the Rhine River. Eisenhower was concerned that giving Montgomery sole command of land operations would halt General George S. Patton Jr.'s U.S. Third Army and essentially would give the armies advancing from southern France not much to do. All this would aggravate already-strained relations among Allied commanders. These were the reasons why Eisenhower did not endorse Montgomery's proposal on 27 August.[17] (A few days later, Eisenhower took direct command of all army groups while Montgomery was promoted to field marshal by the British government.)

A design for a major operation includes a single *operational direction* or axis. Sometimes operational axes run in the same general direction and comprise the individual segments of the strategic axis for a campaign. A design for a major air operation would usually include several tactical axes of attack. The most favorable situation is when one's air base of operations covers a major part of possible approaches to the targets of the attack.

A major joint operation on land, mounted from an exterior position, can be conducted along several mutually supporting tactical thrusts aimed at splitting the enemy's space into several isolated parts. In the major joint operation to liberate the Krajina area on 4–7 August 1995 (Operation Oluja [Storm]), the Croatian ground forces, supported by air forces, attacked over an arc of about 435 miles of the front.[18] Plan of the operation was based on cutting off the rebel forces by almost simultaneous attacks and advance along some 20 major and minor axes of advance (including three thrusts from western Herzegovina).[19]

Operational Idea

All the successful commanders in the modern era based their plans on a certain overarching idea of how to employ their forces to accomplish the assigned military objectives. Napoleon I based his numerous campaign plans on a certain basic idea – destruction of the enemy's army. Field Marshal Moltke, Sr favored detailed planning; but like Napoleon I he was guided by a certain basic idea.[20] The term operational idea (*Operative Idee*) is German in its origin. It was extensively used by the German military in the late nineteenth century and afterward. The Russian theoreticians prior to 1914 preferred that a plan of an operation should rest on an "idea."[21]

The operational idea is the very essence of any operational design. It is an overarching scheme that describes in simple and succinct terms what each subordinate service or force element will do, why, where and when. However, before the final operational idea is formulated and articulated, a large number of elements and sub-elements should be considered in some detail. Obviously, only those elements that are considered critical for accomplishing an operational or strategic objective will become part of the final operational idea.

The operational idea for an offensive or defensive major operation/campaign can be differentiated. A sound operational idea should present the enemy with multidimensional threats that he has little or no chance of countering successfully. It should be creative. It should be ambiguous to the enemy. It should surprise and deceive the enemy, making it difficult for him to counter one's actions. Most important of all, it should be directed at defeating of the enemy's center of gravity.[22] It can be also focused on the weight of main effort (*Schwerpunkt*). For an attacker, the focus should be on the speed of action, while the defender might stress the need for a delaying action. Likewise, the attacker should emphasize deception, while the defender would focus on counter-deception.

There are many examples of highly innovative operational idea as applied by the Germans in their invasion of Denmark and Norway in April 1940 (Case Weserűbung). The German operational idea envisaged simultaneous and multiple thrusts by sea, air, and ground elements to quickly capture Jutland Peninsula and selected major islands and the southern, central, and northern part of Norway under the protective shield of the Luftwaffe. It was the first example in the history of triphibious operations. The operational idea was bold. The Germans took rather high risks to mount such a large-scale effort in the face of the Allied superiority at sea and ability to react quickly to any German landing in Norway. The Germans' multiple simultaneous attacks had the additional advantage of ensuring that a failure of their action at a single or more points would not endanger the ultimate outcome of the campaign.[23] At the same time, the Germans maintained a high degree of operational security, using deception and concealment.[24]

The Allied campaigns in the Pacific in World War II were planned and conducted using some innovative approaches in attacking the enemy's vulnerabilities and avoiding his strengths. In the Solomons and Papua New Guinea campaigns, the Allies selected the key operational positions as the target of attack, leaving the enemy strong points in the rear of their thrusts. This "leapfrogging" (castling move) method was exemplified by the Allied isolation of Rabaul, New Britain, and Truk, Central Carolines, while capturing key points elsewhere in the area.

The U.S./Coalition ground offensive in the Gulf War of 1990–1991 is an example of an innovative operational idea that was highly successful in practice. The plan contemplated the use of deception to mask a frontal attack by U.S. marines and the Arab forces between Minagish and the coast, and the short-range envelopment maneuver of the U.S. 1st Cavalry Division in the west along Wadi al-Batin.

Both actions were aimed at tying down the Iraqi forces in Kuwait, while a massive and wide-ranging envelopment maneuver by land and from the air was underway by the U.S. XVIII Corps and VII Corps. The main thrust of attack along the line Rafha–Hafar al-Batin toward northeast of the Euphrates River was aimed to seize and hold the crossings between Al-Samāwah and An-Nasiryah. Other Coalition forces were tasked to protect the flanks to the west and north so that the attacking force can turn eastward and destroy Iraqi operational reserves deployed between Basra and Kuwait City and then encircle the rest of the Iraqi forces.[25]

In both the Pacific and European theaters in World War II, the Allies used similar and highly predictable operational ideas for their amphibious landings. This made it easier for the enemy to properly deduce Allied intentions. It did not lead to Allied setbacks, mainly because the Allies had enormous superiority in the air, at sea, and on the ground in most of the landings they conducted. In the Pacific War, the Allied major amphibious landings were planned and executed according to a certain repetitive pattern. The Japanese, by closely observing and analyzing U.S. amphibious landings, changed their method of conducting anti amphibious defense from defending the beaches to digging in and establishing several defensive lines farther from the beaches. In that way, the Japanese countered the superior U.S. firepower and maximized their advantages. After U.S. troops landed on Okinawa in April 1945 (Operation Iceberg), the Japanese offered stubborn resistance in the interior of the island. By the time the last resistance on the island ended in late June 1945, the Japanese had lost 110,000 men in combat. The Allies suffered heavy losses too; 49,000, including some 12,500 men killed or missing.[26]

The operational commander and the staff should evaluate relatively large of possible elements of the operational idea, such as selected principles of war, methods of defeating enemy center of gravity, sectors of effort, operational maneuver and tactical fires, operational/strategic deception, branches and sequels operational/strategic sequencing and synchronization, and many others. They should select only those that will ensure success quickly and with the fewest losses for friendly forces. The fewer elements the operational idea has, the better. The *simplicity* of the operational idea is a prerequisite for drafting and executing simple and flexible plans. Because of the characteristics of the physical environment and different forces employed, there are significant differences in the number of elements between the operational idea for land or maritime campaigns, and major land, air and naval operations.

An operational idea should provide for the application of the selected *principles of war*. While some principles of war might be violated with impunity, others should be fully observed. Operational surprise was one of the main factors in the success of many major operations. The Allied landing in southern France on 15 August 1944 (Operation Dragoon) surprised the German defenders because they had expected the Allied invasion to take place near Genoa, not on the Provence coast, in August 1944.[27] The UN forces' landing at Inchon, Korea, in September 1950 was highly successful because it achieved operational surprise. General Douglas MacArthur had only 30 days to plan this complex and highly risky operation. The landing was a complete success and led to the advance of UN forces to the Yalu River.[28]

Some operational ideas have been highly successful although they contradicted some important principles of war. The German invasion of Norway in April 1940 violated every principle of war except for objective, surprise, and security. The invasion, while expected in terms of objectives, found the British and French unprepared in terms of time – and more important, unprepared as well for the method used by the Germans. The Germans moved quickly and occupied key points on the coast and the interior of Norway by a simultaneous employment of naval, air, and ground force elements. Their plan was detailed, the date of attack was kept secret, and the selected axis of attack was unexpected.[29] Likewise, the Japanese, in their campaign to gain control of Southeast Asia in December 1941–April 1942, violated the principles of mass and simplicity but observed the principles of objective, offensive, surprise and security.

The problem facing the stronger and weaker side in defeating the enemy's center of gravity is very different. A stronger side should aim to destroy, annihilate, or neutralize the enemy's center of gravity. In contrast, the weaker side should try to defeat the enemy's center of gravity by weakening, avoiding, or deceiving it. For the attacker, the operational idea should envision direct and/or indirect attack aimed to destroy or neutralize the enemy's center of gravity. In practice, these two methods should be used in combination. Normally, if the attacker has overwhelming forces available, the quickest and most effective approach is to attack a major part of the enemy's forces directly. A *direct attack* means using one's strengths against the enemy's strengths. It often involves a linear, uninterrupted, head-on approach against the enemy's center of gravity.

The enemy's center of gravity can be defeated by destroying or neutralizing the enemy's selected decisive points. These points can be located in the physical proximity of the enemy's center of gravity, or several hundreds or even

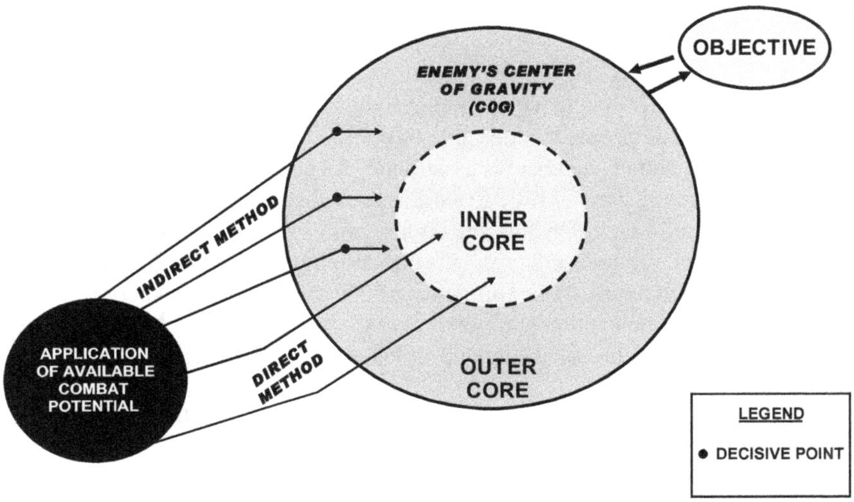

Figure 8.4 Methods of Defeating an Enemy's Center of Gravity

thousand miles away, as in case of the computer networks. A decisive point may have readily identifiable physical features, or it may be composed of imponderables, such as the psychological and mental state of the enemy operational commander.[30]

A decisive point is usually an enemy's critical vulnerability. In some cases, it can be one of the enemy's sources of strength. If such a decisive point is exclusively or predominantly tangible, then it can be subjected to a direct physical attack. Sometimes a decisive point can be degraded, neutralized or destroyed if they are inadequately protected or defended. Attacking intangible decisive points (e.g., the enemy political/military leadership or operational commander) is more complicated and requires some creative thinking on the part of the operational commanders and the staff.

Decisive points can be used as the keys to unlock the enemy's center of gravity. The destruction, annihilation, neutralization, or serious degradation of the selected decisive points can possibly generate a "ripple" (or "cascading") effect on other parts of the outer core of the respective center of gravity. Several decisive points attacked simultaneously or sequentially can synergistically exert leverage on the enemy's center of gravity, separating the mass of his forces from their C4 nodes, critical supply lines, or reinforcements.[31] In some cases, by attacking decisive points, one can neutralize the center of gravity without ever having to attack it directly.[32]

In a case in which the enemy's center of gravity is difficult to attack directly, because either its physical strength or its essential elements are well protected or concealed, the *indirect attack* should be used. This method is often suitable in degrading or neutralizing intangible elements of the enemy's center of gravity. An indirect attack is aimed to identify and then exploit the enemy's critical vulnerabilities and, sometimes, even critical strengths. An enemy's vulnerability is most valuable when it contributes to the elimination or serious degradation of the enemy's center of gravity.[33] Such vulnerabilities are found in the enemy's physical, moral and cybernetic domains.

If the enemy relies extensively on computerized systems for C2, intelligence, theater air/missile defenses, it is possible to indirectly attack the enemy's center of gravity through an attack against his cybernetic-oriented decisive points. Computer networks are in many ways similar to a transportation network. The planners might select a method of attack that focuses on key nodes whose destruction or annihilation would have a ripple effect on the functioning of the network as a whole. Land transportation nodes should not always be destroyed; such an action might be incompatible with one's ultimate objectives (e.g., preserving them for the country's reconstruction) or might be politically, diplomatically, or psychologically unwise or legally suspect.[34]

An indirect attack on the enemy's center of gravity should be conducted from an unexpected direction and/or unexpected time to the enemy. The attacker should avoid the enemy's critical strengths and focus on his critical vulnerabilities. Examples of indirect attacks against the enemy's operational center of gravity are attacking the enemy's flank or deep into his rear, and destroying or neutralizing the enemy's C4 nodes, air defense elements, EW capabilities, elements of his logistical infrastructure and others.

A weaker side can use feint or deception to influence the enemy to move his strongest force – his operational center of gravity – from a certain area and thereby allow other forces potentially attack the enemy's operational center of gravity. In preparing its Sho-1 plan, the Combined Fleet HQ prepared a plan envisaging the use of its much-weakened carrier forces under Vice-Admiral Jisaburō Ozawa to act as bait for Admiral William F. Halsey's TF-38 carrier forces. The Japanese plan worked as anticipated: Halsey made an unsound decision at 1950 on 24 October 1944 to move his three carrier groups north and join another carrier group and then try to destroy Ozawa's force. In the process, he left the San Bernardino Strait unguarded and thereby allowed Kurita's damaged but still powerful heavy surface force to exit the strait and continue its advance undisturbed toward the Leyte Gulf to attack the Allied shipping there.[35]

At the strategic level, morale and public support for war can be weakened by imposing increasing hardship on the enemy's population through the destruction of the installations and facilities necessary for normal everyday life, such as energy plants, electricity grids and transportation systems. A form of indirect attack includes crippling or destroying enemy oil production and refining facilities, maritime or land transportation systems, and defense industrial potential. These and similar actions to weaken the enemy's will to fight usually take a long time to be effective. In addition, oppressive governments are often skillful at manipulating the population to endure protracted punishment, as the examples of Nazi Germany and Japan in World War II illustrate.

In theory, one's sources of power can be applied symmetrically and/or asymmetrically. Each method has its advantages and disadvantages. A *symmetric* application generally aims to avoid the enemy's strength while attacking his vulnerabilities. Obviously, the chances of success are generally greater in acting against an enemy's vulnerabilities than against his strengths. Hence, the weaker side is usually more likely to respond *asymmetrically*. This allows a weaker and less sophisticated enemy to greatly complicate the attacker's problems in achieving its objectives quickly, even when the attacker possesses overwhelming strength in the main categories of power.

One of the major advantages of employing two or more services of one's armed forces is the ability to mount highly successful asymmetric attacks against the enemy's forces. The employment of land-based aircraft against the enemy fleet or merchant shipping is a form of asymmetric attack. Likewise, the employment of ground forces to capture the enemy's basing areas and ports or the use of coastal missile/gun batteries against enemy ships are also forms of asymmetric methods.

The weaker side might use sophisticated and widespread cyber-attacks against the attacker's informational infrastructure or might threaten to use weapons of mass destruction (WMD) and thereby achieve disproportionate results far exceeding its overall combat potential in the conventional categories of power. At the same time, the asymmetric application of available combat potential offers the best opportunity to achieve quick and decisive results, even when the stronger side uses it.

The operational idea should envisage how available combat potential should be applied in terms of time – sequentially, simultaneously, or combined. Which method will be applied depends on the content of the ultimate operational objective and the political limitations imposed on the use of one's military forces. The most decisive and usually quickest results are achieved by the simultaneous application of overwhelming force. The incremental use of power has invariably proven less effective whether it is used against a strong enemy or an inferior one. This would give a resourceful enemy time to adjust and recover from one's blows. It usually prolongs the campaign or major operation and results in unnecessary losses of personnel and materiel. Perhaps nowhere is this more obvious than with air power: it is a blunt instrument that achieves the best results when employed massively from the outset to stun the enemy and make him physically and psychologically unable to resist for too long. Such use of air power by the U.S.-led coalition forces in the Gulf War of 1990–1991 proved to be extremely effective and was the key factor in the subsequent defeat of the Iraqi forces on the ground. Some airpower enthusiasts go too far, insisting that parallel or simultaneous attacks should be conducted in any scenario, that such attacks against the enemy's center of gravity would ensure rapid actions and minimal costs and risk to friendly forces. Parallel attacks would create permanent changes in the situation. In contrast, sequential operations would give the enemy ample opportunity to react. And each time the defender reacts, the attacker is faced with an entirely new set of problems. In contrast, parallel operations are faster, safer, cheaper, and more likely to succeed than sequential operations.[36]

In designing a campaign or major operation, the operational commander and the staff must decide early on which part of the theater or area of operations will be the *sector of main effort* and which will be the sectors of secondary effort. Among other things, the designation of a sector of main effort is critical for focusing one's efforts against the enemy's center of gravity.[37] It also ensures the principle of unity of effort by massing forces to accomplish desired effects.[38] This important decision determines where the strongest force will be deployed and physically concentrated to generate massed effects. In contrast, a *sector of secondary effort* is characterized by shallower operational depth and the assignment of forces with less firepower, fewer reserves, and more limited objectives. The operational depth is that area beyond the enemy's tactical defenses in which both the attacker and the defender can achieve freedom of action.

By designating the sector of main effort, the operational commander directs the staff regarding which friendly forces will receive the greatest attention in planning operational and tactical fires and logistical sustainment. Planning the initial main and supporting major operations in a campaign is directly related to the designation of the sector of main effort and the sectors of secondary effort. Failure to designate the sector of main effort versus those of secondary effort can have fatal consequences, especially in defense.

In a land campaign, the boundaries between the mobile and fixing force and the supporting air forces usually coincide with the sectors of main and secondary effort. In some cases the air force's actual sector of main effort may temporarily be over the sector of secondary effort on the ground, if the aim is to deceive the defender as to where the main offensive thrust will follow.

In a maritime campaign, the sectors of main and secondary efforts are more blurred than those on land. In a campaign aimed to cut off or interdict an enemy's maritime trade, the sector of main effort can encompass a large or small part of an ocean or adjacent sea area where the bulk of one's forces are employed over a relatively long time. The traffic volume and the importance of sea lines of communication for the enemy's war economy war are the principal factors in selecting which sea or ocean area will be a sector of main effort. Normally, the most important sea or ocean routes are also the best protected by the defender. Therefore, the attacker has no choice but to employ his forces in less decisive ways to wear down the enemy's defenses over time. In a campaign to attack or defend maritime trade, the sector of main effort is usually identical for both sides, as was the North Atlantic in the Battle for the Atlantic in 1939–1945.

In an amphibious landing operation, the sector of main effort is usually an island, an archipelago, a strait, or a stretch of the coast where the main operational objective is located. This requirement is observed in most instances, but not all. The operational idea for the Allied invasion of Sicily in July 1943 contained two landing sectors: the Gulf of Noto where the British Eighth Army landed; sector of secondary effort was the 70 miles long beaches of the Gulf of Gela assigned to the U.S. Seventh Army.[39] For some reason, the planners did not designate a single sector of main effort. Consequently, neither army was given priority in terms of combat support and logistics.

Any major error in determining the enemy's proper center of gravity invariably results in a wrongly selected sector of main effort. If the error is corrected in time, the operational commander should *shift the sector of main effort* to another part of the theater to compensate for the error. This shift can be caused by a change of plans. A highly mobile and lethal force allows more flexibility in shifting the sector of main effort. In all cases, a shift in the sector of main effort requires a change in support priorities among one's forces. Thus, flexibility in shifting the main effort should be an integral part of the operational design.[40]

The sectors of effort determine where the *main (or principal) forces* and *supporting forces* should be concentrated or employed. The greatest problem is in finding a balance between the forces deployed in the sector of main effort and those deployed in the sectors of secondary effort. This calls for the operational commander's good judgment and experience.[41]

In land warfare, the forces' size, mix, and mobility depend primarily on the combat potential of the enemy's forces deployed in front of sector of main effort; the characteristics of the operating area; the weather/climate, and time of year. The width and depth of the sector of main effort affect the forces' density. Given a constant number of forces, the smaller the frontage, the higher the forces' density. The sector of main effort should contain multiple lines of operation and adequate communications.[42] In selecting the sector of main effort, the operational commander should assign the best-led, best-trained and most cohesive forces. The sector of main effort would also have the densest concentration of combat potential.[43]

Another factor to consider is which part of the sector of main effort the operational commanders must concentrate a major part of their available combat potential – *point of the main attack* (or *defense*) – to have the best chance of achieving a decisive success in the initial phase of a campaign or major operation. Napoleon I said that the art of warfare rests on a single tenet – concentrate larger mass than the enemy at a single point.[44] The operational commander and planners should keep in mind the characteristics the point of the main attack (or defense) might assume. It may have readily identifiable physical features, or abstract features such as the cohesion of an alliance or the psychological state of the enemy commander, to name a few.[45] The point of the main attack (or defense) is usually selected among decisive points offering an indirect approach for attacking the enemy's center of gravity. By properly selecting the point of the main attack, the attacker can obtain relative superiority over his opponent, even if the overall ratio of forces in the theater is unfavorable. Good tactical and operational intelligence is critical in the selection of the point of the main attack. Effective employment at the point of the main attack requires the preparation of the deception plan.

In land warfare, the point of the main attack is selected from among the geographic or force-oriented decisive points (or a combination of these) within a given sector of the main effort. It is where the enemy's forces are weakest or least densely deployed. A point of the main attack might be the area separating the enemy's major forces. In contrast, point of the main attack (or defense) in the operational idea for a major naval or air operation is often difficult to determine compared to those in land warfare. The problem lies in the inherent fluidity and fast changes in air and naval situations. In a given sector of main effort, points of the main attack (or defense) can change rapidly. The probable exception is an amphibious landing operation in which a sector of the coast allowing the most favorable conditions to indirectly attack the enemy's center of gravity, or a port whose capture is critical for logistical sustainment, is selected as the point of the main attack for the landing force. In a campaign or major operation aimed to interrupt the enemy's, or defend and protect friendly, maritime trade, the selection of the point of the main attack is usually influenced exclusively by the shipping density in a certain sea or ocean area. In other cases, large ports or shipyards and ship-repair facilities whose destruction would cripple maritime trade in a certain area could be selected as points of the main attack. For the defender, such points should obviously be considered the main points of defense.

In attack on the enemy's maritime trade, the point of the main attack is usually a focal area of shipping or a choke point – specifically, approaches to larger ports, channels, narrows, and international straits or canals. Likewise, in defending and protecting maritime trade, the main points of defense should also be focal areas or choke points. In the German campaign to cut off Allied trade in the Atlantic in 1939–1945, points of main attack for the U-boats included the English Channel and its approaches, the Bristol Channel, approaches to New York, and the Strait of Gibraltar.

The operational idea should provide for quick and, if possible, covert *concentration of forces in a sector of main effort* and at the point of the main attack or defense. Concealed and rapid concentration is a critical factor for the successful employment of a numerically inferior force. Despite its obvious advantages, however, concentration makes one's forces vulnerable to enemy fires. This might require dispersal.[46]

The concentration of forces in the sector of the main effort is one of the most critical yet difficult tasks facing any operational commander. There are essentially two options: withholding friendly forces until the enemy's sector of main effort and point of the main attack (or defense) are recognized (a prerequisite for this is having good reconnaissance and sufficient forces for protection), or concentrating adequate forces against the enemy's probable sector of main effort but assigning sufficient forces for protection in the less-threatened sector.[47]

Operational maneuver comprises one of the key parts of a sound operational idea. The outcome of the entire major operation/land campaign might depend on the success or failure of the operational maneuver for the initial major operation. The success of the first phase of the German campaign in the west in May 1940 depended on the combined operational penetration-envelopment maneuver, from Sedan-Dinant to the French Channel coast that the Panzer Group von Kleist conducted. The success of numerous Allied amphibious landing operations was due to successful operational maneuvers. The German success in destroying a large number of Soviet forces in the initial phase of their invasion of Soviet Russia was due to a large number of double-envelopment maneuvers. Likewise, the Soviets conducted a double-envelopment maneuver in encircling and subsequently destroying the German Sixth Army at Stalingrad in November 1942. In southern Russia in February 1943, however, they were unsuccessful in conducting a single-sided operational maneuver aimed at cutting off the German forces south of Stalingrad and in the Caucasus.

In some cases, the initial geostrategic position and the deployment of the enemy preclude, or do not provide an opportunity to incorporate, operational maneuver as part of the operational idea. In such a case, the operational objectives could be accomplished by a series of mutually supporting tactical thrusts, with one's forces operating along converging operations, as the German experience in Poland in September 1939, and Yugoslavia in April 1941 illustrate.

Combat forces conducting an operational maneuver should possess a high degree of mobility and firepower. Modern forces' greater maneuverability and their capabilities for long-range surveillance and fires allow for far more flexible dispositions in the conduct of an operational maneuver than were possible in the past. This is especially true for naval and air elements. On land, a smaller force is usually more mobile than a larger one; however, a small force might not have the necessary combat potential to have an operational impact. Conversely, a large force is usually more powerful but less mobile. Also, if a mobile force is too large, it requires a larger supporting force to protect it. This is particularly true if a force must move over great distances.[48] In planning operational maneuver, the relationship of one's base of operations and/or lines of operation to a given physical objective should be properly evaluated.[49]

In planning an operational maneuver, it is critical to select the most favorable lines of operation for the movement of friendly combat forces. The fewer the lines of operation a mobile force uses, the smaller the spatial distance among force elements will be. While a few lines of operation enhance security and mutual support, they increase the force's vulnerability to enemy attack from the air. Another important consideration is the force-to-space ratio. The larger this ratio, the smaller the space the enemy will have in which to recover. The greater the enemy's ability to recover, the more temporary the effects of an operational maneuver. The planners should select points that are potentially decisive for defeating the enemy's center of gravity or for cutting off the enemy's logistical supply lines.[50]

The operational idea for major operations on land should combine the operational maneuver with tactical fires. These fires can be delivered in the form of close air support, naval surface fire support (NSFS) (formerly naval gunfire support-NGS), and field artillery fires. They are the best means of destroying enemy forces.[51] They facilitate maneuver by fixing the enemy forces and, when necessary, destroying them. Tactical fires are planned to disrupt enemy movement and neutralize enemy fire support, command and control, and sources of logistical sustainment[52]

The operational commanders should do everything possible to provide sufficient *protection for the friendly center of gravity*.[53] In doing so, they should repeat the process of identifying decisive points in the physical, cybernetic and moral domains of friendly forces. This helps the operational commanders to assign sufficient forces and take adequate measures to protect one's decisive points from the enemy's attack.[54] A highly capable but not overly strong force should be assigned to the protection of the friendly center of gravity; otherwise, one could open oneself to a devastating enemy attack. One should also be aware of the danger of overprotecting the wrong center of gravity for friendly forces.[55]

At the strategic level, protection of the friendly center of gravity largely pertains to both the tangible and intangible elements of the strategic center of gravity. Today, not only should the strongest force taking part in the planned campaign be fully protected but also selected military and nonmilitary elements in a theater – the informational structure in particular – should be adequately protected as well. Intangible elements of the strategic center of gravity, such as "will to fight," are not literally protected but are maintained through the successes achieved at the tactical and operational levels. Likewise, the cohesion of the alliance/coalition is preserved through military successes in the field. Another aspect of "protection" of the strategic center of gravity is maintaining solid public support for a war.

Optimally, the operational commander should have sufficient forces to ensure full protection of all the forces deployed in the theater. Yet one's forces are finite. Therefore, the operational commander should balance competing demands for forces to accomplish the principal objective while also having sufficient forces to conduct operational fires and ensure operational protection. In addition to ensuring the protection of operational centers of gravity, the operational commander should assign adequate military resources in protecting the strategic center of gravity. The effectiveness of these measures can be significantly enhanced by conducting strategic and operational deception designed to make it more difficult for the enemy to correctly identify friendly centers of gravity.

At the operational level of war, one option for an operational commander is to disperse friendly forces so that they are not deployed in a relatively small area and thereby represent an easily identifiable operational center of gravity to the enemy. At the same time, these forces should possess sufficient mobility and lethality to defeat the enemy's center of gravity either by physical concentration or by concentrating effects. The latter is a feasible option for protecting the center of gravity of one's air forces or naval forces but is far more difficult to do in the case of one's ground forces. Without taking proper and timely measures to protect the friendly operational center of gravity, the risk of reaching a point of culmination is rather high.

Major operations would be only partially successful or even fail if the commander did not protect a friendly operational center of gravity in the course of pursuing a beaten enemy force. During the British major offensive operation in North Africa (Operation Crusader) in mid-November 1941, Rommel selected the 21st Panzer Division to protect his operational center of gravity. This division and the Italian Ariete armored division had the economy of force and security role to stop or slow down British forces attacking out of Egypt.[56] Yet their efforts failed because the Luftwaffe was unable to prevent Royal Air Force (RAF) strikes from seriously weakening German strength on the ground. Rommel had no choice but to order a general withdrawal from the Tobruk area on 7 December 1941.[57]

A sound operational idea for a major operation should incorporate the concept of *operational/strategic deception.* The chances of a campaign's success are significantly enhanced by having a sound strategic deception plan. An operational deception plan is an integral part of strategic deception. It should be prepared for each phase of a given campaign. In a major operation, tactical deception should be coordinated with the plan of operational deception; any disconnect between these two levels of deception is bound to create difficulties in their execution and can even endanger the success of the entire operation.

A critically important element of a sound operational idea is *sequencing* – the order or arrangement in which the intermediate objectives should be accomplished a given ultimate objective, in the shortest possible time and with the least losses. The events can be arranged to take place consecutively or simultaneously. Sequencing provides a framework within which the synchronization is planned and executed. Proper sequencing should ensure flexibility in the accomplishment of assigned objectives. The sequencing should be used for developing a force list for the forthcoming operation by calculating force composition/mix required to accomplish each intermediate objective. Afterward, the shortfalls are identified by comparing currently available force versus those required for accomplishing a given intermediate objective. Normally, the new task organization should be based on the force list derived from the combat forces' sequencing.

Operational sequencing is both a prerequisite and framework for sound *operational synchronization.* Synchronization is another critical element of an operational idea for a campaign or major operation. Synchronization is both process—arranging or initiating actions in terms of space, time, and purpose—and effect, in generating maximum relative (combat or noncombat) power at a decisive place and time. It should ensure that all elements of the force generate effects that collectively exceed the sum of their individual effects. A soundly conceived and well-executed synchronization plan may allow an inferior force to defeat a superior enemy force.[58] In every instance in which a major operation or campaign failed to accomplish its stated objective, one of the main contributing factors was either a

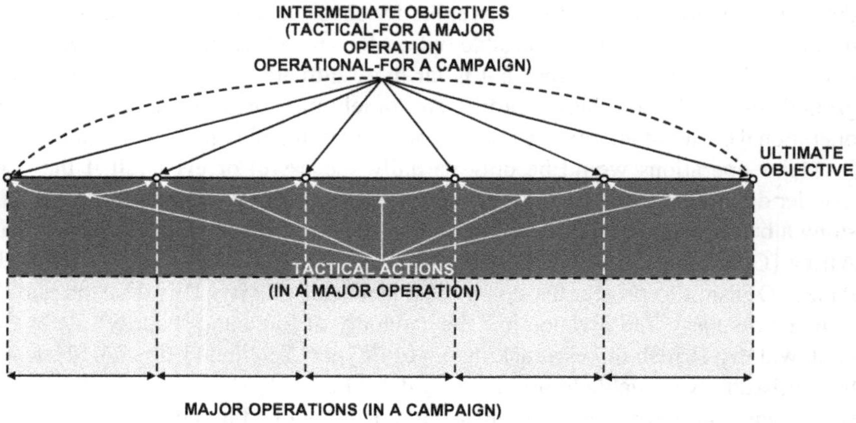

Figure 8.5 Operational Sequencing

poorly conceived or a poorly executed synchronization plan. Without synchronizing all the sources of power, it is difficult, if not impossible, to achieve victory with minimum losses of personnel and materiel.

The operational commander should not be bound to a single option in planning for a campaign or major operation. Sufficient flexibility should be built into the basic plan to preserve one's freedom of action. This is achieved by having *branches.* The operational commanders should anticipate several options for actions planned or in progress. Liddell Hart wrote that a plan, like a tree, must have branches if it is to bear fruit. A plan with a single aim is apt to prove a barren pole. He considered branches as alternate plans.[59]

Sometimes the true meaning of branches is changed to refer to contingencies within a given phase of a campaign or major operations such as change of command relationships or the direction of movement of one's and acceptance or refusal of combat.[60] Branches allow the operational commanders to anticipate future enemy actions that might lead them to drastically modify the basic plan. Such a situation could arise from unexpected action by the enemy, the nonavailability of one's forces, or even a change in the weather or season within the operational area. As conditions change, the operational commanders should be prepared to maintain the continuity of the campaign or major operation by regrouping and preventing their forces from being stopped at an inopportune time.[61]

In designing a campaign, the operational commander needs to plan new or successive major operations – *sequels* – after the preceding major operations have been completed. For a major operation, sequels are major tactical actions. Sequels are often confused with branches. Yet though related, they are not identical in their meaning and purpose.[62] In contrast to branches, sequels follow the completion of a given phase of a campaign or major operation. They are based on the possible outcomes of the preceding major operations or tactical actions. Sequels are the means of allowing the operational commander to exploit tactical or operational success.[63] They are linked to the concept of *successive major operations* in a campaign.

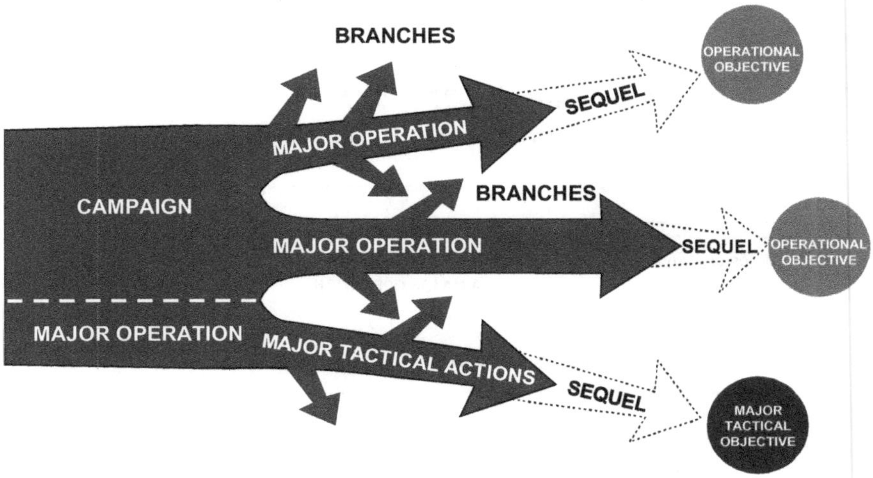

Figure 8.6 Branches and Sequels

A failure to plan sequels, because of a lack of foresight or insufficient forces, has often had serious consequences for the course and even outcome of a campaign or major operation. In their invasion of Soviet Russia in 1941, the Germans failed to prepare any contingency plans or sequels beyond the Dnieper River–Leningrad line. No thought was given to the possibility that the Soviets might not collapse after the initial frontier battles. Hitler apparently believed that after reaching the Dnieper–Leningrad line the campaign would enter into the occupation phase.[64] In contrast, the Soviets invariably planned several successive major operations, often conducted simultaneously and over a large part of the front, especially in the offensive phase of the war on the eastern front in 1943–1945.

Sequels are prepared in a rudimentary form – usually as an outline – in advance, so that subordinate commanders are fully familiar with them. The forces that would carry out a sequel should be identified, and the sequel's objective clearly determined. Individual phases of a campaign or major operation can be viewed as sequels to the basic plan.[65]

The ultimate objective of a campaign or major operation is normally accomplished by spacing one's effort in two or more related *phases* in time and space. In a campaign, each phase may consist of a single or more major operations, while in a major operation they consist of major tactical actions. Phases can be planned to take place simultaneously or sequentially. Their main purpose is to stagger the entire campaign or major operation into manageable parts. Thereby, one's forces would avoid reaching a point of culmination before the assigned ultimate objective has been accomplished. The operational commander should not arbitrarily break down a campaign/major operation into artificial phases; that invariably slows the operational tempo.

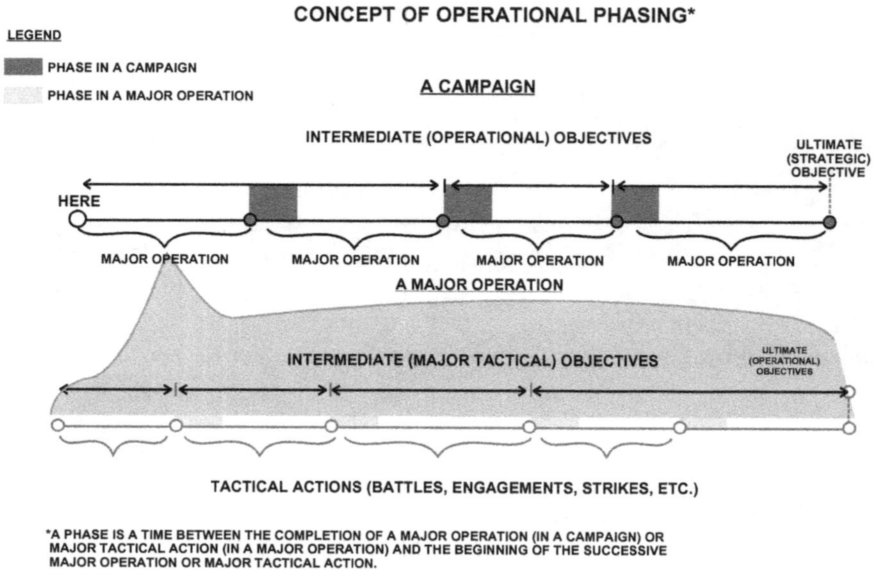

Figure 8.7 Operational Phasing

A change in the phases of a campaign/major operation usually involves changes in the objectives/tasks. These changes – not a timeline – should determine the phasing of a campaign or major operation. Other factors include the distance from the main or intermediate base of operation, logistical resources, and protection of friendly logistical lines of supplies, terrain and weather, to name a few. The individual phases gain importance only in the larger context of the operation.[66] The most critical for success are the links between two consecutive phases. It is necessary to clearly define the conditions to be created for a transition from one phase to another. Ideally, such a transition should be concealed from the enemy. This can be achieved by concurrently intensifying actions of friendly combat forces in a false direction or conducting deception.

In a major operation, phases can be planned to coincide with the anticipated accomplishment of each intermediate major tactical objective. Like a campaign, a major air-land or land operation is rarely completed in a single phase. In contrast, major offensive air and naval operations are often conducted in a single phase. The exceptions to this are major amphibious landing operations and air defense operations, which often encompass two or more phases.

The description of each phase should identify operational or major tactical objectives and provide the operational commander's concept of how and when they must be accomplished. For each phase, the operational commander should estimate the force requirements and supporting actions required from subordinate component commanders. Optimally, each phase of a campaign/major operation should set the stage for the subsequent phase. No sequence of operational phases should be prepared without a clear vision of what the ultimate objective is.[67]

Distinctions between consecutive phases should be made by their separation in time or space, by a difference in the objectives to be accomplished, or by the forces assigned. Each phase should aim at some intermediate objective necessary to achieve the ultimate objective. Phases are planned by identifying transition points between successive major operations or major tactical actions, or by identifying changes in tempo within a particular combat action.

An example of sound operational phasing was the Allied Central Pacific Campaign, November 1943–September 1944. The first phase of the campaign, from the end of the operation to capture the Gilbert Islands (Operation Galvanic) to the start of the next major joint operation to seize control of the Marshalls (Operation Flintlock/Catchpole), lasted from 28 November 1943 to 31 January 1944. The second phase, between the end of the operations to seize the Marshalls and the invasion of the Marianas (Operation Forager), lasted from 23 February to 15 June; and the third and final phase, from the capture of the Marianas to the invasion of the Western Carolines (Operations Stalemate-I/-II), lasted from 10 August to 15 September 1944.

The higher the tempo and the less the anticipated attrition of one's forces, the less need there is to divide a campaign or major operation into several phases. One of the characteristics of major operations today is that they are conducted at a very high intensity; hence, they may not need to be phased. In the Gulf War of 1990–1991, the coalition forces conducted a major ground operation at high intensity. The operation was arbitrarily stopped about 100 hours after it started on 24 February 1991. There were no distinguishable phases in that operation.

The operational commander and the planners should evaluate those elements of the operational idea that most directly affect the *arrival at the culminating point*. Specifically, ultimate and intermediate objectives and their sequencing, phasing, reserves, surprise, deception, and protection of a friendly center of gravity are closely related to the point of culmination. Reaching or overshooting a culminating point can be prevented by determining the right number of intermediate objectives and their sound sequencing. It can be prevented by planning an operational pause after a given intermediate objective has been accomplished but before starting actions to accomplish the next intermediate objective.

The operational commanders should give great attention to the logistical sustainability of their forces, their morale and physical condition, and the limits of their combat power. For the attacker, attrition of enemy's forces and extension of lines of communication may eventually considerably slow down operational tempo so that a pause is necessary to consolidate and regenerate combat potential. For the defender, a point may come when failure to attack will forever forfeit the opportunity. Either situation requires a radical reorientation of operations.

The operational commander's inability or failure to properly sense or anticipate when forces under his command are close to overshooting the point of culmination has often been the primary reason for a severe setback or even a defeat. Sometimes such a failure is compounded by another: misjudgment as to whether the enemy forces are capable of effectively countering one's offensive. This was the case in southern Russia in February–March 1943 when the Soviets failed to properly assess both their own and the enemy's situation. The Soviet High Command (*Stavka*) and General Nikolay K. Vatutin, commander of the Southwestern Front, believed that the Germans were too exhausted (overreached their culminating point) and hence became vulnerable to Soviet attack. The Stavka made a rather serious error in believing that the Germans were unable to effectively reinforce threatened sectors or launch a counteroffensive. They were convinced that the Soviet forces were capable of accomplishing their assigned objectives before being exhausted.[68] As it turned out, Manstein properly assessed the situation of both the Soviets and his own forces, counterattacked, and won a great victory over the much larger enemy forces. In contrast to the Soviet High Command and the Soviet field commander, Manstein properly sensed the point of culmination and went on the counteroffensive at the time when the Soviet forces were close to collapse.

In a campaign or major operation, it is rare that combat actions are conducted uninterruptedly. Periodic slackening, or even the stopping of actions by combat forces – called *operational pauses* – should be planned. An operational pause is much broader in its scope, time, and space than a tactical pause.[69] An operational pause does not mean that one's forces stop all the fighting. It means that a major part of the main forces slackens or even drastically reduces its efforts while the supporting forces (air forces or naval forces) intensify their pressure on the enemy's main forces.

Pauses are highly useful for arranging sound sequencing and synchronization of major operations within a campaign, or tactical actions within a major operation. Their main purpose is to allow sufficient time to regenerate combat potential and avoid arriving at or overreaching the point of culmination. An operational pause can be planned to shift friendly forces from one sector of the main effort to another. If friendly forces are on the defensive, the operational commander may order an operational pause to allow his forces to shift so they can operate along more favorable lines of operation elsewhere.

The operational commander might intentionally and temporarily cede the initiative to the enemy to prepare a major operation in some other part of the theater or to deceive the enemy about his real intentions. An operational pause should be planned by taking the anticipated point of culmination into account, so that the enemy cannot seriously threaten friendly forces while they temporarily reduce the intensity of their actions. The operational commander can use an operational pause to regroup or redeploy, or simply to provide rest for his forces.

Operational pauses may be required when a campaign or major operation temporarily cannot continue because it lacks adequate logistical sustainment, it encounters increasing difficulties in overcoming terrain and weather/climate, the enemy's resistance stiffens unexpectedly, or any combination of these events. If the logistics requirements are inadequate or anticipated attrition is too high, an operational pause should be planned to prevent one's forces from approaching a point of culmination. Other reasons for an operational pause include personnel fatigue, extended and increasingly vulnerable lines of supplies, difficult terrain and severe weather.

The operational commanders should orchestrate actions by friendly forces to accomplish assigned objective before reaching a point of culmination. They can plan for a pause to maintain a relationship to that point that ensures a relative combat power advantage over the enemy.[70] To have any utility, the operational pause should provide the operational commander a tool to avoid the consequences of culmination.[71]

An operational pause offers an advantage only if the combat power can be regenerated faster than that of the enemy. This means that logisticians should be closely involved in determining when and for how long an operational pause should occur. The Germans failed to plan for an operational pause in their invasion of Soviet Russia in 1941. During the war game played prior to that invasion, the German logisticians properly identified the need for an operational pause to revamp the rail transportation system before continuing the advance east of Smolensk. The Germans' overly ambitious plan did not allow for any pause until the ultimate objective of the campaign was reached. As it happened, a combination of strong Soviet resistance, lack of logistical supplies, and harsh winter conditions resulted in the German failure before Moscow in December 1941.[72]

The *operational tempo* refers to the pace of a campaign or major operation toward the accomplishment of its ultimate objective. An operation-size force can move only as fast as its force elements' tactical rate of advance.[73] The operational tempo is understood as the rate or rhythm of activity relative to the enemy's activity in combat. It includes the capability of combat forces to make a successful transition from one posture to another – that is, from offense to defense or vice versa. In practical terms, the tempo is the distance to be covered divided by the amount of time it takes to cover it.[74] Tempo can be measured only in relative terms. A force that acts slowly, but at a rate higher than the enemy, operates in relative terms at a higher tempo. The higher the operational tempo, the better the chance of maintaining the initiative. Operational tempo is directly related to the tactical tempo of subordinate force elements. Tempo depends mainly on the mobility of friendly forces and their tactical rate of advance, quality of combat support, combat service support, surprise, friction, enemy resistance and the attrition rate of friendly forces. Operational support preserves the combat effectiveness of friendly forces in the face of the enemy's resistance. It ensures that friendly forces have sufficient resources to maintain their combat power. Surprise creates conditions that enhance the tactical rate of advance by confusing or slowing down the enemy's response. The primary causes of friction are terrain, weather, light conditions, enemy resistance and poor effectiveness of technical systems.[75]

Operational tempo entails the capacity of forces to transition from one operational posture to another. By increasing and varying the tempo, the operational commander can present threats to which the enemy cannot opportunely and successfully react. High operational tempo is achieved by having a faster decision–action cycle than the enemy. While dictating the tempo, the operational commander should make certain that it can be maintained if possible, for the duration of a campaign or major operation.

The operational commander should find a fine balance between the requirement to slow or stop the actions of friendly forces and the maintenance of a high operational tempo. A faster tempo is achieved by properly sequencing intermediate objectives. The fewer the intermediate objectives and the weaker the enemy resistance, the higher the tempo.

An essential factor in maintaining a high operational tempo is *operational agility* – the ability of one's forces to act or react faster than the enemy. Agility depends on the speed of the decision-making cycle. It can be enhanced by spatially dislocating the enemy's forces, through either maneuver or feints. A high operational tempo can be achieved by conducting major operations simultaneously, rather than in sequence. In addition, the operational tempo is greatly increased by the proper application of the mission command. Operational tempo depends on the ability of part of one's forces to operate at high intensity while the remainder of the forces are regenerating their combat power.

Optimally, an offensive campaign or major operation should maintain a fast pace – *operational momentum* – so that friendly forces obtain all the advantages that accrue from combining speed and mass. The higher the speed and the larger the forces, the higher the momentum of a campaign or major operation. High speed of advance allows rapid concentration, surprise, and exploitation of the physical mass or massed effect. The operational momentum also depends on the length of individual phases and pauses. The longer the phase and pause, the slower the tempo of one's forces.

The operational idea should envisage in broad terms the forces' size and mix and the location of the *operational reserve*. The size of a force is not as important as its mobility and combat potential. The operational commander who does not assign a force to serve as an operational reserve is taking too high a risk. Regardless of the enemy's strength, prudence demands that some force should be committed to be, at short notice, in the area at the time the decision is sought.

The operational design for a campaign or major operation should include ways and procedures to *regenerate combat potential of friendly forces*. Phasing and operational pauses are primarily used to plan for sufficient time to regenerate combat power. After evaluating the enemy's strength, the operational commander can determine *relative combat power* and calculate where friendly forces are in relationship to their culminating point. This information can be used in sequencing tactical actions, in pausing at the most opportune time, or, when on the defense, in designing actions that will lead to the enemy's culmination.[76]

Conclusion

A plan for a campaign should be based on a proper determination of the ultimate and intermediate objectives. The operational commander and the planners should never lose sight of the ultimate strategic or operational objective. One should avoid selecting multiple intermediate operational objectives even when one's forces are superior to the enemy. The time factor cannot be accurately predicted. The accomplishment of multiple objectives usually requires a much longer time than originally anticipated. This can derail the schedule of an entire campaign and ultimately slow the operational tempo or even court disaster from a stronger and more agile enemy.

The design for major operations is sometimes given short shrift, with almost all the attention given to the design for a campaign. However, a campaign cannot be successful if subordinate major operations are improperly designed. In addition, without a design for major operations, the tactical employment of forces cannot be effective. Many elements in the design for a major operation are similar to those for a campaign, with some significant differences due to distinctions in the scale of the ultimate and intermediate objectives.

The operational idea is the very heart of the operational design. The scope of the operational idea for a campaign and the scope of a major operation are different in terms of space–time–force factors. The fewer elements an operational idea has, the less complicated it is and the more quickly it can be executed. None of the elements should be discarded without due analysis of the mission and the situation. The operational idea should be written in a free form; various shortcuts and buzzwords should be avoided. It should express in clear and commonly understood operational terms how the operational commander intends to sequence and synchronize all available forces to accomplish the assigned strategic or operational objective in the shortest possible time and with the fewest friendly losses.

Warfare consists of a series of trade-offs. In each instance, the operational commander and staff should properly balance competing demands for scarce resources while still accomplishing assigned objectives. Designing a campaign or major operation is not a simple job amenable to a few hours of discussion. It requires time, imagination, hard work, and, above all, operational thinking, sound judgment, and common sense on the part of both operational commanders and their staff. The main purpose of operational design is to make this exhaustive effort a coherent one.

Notes

1 Albrecht von Boguslawski, *Heerwesen und Kriegführung* (Berlin: Verlag von R. Eisenschmidt, 1897), 105–6.
2 MCDP 1–2 *Campaigning* (Washington, DC: Department of the Navy, Headquarters United States Marine Corps, August 1, 1997), 35.
3 Burkhart Müller-Hillebrand, *Deutsche und Sowjetrussische Militärische Führung im Zweiten Weltkriege,* ZA 1/56 P-145, Studien der Historical Division Headquarters, United States Army Europe, Foreign Military Studies Branch, BA-MA, 1954, 134–35.
4 Donna Lucchese, *The Relationship of Center of Gravity Analysis, Targeting for Effect, and Measuring Success* (Carlisle Barracks, PA: Army War College, 1998), 3.
5 Robert G. Fix, *Operational Encirclement: Quick Decisive Victory or a Bridge Too Far?* (Fort Leavenworth, KS: School of Advanced Military Studies, U.S. Command and General Staff College, 1993), 21.
6 Martin Blumenson, *United States Army in World War II: The Mediterranean Theater of Operations, Salerno to Cassino* (Washington, DC: Center of Military History, United States Army, CMH Pub 6-3, [1969] 1988), 301, 303, 353.

7 Roy E. Appleman, *United States Army in the Korean War: South to the Naktong, North to the Yalu (June–November 1950)* (Washington, DC: Center of Military History, United States Army, CMH Pub 20-2, [1961] 1992), 499–500; Malcom W. Cagle and Frank A. Manson, *The Sea War in Korea* (Annapolis, MD: Naval Institute Press, 1957), 84.
8 Werner Lange, *Raum, Zeit und Mittel als Faktoren strategischer Zielsetzung. Überlegungen zu Napoleons Feldzug in Russland 1812* (Hamburg: Führungsakademie der Bundeswehr, November 1964), 1.
9 Führungsakademie der Bundeswehr, *Operative Leitlinie für die Führung von Landstreitkräften* (Hamburg: Führungsakademie der Bundeswehr, October 1993), 2.
10 Ajay Singh, "Time: The New Dimension of Warfare," *Joint Force Quarterly* 4 (Winter 1995–96), 56.
11 William W. Mendel and Lamar Tooke, "Operation Logic: Selecting the Center of Gravity," *Military Review* 6 (June 1993), 5–6.
12 Hermann Franke, ed., *Handbuch der neuzeitlichen Wehrwissenschaften*, vol. 1, *Wehrpolitik und Kriegführung* (Berlin/Leipzig: Walter de Gruyter, 1936), 206.
13 Walter A. Vanderbeek, *The Decisive Point. The Key for Victory* (Fort Leavenworth, KS: School of Advanced Military Studies, U.S. Army Command and General Staff College, April 1988), 10.
14 Robert Ross Smith, *United States Army in World War II: The War in the Pacific. The Approach to the Philippines* (Washington, DC: Center of Military History, United States Army, 1948), 2.
15 Headquarters, U.S. Marine Corps, FMFM 1-1, *Campaigning* (Washington, DC: U.S. Government Printing Office, 1990), 50.
16 Kenneth Macksey, *Why the Germans Lose at War: The Myth of German Military Superiority* (London: Greenhill; Harrisburg, PA: Stackpole, 1996), 136.
17 John Ehrman, *History of the Second World War: Grand Strategy*, vol. 5: *August 1943–September 1944* (London: Her Majesty's Stationery Office, 1956), 379–81.
18 Rajko Rakić and Branko Dubravica, *Kratak Pregled Vojnih Djelovanja U Domovinskom Ratu, 1991–1995* (Zagreb: Hrvatsko Vojno Učilište Petar Zrinski, 2009), 259.
19 Rakić and Dubravica, *Kratak Pregled Vojnih Djelovanja u Domovinskom Ratu 1991-1995*, map 3.2, 277.
20 Albrecht von Boguslawski, *Heerwesen und Kriegführung* (Berlin: Verlag von R. Eisenschmidt, 1897), 105–06.
21 The Soviets used the similar term "idea of the operation," as do the Russians today. The content of an operational idea is basically similar to a concept of operations (formerly scheme of maneuver) used by the U.S. and other militaries.
22 Vanderbeek, *The Decisive Point*, 39.
23 Center for Military History, U.S. Army, CMH Pub 104–5, *Terrain Factors in the Russian Campaign* (Washington, DC: U.S. Government Printing Office, 1st printed July 1951, facsimile printing 1982, 1986), 8.
24 Olivier Desarzens, *Nachrichtendienstliche Aspekte der "Weserübung" 1940* (Osnabrück: Biblio Verlag, 1988); T. K. Derry, *The Campaign in Norway* (London: Her Majesty's Stationery Office, 1952), 232.
25 Martin Braun, "Planung und Verlauf der Allierten Streitkräfte Gemeinsamen Landkriegsoperation," in *Der Golfkonflikt. Dokumentation, Analyse und Bewertung aus militärischer Sicht*, ed. Hartmut Zehrer (Herford/Bonn: Verlag E.S. Mittler & Sohn, 1992), 97–98.
26 Roy Appleman, James M. Burns, Russell A. Guegler, and John Stevens, *United States Army in World War II: The War in the Pacific: Okinawa; The Last Battle* (Washington, DC: Center for Military History, United States Army, CMH Pub 5-11, [1948] 1991), 473; some sources cited put U.S. casualties at 65,631 killed or wounded, while the Japanese lost 70,000 men and an equal number of civilians; Robert H. Scales, Jr., "Adaptable Enemies: Achieving Victory by Avoiding Defeat," *Joint Force Quarterly* (Autumn/Winter 1999–2000), 8.

27 Friedrich Ruge, *The Soviets as Naval Opponents 1941–1945* (Annapolis, MD: Naval Institute Press, 1979), 61; Raymond De Belot, *The Struggle for the Mediterranean 1939–1945*, trans. James A. Field, Jr. (Princeton, NJ: Princeton University Press, 1951), 259.
28 Richard Betts, "Strategic Surprise for War Termination: Inchon, Dienbienphu and Tet," in *Strategic Military Surprise Incentives and Opportunities* ed. Klaus Knorr (New Brunswick, NJ: Transaction Books, 1983), 153.
29 Vanderbeek, *The Decisive Point*, 173; Derry, *The Campaign in Norway*, 232.
30 Vanderbeek, *The Decisive Point*, 9.
31 U.K. Army Doctrine, *Operations* (prepublication copy, 1989), 3–12.
32 Gordon M. Wells, *The Center of Gravity Fad: Consequence of the Absence of an Overarching American Theory of War* (Arlington, VA: AUSA's Institute of Land Warfare, March 2001), 4.
33 U.K. Army Doctrine, *Operations*, 3-12.
34 Jeffrey Harley, "Information, Technology, and the Center of Gravity," *Naval War College Review* 50 (Winter 1997), 77.
35 Commander in Chief, U.S. Pacific Fleet and Pacific Ocean Areas, Operations in the Pacific Ocean Areas During the Month of October 1944, RG 38: Records of the Office of the Chief of Naval Operations, National Archives and Records Administration (NARA), College Park, MD., 115; Carl Solberg, *Decision And Dissent. With Halsey at Leyte Gulf* (Annapolis, MD: Naval Institute Press, 1995), 119.
36 John A. Warden III, "Strategy and System Thinking," *Air Power Revue der Schweizer Armee*, no. 3, Addendum to *Allgemeine Schweizerische Militärische Zeitschrift* 12 (December 2004), 21.
37 Peter J. Palmer, *Operational Main Effort and Campaign Planning* (Fort Leavenworth, KS: School of Advanced Military Studies, U.S. Army Command and General Staff College, 1991), 7.
38 John W. Foss, "Command," *Military Review* 5 (May 1990), 4; Peter J. Palmer, *Operational Main Effort and Campaign Planning*, 8.
39 Albert N. Garland et al., *United States Army in World War II: The Mediterranean Theater of Operations, Sicily and the Surrender of Italy* (Washington, DC: Center of Military History, United States Army, CMH Pub 6-2, [1965] 1986), 89.
40 Peter J. Palmer, *Operational Main Effort and Campaign Planning*, 9–10.
41 Foss, "Command," 4; Palmer, *Operational Main Effort and Campaign Planning*, 8.
42 Palmer, *Operational Main Effort and Campaign Planning*, 12.
43 Todd J. Ebel, *Without Coup d'oeil, But with Responsibility: Determining Centers of Gravity and Decisive Points* (Fort Leavenworth, KS: School of Advanced Military Studies, U.S. Army Command and General Staff College, December 17, 1993), 12.
44 Cited in Karl-Heinz Frieser, *Blitzkrieg-Legende. Der Westfeldzug 1940* (Munich: R. Oldenbourg Verlag, 1995), 415.
45 Clayton R. Newell, "What Is Operational Art?" *Military Review* 9 (September 1990), 9.
46 Headquarters, U.S. Marine Corps, FMFM 1, *Warfighting* (Washington, DC: U.S. Government Printing Office, March 6, 1989), 31.
47 Wolfram Hiller, *Vergleich und Bewertung der in der HDv 100/100 (TF/G) und in der Anweisung für Führung und Einsatz 700/108 dargestellten Führungsgrundsätze* (Hamburg: Führungsakademie der Bundeswehr, November 15, 1985), 9.
48 Vanderbeek, *The Decisive Point*, 9–10.
49 Joseph Schroedel, *The Art and Science of Operational Maneuver* (Fort Leavenworth, KS: School of Advanced Military Studies, U.S. Command and General Staff College, 18 May 1988), 17.
50 Schroedel, *The Art and Science of Operational Maneuver*, 17.
51 Hiller, *Vergleich und Bewertung der in der HDv 100/100 (TF/G) und in der Anweisung für Führung und Einsatz 700/108 dargestellten Führungsgrundsätze*, 12.
52 Vanderbeek, *The Decisive Point*, 13–14.

53 Myron J. Griswold, *Considerations in Identifying and Attacking the Enemy's Center of Gravity* (Fort Leavenworth, KS: School of Advanced Military Studies, U.S. Army Command and General Staff College, May 14, 1986), 28.
54 Vanderbeek, *The Decisive Point*, 9.
55 Phillip Kevin Giles and Thomas P. Galvin, *Center of Gravity: Determination, Analysis, and Application* (Carlisle Barracks: *Center for Strategic Leadership, U.S. Army War College*, 31 January 1996), 19.
56 Griswold, *Considerations in Identifying and Attacking the Enemy's Center of Gravity*, 12.
57 Giles and Galvin, *Center of Gravity*, 19.
58 John V. Scudder, *Talk'n Ain't Fight'n: Synchronization and the Joint Task Force Training Process* (Fort Leavenworth, KS: School of Advanced Military Studies, U.S. Army Command and General Staff College, May 1994), 1; Headquarters, Department of the Army, *Operations,* FM 100-5 (Washington, DC: GPO, 14 June 1993), 2-8; Joel J. Snow, "Airland Battle Doctrine Tenets in Opposition," *Military Review* 9 (October 1987), 64–65.
59 Cited in Steven N. Read, *Planning for the Unplannable: Branches, Sequels, and Reserves* (Fort Leavenworth, KS: School of Advanced Military Studies, U.S. Army Command and General Staff College, June 7, 1990), 5.
60 William S. Pennypacker, *Sequels: Thinking About the Future* (Fort Leavenworth, KS: School of Advanced Military Studies, U.S. Army Command and General Staff College, May 7, 1988), 34.
61 Griswold, *Considerations in Identifying and Attacking the Enemy's Center of Gravity*, 14–15.
62 William Morris, ed., *The American Heritage Dictionary of the English Language* (New York, NY: American Heritage, 1973), 1182.
63 Pennypacker, *Sequels*, 4.
64 Adolf Heusinger, *Der Ostfeldzug 1941–1942. Ein operativer Überblick*, T-6, ZA/1 2325, Studien der Historical Division Headquarters, United States Army Europe, Foreign Military Studies Branch, BA-MA, 1947, 11.
65 Headquarters, Department of the Army, FM 100–5, *Operations* (Washington, DC: Headquarters, Department of the Army, June 1993), 6–9, 4–5; L. D. Holder, "A New Day for Operational Art," *Army* 3 (March 1985), 4–5.
66 Headquarters, Department of the Army, FM 5–0 (FM 101–5), *Army Planning and Orders Production* (Washington, DC: U.S. Government Printing Office, January 2005), 1–17.
67 MCDP 1–2 *Campaigning* (Washington, DC: Department of the Navy, Headquarters United States Marine Corps, August 1, 1997), 45–46.
68 John F. Kalb, *A Foundation for Operational Planning: The Concept of Center of Gravity, Decisive Point and the Culminating Point* (Fort Leavenworth, KS: School of Advanced Military Studies, U.S. Army Command and General Staff College, April 21, 1987), 27–28.
69 Donald C. McGraw, Jr., *Operational Pauses: Minimizing the Effects of Culminating in Joint Operations* (Fort Leavenworth, KS: School of Advanced Military Studies, U.S. Army Command and General Staff College, May 9, 1991), 16.
70 David M. Cowan, *The Utility of the Operational Pause in Sequencing Battles to Achieve an Operational Advantage* (Fort Leavenworth, KS: School of Advanced Military Studies, U.S. Army Command and General Staff College, 1987), 22.
71 McGraw, Jr., *Operational Pauses*, 15.
72 John D. Johnson, *Sequencing Operations: Considerations for the Operational Planner* (Fort Leavenworth, KS: School of Advanced Military Studies, U.S. Army Command and General Staff College, May 20, 1991), 21–22.
73 Johnson, *Sequencing Operations: Considerations for the Operational Planner*, 8–9.
74 Johnson, *Sequencing Operations: Considerations for the Operational Planner*, 7–8.
75 Johnson, *Sequencing Operations: Considerations for the Operational Planner*, 9–10.
76 Cowan, *The Utility of the Operational Pause in Sequencing Battles to Achieve an Operational Advantage*, 24.

9 Operational Thinking

One of the principal requirements for success at the operational and strategic levels of command is to think broadly and have a broad vision.[1] Such ability, which the Germans aptly call "operational thinking" (*Operatives Denken*), is not an innate trait of a commander.[2] It is a result of considerable conscious effort on the part of the future operational commanders during their entire professional career. Although operational thinking is one of the most critical factors for success, whether in peacetime or time of war, many operational commanders have remained essentially captives of their narrow tactical perspective. To think tactically is easy; it is an area in which all commanders feel comfortable, because this is what they have done for most of their professional careers. In the past, many operational commanders failed because they were unable or unwilling to raise their perspective above the tactical level. To broaden one's perspective requires considerable mental effort.

The term "operational thinking" is not easy to define concisely, because it encompasses many diverse elements. In broad terms, operational thinking can be described as the commander's ability to think far ahead into the future, have a deep knowledge and understanding of the essential elements of the operational/strategic situation, and ensure that each action is planned and conducted within a given operational/strategic framework. Strategy and operational art deal with broad objectives, and broad objectives require broad vision.[3] The operational commanders must focus on broad objectives that can range from the destruction of the enemy forces in the field to undermining public support for war or the enemy's will to fight.

The need for a higher commander to think broadly has been recognized by many classical military/naval thinkers and practitioners of warfare. The French Marshal Maurice de Saxe wrote that the generals under his command must be

> incompetent indeed if they do not know how to execute this order and to perform the proper maneuvers with their respective divisions. The commander in chief will not be forced to occupy himself with it nor be embarrassed with details, for if he attempts to be a battle sergeant and be everywhere himself, he will resemble the fly in the fable that thought he was driving the coach. He should not be involved with details on the day of battle so when he sees an occasion he can unleash his energies, hasten to the critical point at top

DOI: 10.4324/9781003333685-9

> speed, seize the first troops available, advance them rapidly, and lead them in person.[4]

The Prussian General Gerhard Johann David von Scharnhorst observed that "one must see the whole before seeing its parts. This is really the first rule, and its correctness can be learned from a study of history."[5] Clausewitz) wrote that "small things always depend on great ones – the unimportant on the important, and accidentals on essentials; this must guide our approach."[6] Moltke, Sr, observed "all individual successes achieved through the courage of our troops on the battlefield are useless if not guided by great thoughts and directed by the purpose of the campaign and the war as a whole."[7] He believed that "it is far more important that the high commander retains a clear perspective of the entire state of affairs than that any detail is carried out in a particular way."[8] The commander must make decisions throughout a campaign in conditions that are difficult to predict. Moltke, Sr, believed that

> all successive acts of war are thus not premeditated implementations of some plan but spontaneous actions in response to the military situation of the moment. What is important, in each concrete case, is to see clearly through a mist of uncertainty, assess the facts accurately, guess the unknowns, reach a decision quickly, and then move to carry it out vigorously without letting oneself be sidetracked.[9]

Importance

Operational thinking is one of the principal prerequisites for the commander's judicious use of sources of power in accomplishing strategic or operational objectives. It is especially critical for the operational commanders who command numerically inferior forces. It helps the commander to employ friendly forces in such a way that each action directly or indirectly contributes to the accomplishment of the ultimate strategic or operational objective. This means that the operational commander must have the ability to differentiate between events that are essential for the accomplishment of the ultimate objective and those that are outside the framework of a given campaign/major operation. The commanders who think operationally will focus on the broader elements of the situation instead of on the minutiae. The commanders who are too tactically focused may ultimately achieve operational or even strategic success, but at substantially heavier costs for friendly forces in terms of personnel, materiel, and time than the commanders who think operationally.

Often, a predominant or exclusive focus on tactics leads to defeat at the operational and strategic levels. Field marshal Helmuth von Moltke, Sr, observed that Napoleon I's ultimate defeat was due to his marshals, who with few exceptions were incapable of intelligent cooperation when removed from the commander-in-chief's immediate control.[10] Most of Napoleon I's marshals lacked the required skills to command more than a division. They were unable to follow the workings of Napoleon I's mind and consequently failed to understand the methods that were the expression of his thoughts.[11] Napoleon I reportedly said that Marshal Jean Victor Marie Moreau had no system in either politics or military matters. In his view, Moreau was an excellent soldier, personally very brave and very capable of moving a small army on the battlefield. However, he was an absolute stranger to grand tactics (operational art today). Moreau could not effectively command more than 20,000 men. Napoleon, I had the same view of another of his marshals, Jean-Baptiste Jourdan.[12] Michel Ney was typical of the Napoleonic marshals. He was very courageous and showed great stoicism in the face of danger. Napoleon I called him the "Bravest of the Brave." However, Ney had poor knowledge and understanding of strategy. He was often reckless. Ney lacked the necessary vision to be a successful operational commander.[13]

In July 1813, Napoleon I wrote to Marshal Auguste Frédéric Louis Viesse de Marmont, Duc de Raguse, explaining his plan against the enemy coalition in Germany and asking for his opinion. Marmont wrote in his response: "it is to be feared that on the same day your majesty wins a great victory you will learn that your subordinates have lost two." He knew that his comrades-in-arms, with but a few exceptions, were not capable of commanding armies, however skillful they might be in handling their own army corps. Within four weeks after Marmont sent his response, three of Napoleon's marshals lost the opening battles that led to the Battle of Nations at Leipzig in August 1813 (Nicolas Oudinot was defeated at Großbeeren on 23 August, Etienne MacDonald at Katzbach on 26 August, and Ney at Dannewitz on 6 September 1813). This was another proof that Napoleon I's army was properly controlled only when he was present.[14]

The lack of operational thinking was the main reason for the Allied defeats in Norway and France in 1940 and in Southeast Asia in 1941–1942. The absence of operational thinking was apparent in many UN commanders during the Korean War in 1950–1953. The one exception was General Douglas MacArthur. He conceived and executed a bold and highly successful turning maneuver by landing some 70,000 UN troops about 150 miles into the depth of the enemy's defenses at Inchon in September 1950.

Besides the serious disconnect at the strategy and policy level, the Vietnam War (1965–1975) was essentially conducted at the theater-strategic and tactical levels only; again, operational art was not applied.[15] In contrast, the U.S. performance in the invasion of Panama in December 1989 (Operation Just Cause) and the Gulf War of 1990–1991 showed that the services had learned from their neglect of operational art in the previous decades. Yet, only a few years later and despite ultimate success, NATO stumbled into a war of attrition at the operational level in its air offensive against Serbia in March–June 1999. The lack of operational thinking on the part of the Russian commanders in their invasion of Ukraine in February 2022 led to a highly intense but protracted war.

Attributes

A commander who thinks and acts operationally must have some attributes that are lacking by tactically minded commanders. There is no consensus about these attributes. Yet, an operational commander should display to very high degree the following characteristics: (1) operational perspective; (2) understanding of the mutual relationship and linkage among strategy, operational art and tactics; (3) understanding the true meaning and distinctions among the levels of war; (4) sound balancing operational factors versus a given operational/strategic objective; (5) knowledge and understanding of the operational characteristics of the operating area and military geography of a given theater; (6) understanding main aspects of non-military situation in a given theater; and (7) making sound operational decisions.

It is commonly understood that for the theater commanders to think operationally means, first and foremost, that they have a much broader perspective than tactical commanders. The British Field Marshal William Slim said that to think beyond the limits of immediate combat, a higher commander "must think big."[16]

Based on the scale of a military objective, strategic, operational, and tactical situations, respectively, are differentiated. The larger the scale of the objective, the more complex is the situation. Any military situation consists of physical or tangible and "abstract" or intangible elements. Military elements of a situation coexist with non-military elements (political, diplomatic, geopolitical economic, financial, social, ethnic, religious, linguistic, etc.). Hence, strategic, operational, and tactical perspectives can be differentiated. *The strategic perspective* pertains to a theater of war and *operational perspective* to a single theater of operations. The perspective of tactical commanders is much smaller, because they are focused on planning and executing actions aimed at accomplishing tactical objectives in a combat zone/area of operations.

In purely physical terms, the strategic and operational perspective encompasses the respective theater of war/theater of operations plus an arbitrarily defined area beyond the boundaries of these theaters called in U.S. terms, an area of interest (AOI). The size of AOI depends on the events that potentially could influence the situation within the respective area of responsibility. In terms of factor of force, the theater commanders are not only responsible for the employment of military forces but also must synchronize the use of non-military sources of power within their respective theaters. In terms of the factor of time, the theater commander must have the ability to look several weeks or even months into the future. In contrast, tactical commanders must look no more than two to three days into the future. All considerations of an operational nature are ultimately based, when the initiative has been lost, on appreciation or hypotheses regarding the options the enemy may be expected to take. The greater one's sphere of command, of course, the further ahead one should think.[17] In sequencing and synchronizing the use of their forces, operational commanders must have the ability to focus on the big picture and not be sidetracked by minor or unrelated events.

Theoretically, the building of a mental picture of an operational or strategic situation combined the use of the reductionist or Descartes method [named after the French philosopher and mathematician René Descartes] and the holistic or comprehensive method. That picture cannot be obtained by using only one of these two methods. In the reductionist method the complexities of strategic or operational situations should be reduced to their *essentials*. The holistic or comprehensive method must be used in connecting diverse events/aspects of the situation ("connecting the dots") and then identifying their mutual relationships to derive *patterns* and then from these patterns to project *future trends*.

The penalties for the lack of an operational perspective by the operational commanders can lead to a significant waste of time and sorely needed resources. Often, an exclusive focus on tactics leads to defeat at the operational and strategic levels. The Japanese Combined Fleet, led by Admiral Isoroku Yamamoto, suffered a decisive defeat in the Battle of Midway in June 1942 despite its overwhelming strength over the U.S. Navy forces that took part in that battle. On 3 June, when the first contact was obtained, Admiral Yamamoto's forces were dispersed into ten groups in the central and northern Pacific. Two of his carriers that might have provided the margin needed for victory were sent north to attack Dutch Harbor and seize Attu and Adak in the Aleutians. The Japanese forces in the Midway area were beyond mutual supporting distances. The U.S. Navy would not have achieved that victory regardless of its full knowledge of Japanese intentions and brilliant handling of forces by Admiral Nimitz and his subordinate carrier group commanders if Admiral Yamamoto had kept his forces concentrated and not violated the principles of objectives, mass, and economy of effort. Admiral Yamamoto's perspective was clearly too tactical, and that was probably the single major reason for his defeat.

Operational commanders cannot be highly successful without having full knowledge and understanding of the mutual interrelationships and linkage between strategy and policy on the one hand and strategy, operational art, and tactics on the other. The operational commanders should possess a thorough knowledge and understanding of the theory and practice of operational art.[18] They need to have a keen understanding of the relationship between strategy and policy, on one hand, and operational art and strategy on the other. This consists of mastering all aspects of the employment of large forces and formations to accomplish operational or strategic objectives in the theater. Most of the best operational commanders were well known for their excellent knowledge and understanding of the art of war.

Operational commanders must also fully understand the distinctions among the levels of war and how decisions and actions at one level affect actions at other levels. Moltke, Sr, invariably focused on thoroughly studying the problem at hand to make a sound decision. In his estimate of the situation, he studied the way the enemy commander operated, derived assumptions, and then continuously reviewed various "possibilities" (options).[19] His planning was greatly aided by his intimate knowledge of the terrain, order of battle, intelligence, and knowledge of the mind of the enemy commander.[20]

All successful operational commanders had excellent knowledge and understanding in balancing or harmonizing a given operational/strategic objective with corresponding factors of space, time, and forces. This process is largely art, not a science. Whether the process was successful can be seen only in retrospect.

In contrast to the tactical commander, an operational commander must evaluate the features of the operating area in operational rather than tactical terms. This means, among other things, assessing the effects of terrain, and hydrography/oceanography on the employment of combined arms/branches. The operational commander is more concerned with the effects of climate, rather than weather, on the employment of joint forces in a given part of the theater. The operational commanders must have the ability to deduce operational features of the physical environment in which a campaign/major operation will be conducted. They must not only understand the geometry of the theater but also have a full understanding of the factor of distances in the combat employment of their forces. The operational commanders must appreciate the factor of time. They must know how changes on the ground will affect the employment of both friendly and enemy forces.[21]

Operational warfare is essentially a war on the map. All successful operational commanders have had a solid knowledge of geography; otherwise, the operational commander cannot make an accurate estimate of the feasibility and risk involved in a particular operation.[22] Clausewitz observed that the relationship of war to geography and terrain might be regarded as, if not as the most important, at least the most noticeable peculiarity of military activity. Napoleon I combined a vivid imagination with a formidable capacity for calculating space in terms of time to predict outcomes beyond individual battles. In one case he accurately foresaw the location of decisive encounters several weeks before it occurred.[23] He was once asked how he always divined the intentions of the enemy so accurately. Napoleon I responded, "I did not know beforehand the mistakes the enemy would make which I took advantage of; I simply studied my map." He studied the enemy's possibilities and limitations on the map, more than the enemy did himself. Yet the methods that brought Napoleon I his great successes in central Europe failed him altogether in Russia in 1812, because in that instance he failed to properly evaluate the factors of space and time versus the strategic objective. The British failed to understand the geographic features of South Africa until the Anglo-Boer War (1899–1902) was well underway.[24] Field marshal Erwin Rommel was known for his excellent knowledge of the terrain and sense of orientation. He would reportedly often drive day and night through trackless deserts without maps, using a compass to check the direction.[25] General MacArthur was well known for his excellent knowledge of military geography, which greatly helped him to properly evaluate the factors of space, time and forces in planning and executing his campaigns and major operations.

The operational commanders should also have a broad knowledge of foreign policy, diplomacy, geopolitics, the international economy, ethnicity, religion and other issues that shape the situation in their theater. This was realized by several notable commanders in the past as the example of the Prussian king Frederick the Great shows. He wrote that the "soldier-king"

> Must have an accurate idea of politics in order to be informed of the intention of princes and the forces of states and of their communications; to know the number of troops that princes and their allies can put into the field; and to judge the condition of their finances; knowledge of the country where he must wage war serves as the basis for all strategy; he must be able to imagine himself in the enemy's shoes in order to anticipate all the obstacles that are likely to be placed in his way; above all, he must train his mind to furnish him with a multitude of expedients, ways, and means in case of need; for those

> who are destined for the hard military profession, peace must be a time of meditation, and war the period where one puts his ideas into practice.[26]

The operational commanders should have a thorough knowledge of the theater in which their forces will be employed. They should have deep knowledge of other countries' history, society, and culture so that they are not overly reliant on the advice and judgment of their political advisors. They should study the enemy's country in great depth, concentrating on not only military but also nonmilitary matters. The operational commanders are unlikely to be successful if they do not take the time to know their opposite – the enemy commander; they would likely misunderstand the enemy's intentions and may neglect intelligence indicators.[27] General George S. Patton became impressed with the German military because of his studies of the Prussian and modern German history. He became convinced that there would be another war with Germany. Therefore, he decided to read most of the World War I memoirs of German officers, such as field marshals Paul von Hindenburg and Erich von Falkenhayn, when the memoirs became available in English translation. He read the works of Generals Hans von Seeckt and Erich Ludendorff. Patton's studies were so comprehensive that he read the memoirs of German politicians and works of German philosophers and listened to German music.[28]

The highest art of a commander at any level of command is making timely and sound decisions. The act of exercising command consists of making decisions and ordering their execution.[29] The most important factors in making a decision should be the mission and the situation. The decision can also be based on the task or tasks derived during the decision's execution. Actions conducted on one's initiative also require making a new decision.[30]

The larger the scale of a military objective, the greater the importance of making sound decisions. The number of decisions at the operational and strategic levels is relatively small and the time available to make them is much longer than at the tactical level. The operational commanders should not make tactical decisions, except in rare cases. Doing that would lead to losing so badly needed operational perspective. They must often make operational decisions without having a solid knowledge of the essential elements of the situation.

Thinking operationally means that the operational commanders clearly see how each of their decisions and actions contributes to the accomplishment of the ultimate strategic or operational objective. All the decisions and actions of the operational commander should be made within the given operational or strategic framework; otherwise, they will not contribute to ultimate success and might undermine it. As in a game of chess, the player who views the board as a single interrelated plane of action and each move as a prelude to a series of further moves is more likely to be successful than an opponent who thinks only a single move at a time. The operational commanders should think of how to create opportunities for the employment of their forces while at the same time reducing the enemy's future options.[31] The operational commanders should have the ability to see the situation through the enemy's eyes – what Napoleon I called "seeing the other side of the hill." Largely, this ability is intuitive. Napoleon I and some other successful military leaders had an extraordinary ability, in analyzing to visualize what the enemy's commander would do in countering the movements of their forces.[32]

Because campaigns and major operations are conducted over a much larger area and involve considerably larger and more diverse forces than tactical actions, the operational commander must assess the situation in all its complexity several weeks or even months ahead. The resultant decision will often be made on incomplete and even false information and in a time-space window that is very different from that of the tactical commander. The decisions by the operational commander are more likely hypotheses; because they are based on many assumptions.

The operational commanders usually have more time to make decisions than tactical commanders. Because of steadily compressed time–space relationships, the operational commanders must make decisions based on their instinctive judgment and without the benefit of a careful analysis of the situation, weighing the advantages and disadvantages of each course of action.

The number and importance of the decisions vary for each level of war. The higher the level of war, the fewer the decisions that must be made and the larger the time window for making them. The operational commanders make fewer decisions, but the impact of these decisions is much greater than that of those made at the tactical level. The key to making sound operational decisions is to have an accurate picture of the operational situation. This cannot be obtained by simply collecting vast amounts of tactical data or information. The operational situation should be built as a synthesis of strategic and tactical information pertaining to the operational commander's area of responsibility plus his area of interest. It should encompass both military and nonmilitary aspects of the situation. The operational commander should make his decisions by fully considering the trends in the operational situation several weeks or even months into the future. In contrast, the tactical commander is concerned with the developments of the tactical situation, from a few hours to several days ahead.

Acquiring Operational Thinking

The operational commander's ability to think "operationally" is a result of the influences of many factors. The societal and cultural framework determines to a large degree the character of military institutions, and hence professional education and training. The institutional influences encompass national culture and way of war, service culture/way of warfare and military institutions. They provide an overarching framework within which the future operational commanders and their staff are educated and trained to exercise their authority and responsibilities. Campaigns/major operations cannot be successfully conducted unless operational commanders and their staff have a common *operational outlook* of the fundamentals of operational warfare.

The German general staff system established by Moltke, Sr, provides the best example of the benefits of having a common operational outlook. He imposed a common operational outlook through the sheer force of his personality and intellect and his successes in three wars. The German operational thinking in Schlieffen's era was developed under the nightmare of a two-front war. The aim was first to concentrate a major part of the German army in the west against France and destroy its forces in a decisive major battle (*Schlacht*). Afterward, the German army would be redeployed to the east to defeat Imperial Russia. A war on two fronts would be conducted as two single-front wars in succession. Schlieffen insisted that victory could only be a complete victory – in other words, a war of annihilation.[33] Schlieffen aimed to inculcate common operational thinking through the solution of several operational-strategic tasks conducted during general staff field trips. In contrast to Moltke, Sr, Schlieffen went to extremes in enforcing his views on the value of the envelopment maneuver based on his study of historical examples and in particular Hannibal's victory over the Romans in the battle of Cannae in 216 BC. General Bernhardi, one of the great military thinkers of the Schlieffen era, was also one of Schlieffen's fiercest critics. He feared that applying historical examples to modern conditions had the latent danger of leading to practicable conformism instead of the commander's relying on his own intuition based on the situation at hand.[34]

In the process of inculcating common views on the character of the future war, there is an ever-present danger of stifling creativity on the part of subordinates. Certain views on the impact of technological advances on the relative importance of offense and defense might lead to fatally flawed doctrine and unsound force planning. Prior to 1914, the militaries of all major European powers believed in the need to go on the offensive. As a result, none of the major armies were ready for positional and protracted war, as actually happened on Western Front in 1914–1918. This came to pass despite increasing evidence that the gap between the advantages offered by an offensive and defense had steadily narrowed in the decades preceding the outbreak of hostilities in August 1914. Schlieffen advocated the need to conduct rapid and decisive offensives against Belgium, France, and Russia. Likewise, the French army, and many political leaders, became literally obsessed with the virtue of offensive. The British also resolutely rejected any defensive strategy, despite their bitter experiences in the Anglo-Boer War. The Russians, for their part, embraced the offensive, because their potential enemies, intended to wage a war of "rapid and decisive blows." The British and French military were convinced that superior morale would overcome superior defensive firepower. In short, mind would prevail over materiel, and morale would win over machine guns and barbed wire.[35]

Prior to 1914, most of the German thinking about war, professional education, and training emphasized the value of the factor of speed. In practical terms, this meant a quick estimate of the situation, followed by decision and short orders, and their swift execution. The entire army was so indoctrinated, from the highest general to the simple soldier. Rapid troop movements were required. Decisive battles had to be fought and decided quickly.[36] The entire German army was infused with the cultural attributes of the Great General Staff. This included a systematic approach to all aspects of warfare, and a rigorous selection process that emphasized an officer's intellectual achievement in a professional sense, as well as his performance as a leader.[37] However, the German way of warfare prior to 1914 also had considerable weaknesses. Among other things, the Germans overestimated the value of moral factors such as courage, death in battle, and honor. Perhaps surprisingly, the German officer corps was averse to the importance of technology in modern warfare.[38] In addition, the Germans overemphasized tactics and operations, paid little attention to logistics and neglected intelligence.[39]

In the interwar years, the Soviet theoreticians imposed the cult of the offensive on the Red Army's doctrine and training. During the formative years of the Red Army, rational calculations and organizational concepts were meshed with ideological beliefs on exporting revolution. This predisposed the Soviets to go on the offensive in any future war. This strong bias toward the offensive led to dogmatic beliefs and prevented the Soviets from learning proper lessons either from their own or foreign experiences. By the early 1930s, the cult of the offensive became an article of faith – that if the Soviet Union was attacked, the Red Army would not surrender an inch of Soviet territory to the aggressor. The idea of transferring the war to enemy territory reinforced the principal planning assumption of the Red Army that a significant time interval would separate the initiation of hostilities and the engagement of the main forces of the belligerents. Although the German victories in the West in 1940 challenged such views, the Red Army's leadership continued to hold its erroneous beliefs on the character of the initial phase of a future war.[40]

One should be aware that embracing a common operational outlook has several pitfalls. There is an ever-present danger that education may become indoctrination and end in certain rigidly held beliefs on some key aspects of operational art. Schlieffen dogmatically believed that Hannibal's highly successful envelopment maneuver in the Battle at Cannae was valid everywhere and that one must try to envelop and eventually encircle the enemy in all circumstances.[41] This led to his idea of strategic maneuver in the case of war against France. In contrast, Clausewitz and Moltke, Sr, considered a historical example like the Battle of Cannae an isolated case that could occur only rarely in history.[42]

There should be some consensus within a service and among the services about the true meanings of key *operational terms*; otherwise, it is difficult to write sound service or joint doctrine. This complicates the planning, preparation, and execution of a campaign or major operation. The lack of common terms, the abuse of commonly understood terms, or the use of existing terms interchangeably greatly complicates discussion among theoreticians of various aspects of operational warfare.

B. H. Liddell Hart wrote that there are two types of experiences: direct and indirect. The main *direct influences* in acquiring operational thinking are combat, command of large forces/formations, participation in field trips, planning games/war games and large-scale exercises/maneuvers in peacetime. The most important direct influence in shaping the commander's ability to think broadly is war. Obviously, command of large forces in combat is the best opportunity for broadening the commander's perspective. Opportunities to acquire such direct experience are relatively few because major wars are rare. Direct experiences are inherently limited in forming an adequate foundation either for theory or for its application.[43] The next best method for acquiring operational thinking is to have experience in commanding *large forces/formations* (e.g., numbered/area armies, fleets and air forces) in peacetime. The lack of such experience cannot but have a detrimental effect on the development of future operational commanders. The British General Lord

Kitchener of Khartoum admitted that one of the greatest weaknesses of the British generals in the Anglo-Boer War was that tactical failures were caused by a lack of exercises and ability to command large formations.[44] The main reason for these deficiencies was that the British Army was small and was employed mainly for colonial duties. In the modern era, an overemphasis on insurgency and counterinsurgency warfare, combating terrorism or piracy cannot but have a highly negative effect on the armed forces, especially on the army, to prepare for a high-intensity conventional war.

Some operational commanders did not have experience in handling large numbers of troops but still were highly successful. Moltke, Sr, was a brilliant operational thinker and practitioner even though he did not have experience in commanding large forces; he acquired that knowledge mainly through the study of military history. Yet such commanders are the exception rather than the rule. It is a great advantage for future operational commanders to have practical experience in commanding large forces. One potentially serious consequence of the possible elimination of the intermediate level of command because of the supposed advantages of netting one's forces is that it would drastically reduce opportunities for training future operational commanders in commanding or serving on the staffs of large forces.

Peacetime *exercises/maneuvers* can be extremely useful in training future commanders to conduct a quick estimate of the situation, make sound decisions, and then execute their plans and orders. They are an excellent method of preparing future commanders for properly coordinating the movements and actions of diverse combat arms, and for transmitting messages. A lack of large-scale exercises/maneuvers can be one of the main reasons for setbacks or even defeats on the battlefield.

The participation of future operational commanders in field trips, and planning/war games is a proven and highly effective method of preparing future operational commanders and staff for war. During Moltke, Sr's, era as the chief of the Great General Staff, the Germans started the systematic education and training of future operational commanders. This included *training trips* (*Übungsreisen*) and *tactical and strategic war games*. Between 1858 and 1881, Moltke, Sr, personally conducted training trips, which were used not only to enhance the operational thinking of the German general staff officers but also for testing and refining operational plans prepared for various contingencies. The number of participants in these training trips varied from 20 to 40 staff officers, from the Great General Staff to the troop general staff. After 1871, Moltke, Sr, also included staff officers from the states of Württemberg and Bavaria. Those trips conducted in the coastal area also included German naval officers. The focus of these training trips was the Army Corps level of command. The value of these exercises can hardly be overestimated. The trips were based on the real political-military situation of the time. They were routinely enriched by military historical excursions, especially valuable to junior officers.[45]

Schlieffen, during his time as the chief of the Great General Staff, used training trips for purposes like those of Moltke, Sr General Hans von Seeckt, as the chief of the Reichswehr (1920–1926), greatly increased the number and diversity of exercises. They included sand table, command post, staff, and map exercises, tactical walks, training trips, planning games and war games.[46]

In the interwar years (1919–1939), the Germans used planning games and war games extensively for training commanders at all levels. These games were conducted throughout the training year. They were even conducted during the war; then the actual situation on the front was used as the basis for a game.[47] In the planning games (*Planspiel*), the emphasis was on learning various tactical fundamentals and terms and on making decisions. The planning games were one-sided. In contrast, a war game (*Kriegspiel*) was two-sided. War games were conducted from the strategic to tactical levels of command. The focus in war games was on the assessment of the situation (*Lagebeurteilung*), concise and logical presentation of the main idea (concept of operations, in U.S. terms), and then making the decision on how to accomplish the mission. Another purpose of war games was to provide training in techniques and procedures for writing and issuing orders. War games were used to train commanders at all echelons and as a means of testing new methods and checking some fundamentals of combat. In the German view, the greatest value of planning games and war games was in the mental preparation of the participants and in identifying sources of friction. Wrong measures can be rehearsed in advance during a war game, thereby saving many lives in war.[48]

No exercise or planning game/war game can realistically simulate the effect of friction in real combat or the sense of physical danger. There is always a danger, during a long time of peace, that the future operational commander might either overrate or ignore the effect friction has on the employment of both friendly and enemy forces. Hence, it is critical that future commanders be properly trained in how to overcome the effects of friction.[49]

Indirect influences on the commanders' ability to think operationally include professional education at the service/joint colleges and, perhaps more importantly, their consistent efforts at self-education over the course of their entire career. Operational thinking is only in some very rare cases the result of a commander's inherent predisposition to think big and far ahead of current events. The future operational commanders should study in great depth and width military/naval history, international politics/diplomacy, geopolitics, economics, finances, sociology, ideologies, religions, nationalism and other non-military subjects. Indirect practical experience may be more valuable because it is infinitely wider. The greater value of indirect experience lies in its greater variety and extent.[50]

The *professional education* of future operational commanders and staffers should start early in their careers. Optimally, a process should be in place for identifying officers who think broadly and have the ambition to expand their professional knowledge relatively early in their careers Afterward, such officers should be consistently tracked and groomed for the higher command and staff positions. Lack of solid professional education for future operational commanders and staff is bound to have highly negative consequences on the performance of one's military in a war. One of the major problems the North had in the initial phase of the American

Civil War (1861–1865) was that too many officers lacked professional education and training. Only after General Ulysses S. Grant was appointed commander in chief of all the armies in March 1864 did that highly unfavorable situation began slowly to improve. Yet shortly before the end of the war he still complained to President Abraham Lincoln of the inadequate professional education of many of his subordinates.[51]

The main purpose of the professional education of future operational commanders should be to enhance their mental agility and innovative thinking.[52] Junior officers with the greatest potential should be guided to embark on in-depth studies of military history and the theory of operational art and strategy. Only through mastery of military history and military theory can officers gain the wide frame of reference necessary for sound operational decision-making/planning and directing of campaigns. Education in operational art should be general for many and individualized for a select few. The basic concepts of operational art should be used to promote informed discussion of related topics.[53]

The educational process should encourage the development of initiative, flexibility, decisiveness, and willingness to take responsibility; mistakes should be corrected without condemnation.[54] Stereotyped and universal solutions to a given operational problem should be avoided. Every military situation is unique, and no scenario can possibly encompass all its aspects. Officially approved or "school" solutions should not be used in educating future operational commanders and staffers. Normally there are always several viable solutions to a problem. A critique of a subordinate's actions should focus on the rationale for an action taken. The aim should be to broaden the knowledge, analytical skills and experience of the future operational commanders and staffers. Critiques should be open and frank but not personal or overly harsh. Mistakes made in finding a solution to a military problem should be considered part of the normal process of learning. In other words, the focus should be not on errors and mistakes but on progress made. It is critically important that future operational leaders have a healthy self-confidence and self-esteem (based on real accomplishments); otherwise, it will not be possible to develop the military character traits and skills so necessary for success at the operational level.

Moltke, Sr, emphasized that theoretical knowledge will not of itself lead to victory, but it cannot be ignored. From knowledge to doing is just one step, but from not knowing to doing is a giant step. The best lessons for the future are drawn from one's own experience, but since this may be meager, one must use the study of the military historical experience of others.[55]

History serves as a foundation of education by showing how mankind repeats its errors and what those errors are. It is the story of universal experience.[56] It is inherently broader, deeper and more diverse than the study of any other area of human activity.[57] The study of military/naval history provides the future operational commanders with a core of background knowledge and understanding that allows them to form and reform their vision of the battlefield beyond the realm of their combat experiences.[58] All too often, history is considered the exclusive preserve of professional historians. However, nothing could be further from the truth. Liddell Hart wrote that history is

> the record of man's steps and slips; it shows us that the steps were slow and slight; the slips, quick and abounding. It provides us with the opportunity to profit from the stumbles and tumbles of our forerunners. An awareness of limitations should make us chary of condemning those who made mistakes, but we condemn ourselves if we fail to recognize mistakes.[59]

History serves as a foundation of education by showing how mankind repeats its errors and what those errors are. It is the story of universal experience.[60] The 3,500 plus years of military history is the only academic study that provides the totality of the phenomena of war.[61] A study of past wars is fundamental to preparation for the next, for current military problems cannot be solved without an understanding of the past from which they stem.[62] Napoleon I observed that on the battlefield, what one believed to be a happy inspiration proved to be merely a recollection.[63] Critical study of past wars, and campaigns and major operations in particular, is a major source for developing the operational perspective of future commanders.

Because very few military commanders have the experience of commanding forces at the operational level, the best way to educate them to think operationally is through the study of the successes and failures of great military leaders. Operational commanders should be students of history, not historians – a big difference exists between these two. Napoleon I and Moltke, Sr, were great students of history. Moltke, Sr, had a vast knowledge of natural sciences and philosophy, read in several languages, and wrote extensively on various political and economic issues, such as the Belgian and Polish railways. His travel letters on his visits to Turkey, England, France, Russia and Italy are considered the most-read pieces of German prose.[64] General Patton was also a student of military history. He studied past major operations and campaigns in great depth.[65]

A proper study of military history helps to derive general principles of leadership through a critical reading of the biographies and memoirs of the great captains of the past. It also helps in understanding the reasons for their successes and failures.[66] The knowledge of psychological and moral factors in a war must be based on the study of military history.[67] The greatest utility of military history for officers is that by studying the work of the great captains, they can consider what courses of action have borne success or failure.[68] The study of military history is the best way of impressing on future operational commanders and staffers that warfare is not a science but as was in the past, is today, and will remain in the future largely an art.

A full understanding of the relationship between policymakers and operational commanders can be obtained by studying military and political history. Future operational commanders must fully understand the political purpose of war and strategy before they can start to understand various aspects of operational art. That understanding and knowledge can essentially be acquired only through the critical study of operational aspects of past wars[69]

The study of history provides a broad perspective on events and gives a sense of proportion in relation to time, place, and circumstances.[70] Methods of accomplishing operational or strategic objectives might be obsolete, but the fundamentals of strategy or operational art are essentially the same today as they were in the recent or even distant past. Experience is the raw material of imagination, and history is a great source of experience.[71] The study of history shows how the obstacles and hardships of warfare are magnified by human mental responses. If men know that other men have overcome similar or worse conditions, they may be persuaded that they can do so as well.[72] A more direct inspirational value of military history is the building of esprit de corps.

To have the greatest value, military history must be accurate. Clausewitz aptly observed that military history has value when it always presents the truth, the entire truth and nothing but the truth.[73] The most useful histories of past wars are those written from an operational perspective. Unfortunately, such histories are sorely neglected, and relatively few have been written. To be of any use, military history must be factual and frank. Propagandistic or censored history is very dangerous. In fact, such a history is not history at all, and it can provide no sound lessons or basis for intellectual and professional training. It leads to false conclusions, and it fosters one of the worst failings – self-deception.[74]

Military history should be studied in width, depth, and context. By studying warfare in *width* – that is, over a large timeframe – one can discern and learn the discontinuities in military history. Ideally, the study of military history should focus on the history of the art of war[75] It should not be limited to a certain age or a certain area. Yet one of the dangers in studying military history is in trying to cover too broad a field. It would seem to be much better to cover a limited number of events thoroughly than to give students a superficial picture of the greatest possible number of occurrences.[76] Military history should be studied in great *depth*. This means focusing on a single campaign/major operation and analyzing it thoroughly rather than to superficially focusing on several campaigns/major operations and studying them superficially.

Military history should be studied in a *context*. Wars are violent conflicts between societies; therefore, the nature of societies must be understood. The roots of a victory or defeat must often be sought among a multitude of social, political, economic, religious, and other factors. It is critically important that students of military history understand the political framework for a given war.

Military history must be more than a logical, factual record or account of events. After the facts have been synthesized, they should be evaluated and conclusions made. And then based on those conclusions, some operational lessons learned or identified should be derived; otherwise, there is very little utility in studying military history. Yet there are also pitfalls in the process. Conditions that existed in the past cannot be literally transferred to the present or future. One thing is always unchangeable: all conditions of a given situation can rarely be reproduced.[77] There is a great danger in studying a certain episode from a past campaign/major operation and then trying to elevate lessons learned into a dogma, as Schlieffen did with his study of the battle of Cannae in 216 BC. He reportedly said that military historical studies show how "everything became and will become." Such and similar views of the use of historical case studies as recipes for future actions should not be deduced from the study of military history.[78]

The study of past wars should be methodical and long-term. All sources should be used, from primary documents to official histories, biographies and memoirs, and even historical novels. In studying military history, it is more valuable to focus research and analysis on a single area than on a very broad area.[79]

Views on the importance and value of using selected examples from military history in illustrating certain aspects of operational warfare vary greatly. Clausewitz insisted that ancient historical examples should not be considered but only those from modern military history, insofar as it is properly known and evaluated. In his view, the conditions and ways of warfare in more distant times were different; hence, such examples are of little use in one's own era. The further back one goes, the less useful military history is. Also, with the passage of time such historical examples are bound to lose a mass of essential minor elements and details, and the picture becomes less accurate.[80] However, this is only true regarding tactical and technological lessons deduced from history. The value of the human and psychological aspects of warfare is not limited to any era or historical event; they are timeless – and Clausewitz, for one, might also agree.

Conclusion

The future operational commanders and their staffers must think broadly and beyond the domain of physical combat. Operational thinking is the very foundation of the commander's operational vision: an ability to think far ahead of current events, accurately anticipate the enemy's actions and reactions, make quick but sound decisions, and then energetically execute plans and orders. Operational commanders should fully comprehend both the military and nonmilitary aspects of the situation. The lack of operational thinking not only invariably leads to setbacks in combat but might fatally undermine the prospect of a decisive and quick victory. Many operational commanders have possessed the highest intellect, strength of character, courage, and boldness, and yet failed, simply because they never raised their sights beyond tactics. Tactical excellence has not been in the past, nor will it be in the future, a substitute for operational thinking.

The operational commanders must fully understand the highly dynamic relationship among the three components of the art of war. They need to know and understand the distinctions among the levels of war, and how decisions and actions at one level affect the other levels. They also must properly evaluate the operational features of the physical environment in which a campaign or major operation will be conducted. The outcome of a campaign/major operation depends considerably on the synchronization of the available military and nonmilitary sources of power. Future operational commanders must properly evaluate both the military and nonmilitary aspects of the situation in a theater.

Operational thinking is not an inborn quality. To acquire that broad perspective, future operational commanders must exert great and consistent efforts over most of their career. Military institutions, as well as societal and cultural influences, both provide a framework and shape the commander's operational thinking. The opportunities to acquire an operational perspective in war are relatively rare for any officer. It is even rarer for commanders who hold the operational level of command in peacetime. Hence, indirect experiences, education and training are the principal means of acquiring operational thinking. The more diverse and comprehensive the training, the more opportunities it will present for the future operational commanders to acquire that broad perspective necessary for success at the operational and strategic levels of command. One of the principal sources of indirect experience is the study of military history. Future operational commanders and staffers should read widely and in great depth on all aspects of warfare. The focus should be on the study of campaigns and major operations. The study of history is of little utility for the future operational commander unless he tries to derive from it some operational lessons learned.

Military history is replete with examples of operational commanders who did not fully measure up to the high demands of their office. Only relatively few commanders were both brilliant operational thinkers and practitioners. Many excellent thinkers fell short as practitioners because of defects in character or because they lacked some key personality traits necessary to successfully convert their knowledge of operational art into practice. Other operational commanders were promoted above their true abilities. They did not perform well at the operational level of command because they lacked the necessary education and/or character for such high levels of command. Many operational commanders performed poorly because they were unable or unwilling to think beyond the domain of tactics. Other commanders had a mixed record. They showed a remarkable ability to think and act operationally at some point in their career but then performed poorly either through an inability to delegate authority to their subordinate commanders or by trying to be both operational and tactical commanders. Military history cannot, of course, provide a clear answer as to what are the key prerequisites for successful performance at the operational level of command. However, future operational commanders should be both thinkers and practitioners. They should possess a rare combination of high professional education, diverse training, combat experience, and the great military and personality character traits necessary for success at the operational level of command.

Notes

1 David Jablonsky, "Strategy and the Operational Level of War: Part I," *Parameters* (Spring 1987), 71.

2 Operational thinking is not identical to what the information warfare advocates call "situational awareness" (SA) – a term used in training one's pilots; in its strict definition, situational awareness refers to the degree of accuracy with which one's perception of his current environment mirrors reality; situational awareness does not necessarily also mean an understanding; it is purely a tactical, not operational or strategic, term; the extensive use of the term situational awareness in the U.S. and other militaries is perhaps one of the best proofs of the predominance of a narrow tactical perspective among information warfare advocates.

3 Jablonsky, "Strategy and the Operational Level of War: Part I," 71.

4 Maurice de Saxe, *My Reveries Upon the Art of War*, translated from the French by Thomas R. Phillips, in *Roots of Strategy: A Collection of Military Classics* (Harrisburg, PA: Military Service Publishing, 1940), 295–96.
5 General Johann von Kielmansegg, cited in Norbert Hanisch, *Untersuchen Sie die operativen Ideen Manstein hinsichtlich Schwerpunkt-bildung, Überraschung, Initiative und Handlungsfreiheit an den Beispielen Westfeldzug 1940 (Sichelschnitt-Plan) und Operation Zitadelle* (Hamburg: Führungsakademie der Bundewehr, 15 January 1988), 4.
6 Carl von Clausewitz, *Vom Kriege*, ed. Dr. Werner Hahlweg, 16th ed. (Bonn: Ferdinand Dümmler Verlag, 1952), 874.
7 Hanisch, *Untersuchen Sie die operativen Ideen Manstein hinsichtlich Schwerpunktbildung*, "Motto."
8 Daniel J. Hughes, ed., *Moltke on the Art of War: Selected Writings* (Novato, CA: Presidio, 1993), 184.
9 Gérard Chaliand, *The Art of War in World History: From Antiquity to the Nuclear Age* (Berkeley/Los Angeles/London: University of California Press, 1994), 768.
10 Anonymous, "The Place of Doctrine in War," *Edinburgh Review*, January 1912, 13.
11 Anonymous, "The Place of Doctrine in War," 9–10.
12 *Napoleon on the Art of War*, selected, ed. and trans. Jay Luvaas (New York, NY: Free Press, 1999), 68–69.
13 Hugo von Freytag-Loringhoven, *Die Macht der Persőnlichkeit im Krieg. Studien nach Clausewitz* (Berlin: E. S. Mittler und Sohn, 1911), 30.
14 Anonymous, "The Place of Doctrine in War," 8–9.
15 Edward N. Lutwak, "The Operational Level of War," *International Security* Winter 1980/81, 62.
16 David Jablonsky, "Strategy and the Operational Level of War: Part I," *Parameters* (Spring 1987), 71.
17 Jablonsky, "Strategy and the Operational Level of War: Part I," 72; Erich von Manstein, *Lost Victories* (Novato, CA: Presidio Press, 1982), 409.
18 William Slim, "Higher Command in War," *Royal Air Force Quarterly* 2 (April 1953), 149.
19 Michael D. Krause, "Moltke and the Origins of the Operational Level of War," in *Generalfeldmarschall von Moltke,* ed. Roland G. Foerster (Munich: R. Oldenbourg Verlag, 1991), 145.
20 Krause, "Moltke and the Origins of the Operational Level of War," 149.
21 John E. Schlott, *Operational Vision: The Way Means Reach the End* (Fort Leavenworth, KS: School of Advanced Military Studies, U.S. Army Command and General Staff College, May 12, 1992), 10.
22 Gregory C. Gardner, *Generalship in War: The Principles of Operational Command* (Fort Leavenworth, KS: School of Advanced Military Studies, U.S. Army Command and Staff College, May 4, 1987), 24.
23 Cited in Jablonsky, "Strategy and the Operational Level of War: Part I," 72.
24 Hugo von Freytag-Loringhoven, *Die Macht der Persőnlichkeit im Krieg. Studien nach Clausewitz* (Berlin: E. S. Mittler und Sohn, 1911), 290, 252; von Freytag-Loringhoven, *Die Macht der Persőnlichkeit im Krieg. Studien nach Clausewitz*, 151.
25 Alfred Gause, "Command Techniques Employed by Field Marshal Rommel in Africa," *Armor* 4 (July–August 1958), 23.
26 Cited in Armstrong Starkey, *War in the Age of Enlightenment 1700–1789* (Westport, CT: Praeger, 2003), 48.
27 Gardner, *Generalship in War*, 22–23.
28 Roger H. Nye, "Whence Patton's Military Genius," *Parameters* Winter 1991–92, 71.
29 Martin Blumenson, "Essence of Command: Competence, Iron Soul," *Army* 3 (March 1993), 42.
30 Lothar Rendulic, *Grundlagen militärischer Führung* (Herford/Bonn: Maximilian Verlag, 1967), 50.

31 Robert W. RisCassi, "Doctrine for Joint Operations in a Combined Environment: A Necessity," *Military Review* (June 1993), 24.
32 Martin Blumenson and James L. Stokesbury, *Masters of the Art of Command* (Boston, MA: Houghton Mifflin, 1975), 3.
33 Karl-Heinz Frieser, *Blitzkrieg-Legende. Der Westfeldzug 1940* (Munich: R. Oldenbourg Verlag, 1995), 416.
34 Volker Wieker, *Operation, Operative Führung zwischen Taktik und Strategie – Bedeutung und Entwicklung des Begriffes seit Moltke d. Ä. – Bearbeitung aus der Sicht der Landstreitkräfteführung* (Hamburg: Führungsakademie der Bundeswehr, 12 December 1988), 14.
35 Stephen van Evera, "The Cult of the Offensive and the Origins of the First World War," in *Offense, Defense, and War*, ed. Michael E. Brown et al. (Cambridge, MA: MIT Press, 2004), 69–73.
36 Günther Blumentritt, *Das alte deutsche Heer von 1914 und das neue deutsche Heere von 1939 (eine vergleichende Urteil)*, B-296, ZA/1 646, Studien der Historical Division Headquarters, United States Army Europe, Foreign Military Studies Branch, Bundesarchiv-Militärarchiv (BA-MA), Freiburg, i.Br., January 1947, 21–22.
37 Williamson Murray, "Does Military Culture Matter," in *America the Vulnerable: Our Military Problems and How to Fix Them*, ed. John Lehman and Harvey Sicherman (Philadelphia, PA: Foreign Policy Research Institute, 2002), 139.
38 Freiherrn von Freytag-Loringhoven, *Die Verwertung Kriegsgeschichtlicher Erfahrungen* (Berlin: E. S. Mittler & Sohn, 1925), 24; Günther Blumentritt, *Das alte deutsche Heer von 1914 und das neue deutsche Heere von 1939*, January 1947, 28.
39 Murray, "Does Military Culture Matter," 136.
40 Cynthia A. Roberts, "Planning for War: The Red Army and the Catastrophe of 1941," *Europe-Asia Studies* 47, no. 8 (1995), 1321.
41 Militärgeschichtliche Forschungsamt, ed., *Development, Planning and Realization of Operational Conceptions in World Wars I and II* (Herford and Bonn: Verlag E. S. Mittler & Sohn, 1989), 22.
42 Günter Roth, "Operatives Denken bei Schlieffen und Manstein," in *Ausgewählte Operationen und ihre militärhistorischen Grundlagen*; Series *Operatives Denken und Handeln in deutschen Streitkräfte*, published under the auspices of Militärgeschichtlichen Forschungsamt, ed. Hans-Martin Ottmer and Heiger Ostertag (Bonn/Herford: Verlag E. S. Mittler & Sohn, 1993), 289–90.
43 B. H. Liddell Hart, *Strategy* (New York, NY: F.A. Praeger, 2nd printing 1954), 23–24.
44 von Freytag-Loringhoven, *Die Macht der Persőnlichkeit im Krieg. Studien nach Clausewitz*, 83.
45 Wilhelm Meier-Dőrnberg, "Moltke und die taktische-operative Ausbildung im preussisch-deutschen Heer," in *Generalfeldmarschall von Moltke. Bedeutung und Wirkung*, ed. Roland G. Foerster (Munich: R. Oldenbourg Verlag, 1991), 46.
46 Vad, "Operative Führung: Grundlagen, Merkmale und Perspektiven," 130.
47 Rudolf Hofmann et al., *War Games*, MS # P-094, ZA 1/2014, Studien der Historical Division Headquarters, United States Army Europe, Foreign Military Studies Branch, BA-MA, 1952, 6.
48 Rudolf Hofmann et al., *War Games*, MS # P-094, ZA 1/2014, 6–7.
49 von Freytag-Loringhoven, *Die Macht der Persőnlichkeit im Krieg. Studien nach Clausewitz*, 82.
50 Liddell Hart, *Strategy*, 23–24.
51 von Freytag-Loringhoven, *Die Macht der Persőnlichkeit im Krieg. Studien nach Clausewitz*, 85–86.
52 Hermann Foertsch, *Auswahl, Erziehung und Ausbildung zum Generalstabsoffizier. Eine kritische Betrachtung der deutschen Verhältnisse vor dem 2. Weltkrieg*, MS # P-031b, ZA/1 1856, Studien der Historical Division Headquarters, United States Army Europe,

Foreign Military Studies Branch, BA-MA, 30; Wilhelm Speidel, *Der Weg zum Generalstab, Auswahl, Erziehung u. Ausbildung des deutschen Generalstabsoffiziere, German General Staff, Project #6: Training and Development of German General Staff Officers, P-031a/26, vol. 26: Part II, ZA/1 1875,* Foreign Military Studies, Historical Division, Headquarters U.S. Army, Europe, BA-MA, 1950, 31.

53 L. D. Holder, "Educating and Training for Theater Warfare," *Military Review* (September 1990), 87, 89, 91.

54 Charles G. Sutten, "Command and Control at the Operational Level," *Parameters* (Winter 1986), 19.

55 Krause, "Moltke and the Origins of the Operational Level of War," 155.

56 B. H. Liddell Hart, *Why Don't We Learn from History* (London: Allen & Unwin, 1944), 7.

57 Herbert Richmond, *National Policy and Naval Strength and Other Essays* (London/New York/Toronto: Longmans, Green, 1934), 279.

58 John McDonald, *General Walter Krueger: A Case Study in Operational Command* (Fort Leavenworth, KS: School of Advanced Military Studies, U.S. Army Command and General Staff College, May 5, 1989), 11.

59 Liddell Hart, *Why Don't We Learn from History*, 7.

60 Liddell Hart, *Why Don't We Learn from History*, 7.

61 Andreas Broicher, "Wie alles gekommen ist. Gedanken zur Bildung militärischer Führer auf dem Gebiet der Militärgeschichte," *Truppenpraxis* 2 (February 1995), 132.

62 Paul M. Robinett, "History and the Military Profession," *Military Review* 5 (May 1954), 17.

63 Richmond, *National Policy and Naval Strength and Other Essays*, 289.

64 Roland G. Foerster, *Generalfeldmarschall von Moltke: Bedeutung und Wirkung* (Munich: R. Oldenbourg Verlag, 1991), xi.

65 Nye, "Whence Patton's Military Genius," 71.

66 Robinett, "History and the Military Profession," 21.

67 Ebbe Gyllenstierna, "What Can the Study of the Art of War Teach Us?" *Ny Militar Tidscrift*, no. 2–3, 1956; trans. *Military Review* (August 1956), 104.

68 Ralph L. Allen, "Piercing the Veil of Operational Art," *Parameters*, no. 4 (Winter 1986), 26.

69 Robinett, "History and the Military Profession," 22.

70 James A. Huston, "The Uses of History," *Military Review* (June 1957), 27.

71 Huston, "The Uses of History," 28.

72 Huston, "The Uses of History," 27.

73 Max Kemmerich, "Gedanken über die eigene Kriegsgeschichtliche Weiterbildung der Offiziere nach Verlassen der Kriegsakademie," Freiburg, i.Br.; RH 16/v. 95, Bundesarchiv-Militärarchiv (BA-MA), Freiburg i.Br., May 1939, 7.

74 Robinett, "History and the Military Profession," 22.

75 Gyllenstierna, "What Can the Study of the Art of War Teach Us?" 105.

76 Gyllenstierna, "What Can the Study of the Art of War Teach Us?" 104.

77 Huston, "The Uses of History," 26.

78 Broicher, "Wie alles gekommen ist, 132.

79 Kemmerich, "Gedanken über die eigene Kriegsgeschichtliche Weiterbildung der Offiziere nach Verlassen der Kriegsakademie," 10.

80 Jehuda L. Wallach, *The Dogma of the Battle of Annihilation. The Theories of Clausewitz and Schlieffen and Their Impact on the German Conduct of Two World Wars* (Westport, CT/London: Greenwood Press, 1986), 74.

10 Tenets

Success in combat is considerably enhanced when the operational commanders use certain general truths or tenets in the employment of their forces. This term is used here instead of other terms such as maxims, axioms or principles, because it is broader and less rigid. In contrast to principles of war, application of tenets would not lead to violation of other tenets. The French Marshal Maurice de Saxe was critical for blindly following maxims, without any examination of the principles on which they were founded. He observed: "our present practice is nothing more than a passive compliance with received customs to the grounds of which we are absolutely strangers."[1] Moltke, Sr warned of the danger in conducting war by rules. For him, war was an art that could be learned only empirically. There are no general norms in the conduct of war.[2] Rear Admiral Alfred T. Mahan, in referring to the principles of war, wrote that the tendency to accept principles without hesitation and often with too little reflection is a source of the greatest danger. In his view, once "embedded in naval instructions or army field regulations these so-called principles bound the minds of officers in unbreakable bonds."[3] However, ignoring or poorly applying these tenets has been the cause of many failures.[4]

The tenets of operational leadership in combat are derived from their application in many wars of the past eras and those fought more recently. Yet regardless of national and service differences in the conduct of operational warfare, experience has shown that violating or failing to observe certain tenets of operational leadership greatly complicates the planning and execution of a campaign/major operation and often leads to its failure.

The most important tenets of operational leadership in combat are as follows:

- Absolute primacy of policy and strategy.
- Firm and unwavering focus on the objective.
- Obtaining & maintaining freedom of action.
- Exercising the initiative.
- Using overwhelming power.
- High speed of action.
- Taking high but prudent risks.
- Achieving surprise.
- Creativeness.

DOI: 10.4324/9781003333685-10

Absolute Primacy of Policy and Strategy

The employment of national/coalition military sources is dictated by policy and strategy. Operational commanders must in all circumstances accept the absolute primacy of policy and strategy in planning their campaigns/major operations. Policy dominates strategy, and strategy dominates operational art and tactics. It determines the objectives of national or alliance/coalition strategy and military strategy. Policy should never demand what is militarily impossible.

Political leadership should allocate adequate resources for accomplishing the assigned military strategic objectives. It determines political, diplomatic, legal, and other limitations on the employment of one's military forces.

In determining strategic objectives, the political leadership should ask for the advice of the military leadership. At the same time, the highest military leadership should have sufficient freedom of action to accomplish assigned strategic objectives. The military leadership has the responsibility to point out to the political leaders clearly and forcefully the possibilities, risks, boundaries, and consequences of the available courses of action.[5]

The military strategy should create conditions for the application of operational art, developing strategic concepts, determining political and military strategic objectives, and determining the framework for the strategic employment of national military power. The operational commander should be skillful in converting political objectives into achievable military-/theater-strategic objectives.[6] When the primacy of policy over the military is well established, the military leadership can act only within the framework determined by national policy and strategy. The greater the limitations on the military leadership's freedom of action, the larger the combat potential (not necessarily the larger the force) that is required to accomplish the stated political objectives.

The employment of military instruments should be well thought out and integrated with other sources of national power (political, diplomatic, economic, financial, informational, etc.).[7] The military and political leadership should work together to determine specific militarily achievable strategic objectives. The responsibility of the highest military leaders is to make sure that the political decision-makers assign clear and realistic military-strategic objectives – means and ends must be in harmony.

Firm and Unwavering Focus on the Objective

The operational commander should have a firm and unwavering focus on the accomplishment of the ultimate objective of a campaign/major operation. All the efforts of the operational commander should be devoted to that end.[8] The objective, once selected, should be adhered to until accomplished. Yet the initially selected objective should be changed, modified or even abandoned if the situation radically changes so that the objective cannot be accomplished with the resources on hand or becoming available. The operational commanders should realize that there is always more than one way to accomplish the objective.[9] They should pursue the ultimate objective with all determination. They should not be distracted by some relatively insignificant event and shift or even abandon their original objective. Often in a dynamic combat situation, rigid adherence to the original objective can lead to a major setback or even the failure of the entire campaign or major operation.

In a high-intensity conventional war, the main objective should be to destroy or annihilate the enemy forces, not to capture or retain territory, the enemy's capital, or some economically important area. The aim should be to impose one's will on the enemy and to break his combat power.[10] Single-mindedness in pursuing one's objective can sometimes adversely affect the operational commander's creativity and reduce the range and the number of his options.[11] The destruction or annihilation of the enemy usually should be the main objective of a war. Many roads lead to that objective. A campaign or major operation should revolve around a simple operational idea, and everything should be subordinated to it.

There is no greater mistake than to formulate and pursue several operational or strategic objectives simultaneously. The pursuit of multiple objectives can often unhinge the entire balance of the factors of time, space, and forces. Such a course of action can be taken only if one's forces possess overwhelming power against any conceivable combination of the enemy forces. Hitler constantly frittered away Germany's scarce resources because he pursued several strategic or operational objectives simultaneously and often in several theaters of war. He never grasped that one can never be too strong at the crucial point – that one may even have to dispense with less vital fronts or accept the risk of radically weakening them to achieve a decisive objective. Hitler also could not bring himself to stake everything on success. He was neither able nor willing to see what action would be necessary to compensate for the unfavorable turn events then took.[12]

A prudent operational commander should try to have an *alternate objective* in case the main objective cannot be accomplished. The famous British military theoretician B. H. Liddell Hart wrote that by having an alternate objective the commander "will put the enemy on the horn of dilemma, which goes far to assure the chance of gaining one objective at least, whichever he guards least, and may enable you to gain one after the other." An alternate objective will also allow one

> to keep the opportunity of gaining an objective while the single objective, unless the enemy is helplessly inferior, means the certainty that you will not gain it. Once the enemy is no longer uncertain as to your aim, there is no more common mistake than to confuse a single line of operations, which is usually wise, with a single objective, which is usually futile.[13]

Obtaining & Maintaining Freedom of Action

One of the most, critical factors for the operational commander's success is obtaining and maintaining freedom of action.[14] Without freedom to act, the commander's ability to exercise the initiative is lost.[15] The area of the operational commander's actions should be at least large enough to always give him several options from which to choose. Conversely, the operational commander should do everything possible to reduce the enemy's freedom to act from the very outset, and to derail the enemy's plans.[16]

Ideally, the operational commander should have a large freedom to act prior to, and during the accomplishment of military-/theater-strategic and operational objectives. In practice, freedom of action is not absolute but is subject to certain military and non-military limitations. These limitations dictate whether the operational commanders have a larger or smaller room within which they can operate freely. The more limitations on the operational commander's freedom of action, the fewer the means and ways the highest political leadership will have for accomplishing their stated political objectives.[17]

The operational commanders should allow their subordinates freedom to act, as long as that does not adversely affect their overall intent. The U.S. Admiral Ernest J. King, Chief of Naval Operations (CNO)/Commander-in-Chief, United States Fleet (COMINCH), generally did not interfere in the responsibilities of subordinate flag officers. He disciplined himself not to press his combat commanders for information while the operation was in progress. King would not send communications; as he put it, he would not interfere, because they needed all the communication, they could use to coordinate their work effectively.[18] Admiral Nimitz left subordinate commanders alone, believing that looking over their shoulders would only inhibit them. At the same time, he made recommendations via radio messages if it appeared that they were overlooking opportunities for success. Nimitz did not hesitate to intervene and, if necessary, fire a subordinate commander who did not perform well.[19]

The assigned objectives are the most critical factors in determining the boundaries within which the operational commander is allowed to act freely. The more general the objective, the fewer the limitations the higher commander gives, and therefore the greater the freedom of action for the subordinate commanders. Scaling down the original strategic or operational objective or changing the sequence of objectives may result in greater freedom to act for the operational commander. The use of mission vice directive orders to subordinate commanders will enlarge their freedom to act.

To allow freedom of action for subordinates, the operational commanders should first obtain maximum freedom of action for themselves from the top political leadership. Their freedom to act is ultimately built on the trust of the government. Perhaps the greatest threat to the operational commanders' freedom of action is the political leadership's ability, through information technology, to unnecessarily interfere in their decisions and actions. The political leadership should resist such tendencies and make every effort to ensure that operational commanders maintain an appropriate degree of authority and flexibility to respond quickly to changes in the operational or strategic situation.

The larger the factor of space, the larger the freedom of action for an operational commander. A large space gives the operational commander the possibility of fully using the mobility, range, and lethality of subordinate forces. Freedom of action is greatly enhanced by creating an operational reserve. Other methods include withdrawing friendly forces from untenable positions so that the operational commander can release or regroup forces and using the factor of space purposefully.[20] Field Marshal Erich von Manstein obtained operational freedom of action prior to his brilliant counterstrike in southern Russia in February–March 1943 by an orderly withdrawal from the Donets River near Rostov to the lines behind the Mius River. That move shortened his front from 248 to 112 miles and freed the 1st and 4th Panzer armies for mobile operations. The 4th Panzer Army was then redeployed from the right flank of Manstein's Army Group Don to the left flank. Not only was the new front line successfully defended but Manstein also inflicted a major defeat on the advancing and numerically superior Soviet forces and recaptured Kharkiv in March 1943.[21]

Generally, the larger and more capable the forces at hand, the larger the space the operational commander can use to act freely and impose one's will on the enemy. The availability of reinforcements or operational/strategic reserves can significantly increase the operational commander's freedom of action. The operational commander also must have sufficient freedom of action to exploit opportunities or to meet threats that endanger the accomplishment of the assigned mission. This requires applying the principle of economy of effort and having operational reserves.[22]

Too-rigid C2 invariably limits the commander's freedom to act. Likewise, too many or too few command echelons reduce the operational commander's freedom of action. A very poor command climate can reduce freedom of action for subordinate commanders. Thus, it is critical that the operational commander establishes a command structure that clearly defines overall command responsibilities Afterward, the operational commander should give subordinates sufficient authority to discharge their responsibilities.

Accurate and, above all, relevant and timely intelligence is the key element for making a timely and sound decision. Inadequate logistical support and sustainment are probably among the most critical factors that reduce the operational commander's freedom to act. The need to provide theaterwide protection of military infrastructure and the civilian population reduces the forces available for force protection and for protection of friendly center of gravity.

Prescriptive service/joint doctrine invariably limits the operational commander's freedom to act, whereas descriptive doctrine increases it. A sound service/joint doctrine should not prescribe the number and names of phases in a campaign, deployment and logistical support and sustainment of combat forces, and organizational options. It should provide only broad guidance, leaving details to the operational commander; otherwise, his freedom of action will be further reduced, with all the attendant dangers to his performance in combat.

The shorter the decision cycle, the more time is available for the commander and his subordinates to act. Freedom to act is generally enlarged by having a larger and more diverse number of highly advanced weapons and equipment. The greater the range and lethality of weapons, the greater the freedom to act. However, technology can also be a limiting factor. The use of some weapons (WMD, dumb land mines, cluster bombs, napalm bombs, antipersonnel ammo, etc.), while militarily effective, is likely to be politically and morally prohibitive for many militaries. Sometimes the political leadership retains tight control even over the use of very advanced conventional weapons, and this considerably reduces the operational commander's freedom of action.

Exercising the Initiative

Freedom of action creates the conditions for exercising *initiative* on the part of subordinate commanders. The more freedom of action given to subordinate commanders, the more room they would have to exercise initiative; and the more initiative is exercised, the more freedom of action would be created for the operational commander. Initiative is the ability to set or change the terms of combat through action. It requires that commanders maintain an aggressive spirit. It means that, when an opportunity presents itself, the commander may depart from planned actions to speed up mission accomplishment. To do this without jeopardizing the higher plan, the commander must understand and act within the intent of the commander two levels up. This implies both a degree of independence and risk.

Field Marshal Moltke, Sr, wrote

> In general, in doubtful cases and in unclear circumstances, as so often exist in war, it will be more advisable to proceed actively and to maintain one's own initiative than to expect the law from the enemy. The leaders of the individual branches of the army must adhere to the old rule of marching in the direction of the thunder of cannon.[23]

Field Marshal Manstein observed that the operational commander should always conduct operations flexibly and resourcefully and should give all possible scope to the initiative and self-sufficiency of subordinate commanders at all levels.[24]

The initiative can be achieved through the agility of C2 and through the decisive exploitation of operational surprises and preemptive actions.[25] By exercising the initiative, the operational commander can dictate the terms of combat to the enemy.[26]

The German army's *Troop Command and Control* (*Truppenführung*) of 1933 emphasized that no plan, no matter how detailed, can foresee every contingency, development, vulnerability, or opportunity that will arise in a battle. In fact, a plan can become so detailed that it has the reverse effect of limiting or restraining initiative.[27] The Germans highlighted that what they called "the emptiness" of the battlefield demands fighters who think and act independently and who master and decisively exploit each situation. Mobility allows a better and deeper exploitation of space, but at the same time it requires linkage to the factor of time. This exploitation of space leads to a greater emptiness of the battlefield, and the emptiness in turn requires more initiative on the part of the commander. In unclear situations the commander must ensure freedom of action for their subordinates.[28]

Agility is the first prerequisite for seizing and holding the initiative. It is one's ability to think and act faster than the enemy. It involves mental, C2, and organizational abilities to adjust rapidly and to use the situation, terrain, and weather to defeat the enemy. Agility derives from a keen sense of what is happening in combat. It means that the operational commander should have the ability to shift quickly from one situation to the next. The operational commander should have the physical and mental ability to always have more options than the enemy has.[29]

Using Overwhelming Power

The key prerequisite for success in combat is to be stronger than the enemy at the decisive point and to use speed, surprise and deception.[30] Moltke, Sr, advised to use in a major battle whatever forces are available. No one can have too many forces at a place and not too many chances for victory.[31] In all circumstances one should use all the sources of power available or becoming available. One cannot have too great a chance of victory.[32] One is never too strong in war if the aim is to achieve a quick and decisive victory.[33] The mass of one's forces should be brought to the decisive point.[34] The commander who tries to be strong everywhere or who wastes his forces on secondary missions acts contrary to this basic rule.[35]

Today, a force does not necessarily need to be physically massed, but one should create a mass effect at a decisive place and time. It is the successful application of overwhelming military force that will seize the initiative from the enemy and defeat him. Maximum combat potential or combat power at the decisive time and place can be achieved only with the proper arrangement of actions by all available forces. Clausewitz observed that achieving strength at the decisive point depends on the strength of the army and the skill with which this strength is employed. The aim, then, is to take the field in the greatest possible strength, either to get the upper hand or at least to make sure the enemy does not.[36] One of the reasons for the Germans' success in their invasion of France in May 1940 was that in the sector of main effort in the Ardennes they concentrated 42 divisions against only nine poorly equipped and trained French divisions.[37]

One of the most critical elements in the employment of combat forces in a major operation or campaign is the selection of the proper sector of main effort and point of the main attack (defense). This is the operational commander's main means of ensuring decisive action. Surprise is one of the most important factors in this regard.[38] Concentration of combat forces in the sector of main effort and the point of the main attack or defense is the most important, but also the most difficult, task. It requires the operational commander's willingness to take high risks to weaken forces in other sectors.[39] All available forces should be concentrated against the part of the enemy force where the decision is being sought.[40] Concentration of forces in the sector of main effort/point of the main attack (defense) is a critical prerequisite for the success in combat.[41]

The operational commanders should not arbitrarily decide what force size and mix should be employed to accomplish the assigned operational or strategic objective. There is simply no such thing as being strong enough.[42] In a war one should fight not a fair fight but an unfair fight. Napoleon I always strove for the annihilation of his enemies. An overwhelming victory could come only by using one's overwhelming strength. Admiral Nelson said that only numbers can annihilate. Neither Napoleon I nor Nelson thought in terms of total strength superiority. What they aimed at was to employ their available forces so that they could fall in overwhelming force on a portion of the enemy and, having defeated it, do the same to some other part.[43]

The Austrian Archduke Charles observed that

> in war, the predominant power is victorious, whether by majority, art, or bravery. In order to achieve a decisive superiority with the most probable success, the greatest possible mass of forces must be mobilized and used united on the decisive point, so that as much as the mutual relationship permits, one's own strength is increased, and the enemy's is paralyzed.[44]

He insisted that the main art of a general was

> to judge correctly what was the moment, what was the point at which such decisive blows can be delivered with the greatest possible probability of a happy success. It is only by superiority on such a point that such a decisive blow can be struck.[45]

A weaker force can become stronger at the decisive point through speed, mobility, the use of darkness, terrain, surprise and deception.[46] In practice, this means to "mass" one's forces in the sector of main effort and assign relatively weak forces in the sectors of secondary effort.[47] Today, one needs not necessarily to physically mass a force but to have a mass effect created at a decisive place and time. It is the application of overwhelming military force that will seize the initiative from the enemy and, when applied successfully, defeat him. Maximum combat potential or combat power at the decisive time and place can be achieved only with the proper arrangement of actions by all available sources of power.

All available forces should be concentrated against the part of the enemy at which the decision is being sought.[48] It is a big mistake to believe that new technologies can compensate for a lack of numbers. A fight against overwhelming odds can be conducted only if superiority in manpower and material is being concentrated against one part of the enemy.[49] The British general and theoretician Henry H. E. Lloyd observed that it is an axiom that "you ought to bring as many men into action at once as possible."[50]

High Speed of Action

Speed and surprise have been among the most important tenets of warfare in all eras. Inferiority in numbers and weapons should be rectified through higher speed of action. All highly successful commanders, like Alexander the Great, Hannibal, Julius Caesar, Genghis Khan, Frederick the Great, Napoleon I and Moltke, Sr, to name a few, emphasized the critical role of speed of action as the main prerequisite for success. Caesar's campaigns were based not on numerical superiority but on celerity and audacity. By surprising his opponent, he caught him unguarded and so thoroughly rattled that he either refused to fight or was morally half-broken before the engagement took place. Caesar reportedly said to his legions: "I consider rapidity of movement the best substitute for all these things," and "the most potent thing in war is the unexpected."[51] Frederick the Great paid great attention to the factor of speed in a tactical and operational sense. This was the reason for many of his major victories over the Austrians. Archduke Charles wrote that

> success of which depends chiefly on the speed of decision and execution, and which can be applied according to the proportion of friendly forces and situation in which one finds oneself, as a march to the point which the enemy wants to reach in order to prevent him, an attack while he is in motion, and advance on his line of communications, whereby one's own (line of communications) must not be endangered.[52]

Napoleon I insisted that waging war energetically and with severity was the only way to make it shorter.[53] He believed that the strength of an army, like power in mechanics, is estimated by multiplying mass by rapidity: a rapid march augments the morale of an army and increases all the chances of victory.[54] Napoleon I emphasized speed to compensate for what he may have lacked in numbers of troops. The rapidity of his marches allowed him to concentrate the larger force on the eve of a battle even if his opponent had overall numerical superiority. Prior to the battle of Austerlitz (Slavkov today) in December 1805, divisions under Marshal Louis-Nicolas Davout advanced 87 miles in just 48 hours to join Napoleon I on the battlefield. The high speed of movement and high tempo were hallmarks of Napoleon I's style of warfare.[55] Clausewitz wrote that the second principle is

> the rapid use of your forces; any unnecessary expenditure of time, every unnecessary detour is waste of strength and thus abhorrent to strategic thought. It is still more important to remember that almost the only advantage of the attack rests on its initial attack. Speed and impetus are its strongest elements and are usually indispensable if we are to defeat the enemy. Thus, theory demands the shortest roads to the goal.[56]

The high speed of action should not be carried to the extreme as Field Marshal Alfred von Schlieffen did. He overemphasized the importance of speed in the conduct of major operations and campaigns.[57] Schlieffen's plan (actually a memorandum) of 1905 for a campaign against France and Belgium envisaged defeating a large French army quickly in a single blow and then shifting the weight of effort east against Russia. The campaign would consist of a single phase; hence, no operational pause was planned.[58] General Friedrich von Bernhardi, a well-known German theoretician and Schlieffen's contemporary, was highly critical of Schlieffen's excessive focus on the value of the double envelopment and his underestimation of the penetration maneuver. Another problem was that Schlieffen rejected the idea of having reserves behind the front line. This, combined with the insufficient maneuverability of the German forces, led to difficulties in meeting unexpected changes in the situation and exploiting sudden favorable opportunities during the plan's execution in August 1914.[59]

High speed of action allows the rapid concentration of one's strength against enemy vulnerabilities. This should be done repeatedly, so that by the time the enemy reacts to one action, another has already taken its place, disrupted the enemy's plans and led to late, uncoordinated, and piecemeal enemy responses. It is this process of successive concentration against locally weaker or unprepared enemy forces that enables smaller forces to disorient, fragment, and eventually defeat a much larger hostile force.[60] Mobility, flexible task organization, timely intelligence, and rapid decision-making enable the operational commander to cause the enemy to react; thus, the commander can set the terms of the battle or major operation. Agility can be attained by achieving surprise, the availability or creation of an operational reserve, the trading of space for time, and the operational employment of air power or naval forces. Surprise is a decisive factor in success. However, actions based on surprise are successful only if the enemy is given no time to take effective counteraction.[61]

Taking High but Prudent Risks

The higher the command echelon, the larger the area of uncertainty, and the higher the risks the commander should take. Despite all the advances in information technologies, there will always be a rather large area of uncertainty in any given situation. The operational commanders would rarely possess complete knowledge of the intangible factors in a situation. Moreover, very often they would be forced to make operational decisions without waiting for complete information on all essential elements of the situation. Operational decision-making is inherently based on taking high but prudent risks. The uncertainties regarding the enemy's intentions are much greater at the strategic and operational levels than at the tactical level.[62] The consequences of a failure at the operational level are much more severe than at the tactical level, and they cannot be overcome easily, if at all. At the same time, however, the potential gain is much greater at the operational and strategic levels of command than at the tactical level.

A willingness to take calculated risks has distinguished all the great military leaders of the past. The attempt to fight a safe battle without taking risks has never been successful. The doctrine that nothing must be left to chance has never resulted in a decisive victory. Admiral Nelson was always taking risks. Those who take big risks in a war nearly always seem to have luck on their side.[63] No great successes in a war can be achieved without taking calculated risks. At the same time, there is also no rule for weighing the risks to be taken. This is only a partially quantifiable problem. It cannot be resolved by technology, no matter how advanced. Taking a calculated risk depends on the character of the operational commander and his/her artistic activity.[64] It also depends greatly on the operational commander's experience and professional judgment.

In developing a plan, the operational commanders must assess the risk involved in the operation and then must make a decision to execute the plan despite the risk.[65] Willingness to take prudent risks means making operational decisions in varying degrees of uncertainty. Such decisions are critical for success, especially when the operational commander's forces are weaker than those of the enemy. They are not gambles, but carefully made, calculated decisions. An operational commander often achieves success by taking reasonable risks. In general, there is never enough time or enough resources, and most choices involve some risks. There are no certainties in war.[66]

Without calculated risks based on the best available intelligence and estimates of friendly capabilities and enemy intentions, the operational commander will rarely be able to apply combat potential in a coherent manner and faster than the enemy's ability to react. On the other hand, even if outnumbered and outgunned, by judiciously combining risk with flexibility and appropriate economy of effort, the operational commander can generate mass, momentum, and superiority at the decisive point and time. Barring some major intelligence successes, the operational commanders cannot realistically hope to operate completely within their counterpart's decision-implementation process. If they wait for a complete picture of the situation, they will lose the opportunity to seize and maintain the initiative and, thus, to dictate the direction and tempo of operations.

In retrospect, Admiral Ernest J. King, CINC U.S. Fleet (COMINCH)/Chief of Naval Operations (CNO), made a highly risky but sound decision by directing Admiral Chester W. Nimitz, CINC Pacific Ocean Area (CINCPOA)/CINC Pacific Fleet (CINCPac), in the aftermath of the battle of Midway in June 1942, to prepare to seize Tulagi in the Solomons by amphibious assault. He, in fact, intruded into MacArthur's area of responsibility without consulting the Joint Chiefs of Staff (JCS) or President Roosevelt. On 25 June, he presented the JCS with a fait accompli and then asked boldly for concurrence that Nimitz should attack Tulagi.[67] Subsequently, the decision was made to make the main landing on Guadalcanal (Operation Watchtower).

In the initial phase of the Korean War, General Douglas MacArthur's actions were instrumental in saving South Korea from falling under communist rule. After initial setbacks, MacArthur proposed to the JCS in Washington, D.C., a bold plan to land UN forces at Inchon, deep in the rear of the North Korean army. The plan was hotly opposed by many high officials in Washington because of the extremely high risks involved. In arguing for his plan to the JCS, he said he was firmly convinced that an early and strong effort behind the enemy's front would sever his main lines of communications and enable U.S. forces to deliver a decisive and crushing blow; the alternative was a frontal attack from the Naktong line that could only result in a protracted and expensive campaign to slowly drive the enemy north of the 38th Parallel.[68] In the end, MacArthur's views prevailed; the amphibious landing at Inchon, some 150 miles behind the North Korean forces then besieging the Pusan perimeter, was carried out on 15 September 1950. The landing was brilliantly executed, and it quickly led to the collapse of the North Korean forces. After this huge success, MacArthur seemed to lose his sure hand and overreached during the pursuit of the retreating North Korean forces north of the 38th Parallel. In Korea, he at times displayed his old strategic brilliance but at other times either hesitated to act or acted recklessly.[69] In the end, his military career ended abruptly because of his open defiance of President Harry Truman. He was dismissed from all his posts in April 1951.

Achieving Surprise

Surprise is one of the critical requirements for achieving victory. The higher the level of war, the more important this tenet is and the more difficult it is to achieve it. Surprise is one of the key force multipliers. Its effect on the enemy's morale can be substantial. Surprise consists of actions unexpected by the enemy that enhance the possibility of achieving success in combat. It consists of selecting the time, procedures and methods of combat employment to make it possible to strike when the enemy is the least prepared to ward off a strike, thereby paralyzing his will to resist. The best way to achieve a surprise is to try to put yourself in the enemy's state of mind and think about what course he is least likely to foresee and forestall. Normally, the surest way to succeed in war is to choose the course with the least expectation. Surprise achieved at the beginning of a major operation or campaign may prove decisive for its outcome. It creates the effect of numerical superiority, and without successful surprise, no superiority at the decisive point can be attained. The superiority of numbers is often a prerequisite for victory. However, absolute superiority everywhere cannot usually be achieved and hence should not be sought. What matters most is superiority at the point of the main attack. But to possess superior force at the point of attack, the enemy must be found unprepared. The value of surprise depends on the ability to deceive the enemy so completely that it upsets his mental balance at a critical moment.

Surprise is a temporary factor. It creates the opportunity for inflicting considerable losses on the enemy within a relatively short time and for disorganizing and disrupting his control of forces and his plans. It should ensure the execution of an operation that is unexpected for the enemy in terms of time and space and methods of application of combat power. The Japanese achieved both strategic and tactical surprise in their attack against the Russian naval base at Port Arthur on the night of 8 February 1904. The formal declaration of the war came two days later. Hence, the Japanese seized both the military and psychological initiative. They landed troops in Korea and marched to Manchuria to cut off the Russian communications.

The German invasion of Denmark and Norway in April 1940 was not unexpected. Yet the British and the French were caught unprepared to effectively respond. The Germans moved rapidly and captured key points on the coast and the interior of Norway. Their plan was detailed, the date of the attack was unexpected, the selected axis of attack was unexpected, and the operational concept was highly novel – in fact, it had never been tried before. At the same time, the Germans maintained a high degree of operational security, using deception and camouflage.[70]

The main elements of surprise are secrecy, speed, and concealment. A force that can appear before anyone has thought its presence to be possible stands an excellent chance of achieving surprise. If a rapid movement is to have its surprising effect, it must be accompanied by secrecy. Too much secrecy may, however, lead to insufficient cooperation and problems of morale among various parts of a force, because of ignorance of the forthcoming operation. Secrecy is therefore to some extent a double-edged weapon. Concealment can often be found in the obvious.

In some cases, the most direct approach can become the least expected of the enemy. The art of secrecy lies in being so open about most things that the few things that matter are not even suspected to exist.

Deception is related to surprise, but there is a subtle difference between these two. To surprise your opponent, it is not always necessary to deceive or mislead him. It is often preferable to use deception to achieve surprise. Deception can be achieved either by deceiving the enemy as to the existence, location, or size of the counterstroke or its purpose and thrust.

One of the main prerequisites of attaining surprise is accuracy, reliability and above all timely and relevant intelligence on the enemy's capabilities, intentions, plans and dispositions. In most military expeditions intelligence makes the difference between success and failure. There is little doubt that the Germans achieved strategic surprise in their invasion of Norway, thanks largely to their excellent operational intelligence.[71] Lack of intelligence led the Royal Navy not to sail out and strike German transports while they were proceeding across the Skagerrak to southern Norway. The British were taken by surprise not only by the German plans and the effectiveness of their execution but also on matters of topography and local practice, about which they had had ample opportunity to acquire information. There was sheer miscalculation, or failure of imagination, by the British. The likelihood that the Germans would use their heavy ships as they did was not foreseen. The daring and performance of Germans of all three services were underestimated, and that of the British troops were exaggerated.[72]

In another example, the Germans had expected the Allied invasion to take place near Genoa, not on the Provence coast, in August 1944. They were surprised both by the time and place of the operation.[73] The UN forces' landing at Inchon, Korea, in September 1950 was an example of successful operational surprise. General MacArthur had only thirty days to plan this complex and highly risky operation. The landing was a complete success, and it led UN forces to advance all the way to the Yalu River.[74]

The operational commanders should strive to achieve operational and tactical surprise to ensure success in the sector of the main effort and the point of the main attack (defense). They should control space with inferior forces by flexible composition of mobile forces and concentration at, or a shift to, the sector of main effort or point of the main attack (defense).[75] It is rarely a good thing to be invariably on the offensive or to always aim to envelop the enemy's flanks; then the operational commander acts too predictably. It stands to reason that in numerous instances the locale and other advantages of the defensive are so great as to surpass those of the attack, or that the frontal attack is more advantageous than envelopment.[76]

Napoleon I said that one should never do what one's enemy wishes, for the very reason that he does wish it.[77] The operational commanders should try to achieve surprise by attacking the enemy at unexpected times and places. They should never give the enemy enough time to prepare and conduct countermeasures. They should also always be prepared for the possibility that the enemy might achieve surprise. Deception, cover, and concealment are often key prerequisites for achieving surprise.[78]

Creativeness

The operational commanders must possess a high degree of creativity in their thinking and a readiness to take risks.[79] Creative thinking and mental agility refer to the operational commander's ability to see the whole picture from its individual parts. This also requires a mastery of combat with combined arms, exploitation of the possibilities of information warfare, and, in the employment of multinational forces, "combinedness" – knowledge of multinational structures and procedures, and understanding of other nations' culture, history, traditions, and society.[80]

The operational commanders should always be both bold and innovative; they should not use forces in a traditional manner. This means not being fixated on the mechanical or schematic employment of one's combat forces. The operational commander should never act according to a stereotyped scheme.[81] He should surprise the enemy. Liddell Hart aptly advised that the commander should choose the lines of least expectation. The commander should try to put himself in the enemy's shoes and think what course is the least likely the enemy will foresee or forestall.[82] Nothing is more disconcerting than facing a situation for which there is no answer. Whenever possible, the operational commander should use friendly forces asymmetrically and plan for operational/strategic deception. To preserve the versatility and variability of decisions, the operational commander should never act according to conventional views and preconceived notions. The art of warfare rests on the freest application of its fundamentals under constantly changing conditions.[83]

The U.S. concept of bypassing enemy strongpoints in the Solomons and along the Papua New Guinea coast is an excellent example of a creativeness for a maritime campaign. Instead of capturing many Japanese strong points, the Allies selected and then captured only a relatively few operationally significant positions. They left many Japanese strong points in the rear of their thrusts. Moreover, the Japanese rarely correctly guessed the real target of the next Allied move. This strategy was exemplified by the Allied isolation of Rabaul while capturing key points elsewhere in the area. In the Hollandia operation, the Japanese did not have a clear idea where the next blow would fall. Moreover, their strength was too small to enable them adequately to defend more than one or two places. They had expected that Wewak would be one of the sites selected by the Americans for an assault from the sea. The Japanese garrison in the proximity of Wewak was some fifteen thousand men strong. The Allied attack objective was Aitape and Humboldt Bay, where the Japanese kept only a small force. Thus, the much stronger garrison at Wewak found itself bypassed.[84]

Conclusion

The tenets of operational leadership should not be considered dogma. Unlike with the principles of war, observing all the tenets of operational leadership does not mean the violation of some of them. In general, the fewer the tenets the better. Tenets of operational leadership should be based on logic, common sense, and lessons learned. Because of the differences in national and service ways of warfare, there will never be a commonly accepted list of these tenets. However, each military and each service should reach a consensus on the tenets of operational leadership; otherwise, it is hard to see how a common operational outlook can be acquired to provide a common basis for education and training.

Because war is a political act, the primacy of policy and strategy should be ensured. Yet, this should not lead to an imbalance in which policy reigns absolute, making it exceedingly difficult or even impossible to use military forces effectively and decisively. The operational commander should always be focused firmly on the ultimate objective of the campaign or major operation. At the same time, having constancy of purpose does not mean pursuing an objective that is clearly unattainable with the military means currently available or becoming available. The means and the ends should be roughly in harmony.

Operational commanders should first have as much freedom as possible to act within the framework set by policy and strategy. Next, they should obtain and then maintain and enlarge their freedom to act. This is principally accomplished when the operational commanders and their subordinate commanders exercise initiative. The operational commander's intent is the key to ensuring sufficient freedom of action for subordinate commanders, who then use the initiative as the means of obtaining and enlarging their freedom to act. The result of this interplay is increased freedom of action throughout the chain of command.

Proper selection of the sector of main effort and the point of the main attack (defense) and the concentration of combat forces is a prerequisite for achieving operational or strategic success. The operational commanders should try to surprise the enemy as to place and/or time of employing their forces. Deception is one of the most effective methods of achieving surprise. Operational commanders should be willing to take high but prudent risks. The best way to use military force is to use it overwhelmingly to crush the enemy. Operational commanders cannot be too strong in pursuing their assigned objectives. They should be bold and innovative. The worst mistake is to act in a highly predictable and conventional way. Creativeness and innovation are not only desirable but also necessary for the operational commander who wants to achieve swift and decisive victories.

Notes

1 B. H. Liddell Hart, *Great Captains Unveiled* (Boston: Little, Brown, 1927), 41.
2 Helmuth von Moltke Sr., *Ausgewählte Werke*, vol. 1, *Feldherr und Kriegslehrmeister* (Berlin: Verlag von Reimar Hobbing, 1925), 242.
3 William D. Puleston, *Mahan: The Life and Works of Captain Alfred Thayer Mahan, U.S.N.* (New Haven, CT: Yale University Press, 1939), 297–98.
4 Arbeitspapier, *Operative Führung* (Hamburg: Führungsakademie der Bundeswehr, 1992), 17.
5 Arbeitspapier, *Operative Führung*, 2.
6 William W. Mendel and Lamar Tooke, "Operation Logic: Selecting the Center of Gravity," *Military Review* 6 (June 1993), 3.
7 Clayton R. Newell, "What Is Operational Art?" *Military Review* 9 (September 1990), 28.
8 Wolfram Hiller, *Vergleich und Bewertung der in der HDv 100/100 (TF/G) und in der Anweisung für Führung und Einsatz 700/108 dargestellten Führungsgrundsätze*

(Hamburg: Führungsakademie der Bundeswehr, 15 November 1985), 5; Wilhelm Stanger, *Grundzüge der Lehre von der Strategie. Studienbehelf für die K.K. Kriegsschule, vol. 1: Theorie mit kürzeren Beispielen* (Wien: Verlag der K.K. Kriegschule, 1884), 13–14.

9 B. H. Liddell Hart, *Strategy* (New York, NY: F.A. Praeger, 2nd printing, 1954), 348.

10 Stanger, *Grundzüge der Lehre von der Strategie*, 6.

11 Stanger, *Grundzüge der Lehre von der Strategie*, 6.

12 Erich von Manstein, *Lost Victories*, ed. and trans. Anthony G. Powell (Chicago, IL: Henry Regnery, 1958), 276.

13 Liddell Hart, *Strategy*, 348–49.

14 Norbert Hanisch, *Untersuchen Sie die operativen Ideen Manstein hinsichtlich Schwerpunktbildung, Überraschung, Initiative und Handlungsfreiheit an den Beispielen Westfeldzug 1940 (Sichelschnitt-Plan) und Operation Zitadelle* (Hamburg: Führungsakademie der Bundeswehr, 15 January 1988), 2.

15 Christian Leggemann, *Die taktisch/operativen Ideen Mansteins hinsichtlich Schwerpunktbildung, Überraschung, Initiative und Handlungsfreiheit* (Hamburg: Führungsakademie der Bundeswehr, January 4, 1989), 2.

16 Arbeitspapier, *Operative Führung*, 18.

17 Eberhard Fuhr, *Die Handlungsfreiheit der militärischen Führung-Mőglichkeiten und Grenzen aufgrund des heutigen Kriegsbildes* (Hamburg: Führungsakademie der Bundeswehr, February 16, 1968), 3.

18 Thomas B. Buell, *Master of Sea Power: A Biography of Fleet Admiral Ernest J. King*, 1st ed. (Boston/Toronto: Little, Brown, 1980), 199.

19 Richard P. Harrison, *Fleet Admiral Chester W. Nimitz: The Right Man for the Times* (Toronto: Canadian Forces College, Advanced Military Studies Course 2, December 1999), 7; Buell, *The Quiet Warrior*, 173.

20 Eike Middeldorf, *Führung und Gefecht: Grundriss der Taktik* (Frankfurt, a.M: Bernard & Graefe, Verlag für Wehrwesen, 1968), 6; Hanisch, *Untersuchen Sie die operativen Ideen Manstein hinsichtlich Schwerpunktbildung*, 29.

21 Roland G. Foerster, ed., *Gezeitenwechsel im Zweiten Weltkrieg: Die Schlachten von Char'kov und Kursk im Frühjahr und Sommer 1943 in operativer Anlage, Verlauf und politischer Bedeutung* (Hamburg/Berlin/Bonn: Verlag E. S. Mittler & Sohn, 1996), 104–05; Erich von Manstein, *Verlorene Siege*, 13th–14th ed. (Bonn: Bernard & Graefe, 1983), 460–61.

22 Arbeitspapier, *Operative Führung*, 18.

23 "Aus den Verordnungen für die höheren Truppenführer vom 24 June 1869," *Moltkes Taktisch-strategische Aufsätze aus dem Jahren 1857 bis 1871* in *Moltkes Militärische Werke, II. Die Thaetigkeit als Chef des Generalstabes der Armee im Frieden, Zweiter Theil* (Berlin: Ernst Siegfried Mittler und Sohn, 1900), 207.

24 von Manstein, *Lost Victories*, 382.

25 Hanisch, *Untersuchen Sie die operativen Ideen Manstein hinsichtlich Schwerpunktbildung*, 29.

26 Robert W. RisCassi, "Doctrine for Joint Operations in a Combined Environment: A Necessity," *Military Review* (June 1993), 23.

27 RisCassi, "Doctrine for Joint Operations in a Combined Environment," 23.

28 Hartmut Foertsch, *Entwicklung militärischer Führungsgrundsätze des deutschen Heeres im Zeitraum 1921–1934- dargestellt an den Vorschriften 'Führung und Gefecht der verbundenen Waffen' (D.V. Pl. NR. 487) und 'Truppenführung' (HDv 300/1)* (Hamburg: Führungsakademie der Bundeswehr, December 1968), 10.

29 RisCassi, "Doctrine for Joint Operations in a Combined Environment," 23.

30 Middeldorf, *Führung und Gefecht: Grundriss der Taktik*, 51.

31 "Aus den Verordnungen für die hőheren Truppenführer vom 24 June 1869," *Moltkes Taktisch-strategische Aufsätze aus dem Jahren 1857 bis 1871*, 211.

32 Helmuth von Moltke Sr., *Moltke on the Art of War: Selected Writings*, ed. and trans. Daniel J. Hughes and Harry Bell (Novato, CA: Presidio, 1995), 219.

33 Stanger, *Grundzüge der Lehre von der Strategie*, 15.
34 Stanger, *Grundzüge der Lehre von der Strategie*, 18.
35 Bruce Condell and David T. Zabecki, eds., *On the German Art of War: Truppenfuehrung* (Boulder, CO: Lynne Rienner, 2001), 22.
36 von Clausewitz, *On War*, 195–96.
37 Eliot A. Cohen and John Gooch, *Military Misfortunes, The Anatomy of Failure in War*, 1st ed. (New York: Vintage, 1991), 201.
38 Gustav Däniker, *Raum, Kraft Und Zeit In Der Militärischen Kriegführung* (Frauenfeld, Switzerland: Verlag von Huber, 1944), 30–31.
39 Friedrich von Bőtticher, "Feldherrnturm und Generalstab: Grundsätze und Geist," *Wehrkunde* 9 (September 1964), 455.
40 Friedrich von Bőtticher *The Art of War: Principles of the German General Staff in the Light of Our Time*, P-100, ZA/1 2019, Studien der Historical Division Headquarters, United States Army Europe, Foreign Military Branch, BA-MA, May 1951, 9.
41 von Bőtticher *The Art of War*, 10.
42 Arbeitspapier, *Operative Führung*, 17.
43 Russell Grenfell, *The Art of the Admiral* (London: Faber and Faber, 1937), 179.
44 Freiherr von Waldstätten, ed., *Erzherzog Karl Ausgewählte militärische Schriften*, (Berlin: Richard Wilhelmi, 1882), from his *Grundsätze der Strategie*, 77.
45 von Waldstätten, *Erzherzog Karl Ausgewählte militärische Schriften*, 16.
46 Condell and Zabecki, *On the German Art of War*, 22.
47 Arbeitspapier, *Operative Führung*, 18.
48 Friedrich von Bőtticher, *The Art of War: Principles of the German General Staff in the Light of Our Time*, 9.
49 von Bőtticher, *The Art of War*, 12.
50 Patrick J. Speelman, *Henry Lloyd and the Military Enlightenment of Eighteenth-Century Europe* (Westport, CT: Greenwood, 2002), 127.
51 J. F. C. Fuller, *Julius Caesar: Man, Soldier, and Tyrant* (New York, NY: Da Capo Press reprint of 1965 ed.), 321.
52 von Waldstätten, *Erzherzog Karl Ausgewählte militärische Schriften*, 69.
53 William Hawkins, "Iraq Shows This Is No Time to Go Light," *Defense News*, July 7, 2003, 29.
54 David G. Chandler, *The Military Maxims of Napoleon,* trans. from the French by George C. D'Aguilar (London: Greenhill, 1994), 58.
55 James N. Wasson, *Innovator or Imitator: Napoleon's Operational Concepts and the Legacies of Bourcet and Guibert* (Fort Leavenworth, KS: School of Advanced Military Studies, United States Army Command and General Staff College, May 1998), 25.
56 Michael I. Handel, *Masters of War: Classical Strategic Thought,* 3rd rev. and expanded ed. (London: Frank Cass, 2001), 156.
57 David S. Alberts, John J. Garstka, and Frederick P. Stein, *Network-Centric Warfare: Developing and Leveraging Information Superiority*, 2nd rev. ed. (Washington, DC: DoD C4ISR Cooperative Research Program, August 1999), 165.
58 Emanuel Kiliani, "Die Operationslehre des Grafen Schlieffen und ihre deutschen Gegnern (II. Teil 2)," *Wehrkunde* 3 (March), 1961, 134–35.
59 Kiliani, "Die Operationslehre des Grafen Schlieffen und ihre deutschen Gegnern (II. Teil 2)," 135.
60 John T. Nelsen, *Where to Go from Here? Considerations for the Formal Adoption of Auftragtaktik by the US Army* (Fort Leavenworth, KS: School of Advanced Military Studies, U.S. Army Command and General Staff College, December 1986), 15–16.
61 Condell and Zabecki, *On the German Art of War*, 22.
62 Arbeitspapier, *Operative Führung*, 18.
63 Grenfell, *The Art of the Admiral*, 134–35.
64 Johann Adolf Graf von Kielmansegg in Hanisch, *Untersuchen Sie die operativen Ideen Manstein hinsichtlich Schwerpunktbildung*, 6.

65 Gregory C. Gardner, *Generalship in War: The Principles of Operational Command* (Fort Leavenworth, KS: School of Advanced Military Studies, U.S. Army Command and Staff College, May 4, 1987), 21.
66 Headquarters, Department of the Army, FM 22–103, *Leadership and Command at Senior Levels* (Washington, DC: U.S. Government Printing Office, June 21, 1987), 33.
67 Buell, *Master of Sea Power*, 200.
68 D. Clayton James with Anne Sharp Wells, *Refighting the Last War: Command and Crisis in Korea 1950–1953* (New York, NY: Free Press, 1993), 162.
69 Clayton James with Anne Sharp Wells, *Refighting the Last War*, 39.
70 Olivier Desarzens, *Nachrichtendienstliche Aspekte der "Weserűbung" 1940* (Osnabrűck: Biblio Verlag, 1988), 173; K. Derry, *The Campaign in Norway* (London: Her Majesty Stationery Office, 1952), 232.
71 *The German Northern Theater of Operations 1940–1945* (Washington, DC: Center for Military History, United States Army, CMH-Pub 104, 23, 1989), pp. 43–44.
72 J.R.M. Butler, *Grand Strategy*, vol. II: *September 1939-June 1941* (London: Her Majesty's Stationery Office, 1957), pp. 147, 149.
73 Friedrich Ruge, *The Soviets as Naval Opponents 1941–1945* (Annapolis, MD: Naval Institute Press, 1979), 61; Raymond De Belot, *The Struggle for the Mediterranean 1939–1945* trans. James A. Field Jr. (Princeton, NJ: Princeton University Press, 1951), 259.
74 Richard Betts, "Strategic Surprise for War Termination: Inchon, Dienbienphu and Tet," in *Strategic Military Surprise Incentives and Opportunities*, ed. Klaus Knorr (New Brunswick, NJ: Transaction Books, 1983), 153.
75 Erich Vad, "Operative Führung: Grundlagen, Merkmale und Perspektiven," *Österreichische Militärische Zeitschrift*, no. 2 (March–April 1998), 132.
76 Friedrich von Bernhardi, *On War of To-Day, vol. 2, Combat and Conduct of War*, trans. Karl von Donat (New York, NY: Dodd, Mead, 1914), 413–14.
77 *Napoleon on the Art of War*, selected, ed. and trans. Jay Luvaas (New York, NY: Free Press, 1999), 130; Stanger, *Grundzüge der Lehre von der Strategie*, 16.
78 Middeldorf, *Führung und Gefecht: Grundriss der Taktik*, 52.
79 Arbeitspapier, *Operative Führung*, 18.
80 Vad, "Operative Führung: Grundlagen, Merkmale und Perspektiven," 132.
81 Arbeitspapier, *Operative Führung*, 18.
82 Liddell Hart, *Strategy*, 348.
83 von Bernhardi, *On War of To-Day*, vol. 2, 413.
84 Stephen W. Roskill, *The War at Sea 1939–1945, vol. III: The Offensive, Part I, 1st June-31st May 1944* (London: Her Majesty's Stationary Office, 1960), 341.

Select Bibliography

Official Histories

Appleman, Roy E. *United States Army in the Korean War, South to the Naktong, North to the Yalu (June–November 1950).* Washington, DC: Center of Military History, United States Army, CMH Pub 20–2, (1961) 1992.

Appleman, Roy E., James M. Burns, Russell A. Gugeler, and John Stevens. *The War in the Pacific, United States Army in World War II.* Vol. 2: *Okinawa: The Last Battle.* Washington, DC: Historical Divisions Department of the Army, 1948.

Blumenson, Martin. *United States Army in World War II.* Vol. 11*: Salerno to Cassino, The Mediterranean Theater of Operations.* Washington, DC: Office of the Chief of Military History, United States Army; U.S. Government Printing Office, 1969.

Boog, Horst, Jürgen Förster, Joachim Hoffmann, Ernst Klink, Rolf-Dieter Müller, and Gerd R. Ueberschär, eds. *Das Deutsche Reich Und Der Zweite Weltkrieg Der Angriff Auf Die Sowjetunion.* Vol. 4. Stuttgart: Deutsche Verlags-Anstalt, 1983.

Boog, Horst, Werner Rahn, Reinhard Stumpf, and Bernd Wegner, eds. *Das Deutsche Reich und der Zweite Weltkrieg, Der Globale Krieg. Die Ausweitung zum Weltkrieg und der Wechsel der Initiative 1941–1943.* Vol. 6: *Das Deutsche Reich und der Zweite Weltkrieg.* Stuttgart: Deutsche Verlags-Anstalt, 1990.

Cagle, Malcolm W., and Frank A. Manson. *The Sea War in Korea.* Annapolis, MD: United States Naval Institute, 1957.

Crowl, Philip A. *Campaign in the Marianas, United States Army in World War II: The War in the Pacific.* Washington, DC: Center of Military History, United States Army, (1960) 1989.

Crowl, Philip A., and Edmund G. Love. *United States Army in World War II: The War in the Pacific.* Vol. 3*: Seizure of the Gilberts and Marshalls.* Washington, DC: Center of Military History, United States Army, 1989.

Derry, T. K. *The Campaign in Norway.* London: Her Majesty's Stationery Office, 1952.

Freedman, Lawrence. *The Official History of the Falklands Campaign.* Vol. 2*: War and Diplomacy.* London/New York: Routledge, Taylor & Francis Group, first printed 2005; reprinted 2006.

Garland, Albert N., Howard McGaw Smyth, and Martin Blumenson. *United States Army in World War II: The Mediterranean Theater of Operations; Sicily and Surrender of Italy.* Washington, DC: Office of the Chief of Military History, Department of the Army, 1965.

Greenfield, Kent Roberts, ed. *Command Decisions.* Washington, DC: U.S. Government Printing Office, 1990.

Harrison, Gordon A. *United States Army in World War II: The European Theater of Operations. Cross-Channel Attack.* Washington, DC: Center of Military History, United States Army, 1989.

Keaney, Thomas A., and Eliot A. Cohen. *Gulf War: Air Power Survey Summary Report.* Washington, DC: U.S. Government Printing Office, 1993.

Marolda, Edward J., and Robert J. Schneller. *Shield and Sword: The United States Navy and the Persian Gulf War.* Washington, DC: Naval Historical Center, 1998.

Molony, C. J. C., F. C. Flynn, H. L. Davies, T. P. Gleaves. *History of Second World War.* Vol. 5: *The Mediterranean and Middle East. The Campaign in Sicily 1943 and The Campaign in Italy, 3rd September 1943 to 31st March 1944.* London: Her Majesty's Stationery Office, 1973.

Morison, Samuel E. *History of United States Naval Operations in World War II.* Vol. 1: *The Battle of the Atlantic, September 1939–May 1943.* Boston: Little, Brown and Company, reprint 1984.

Morison, Samuel E. *History of United States Naval Operations in World War II.* Vol. 3: *The Rising Sun in the Pacific.* Boston: Little, Brown and Company, reprint 1984.

Morison, Samuel E. *History of United States Naval Operations in World War II.* Vol. 4: *Coral Sea, Midway, and Submarine Actions, May 1942–August 1942.* Boston: Little, Brown and Company, reprint 1984.

Morison, Samuel E. *History of United States Naval Operations in World War II.* Vol. 5: *The Struggle for Guadalcanal, August 1942–February 1943.* Boston: Little, Brown and Company, 1951.

Morison, Samuel E. *History of United States Naval Operations in World War II.* Vol. 8: *New Guinea and the Marianas, March 1944–August 1944.* Boston: Little, Brown and Company, 1953.

Morison, Samuel E. *History of United States Naval Operations in World War II.* Vol. 9: *Sicily-Salerno-Anzio: January 1943–June 1944.* Boston: Little, Brown and Company, 1st printed 1954; reprinted 1984.

Morison, Samuel E. *History of United States Naval Operations in World War II.* Vol. 12: *Leyte, June 1944–January 1945.* Boston: Little, Brown and Company, reprint 1984.

Morison, Samuel E. *The Two Ocean War: A Short History of the United States Navy in the Second World War.* Boston: Little, Brown and Company, 1963.

Morton, Louis. *United States Army in World War II. The War in the Pacific. Strategy and Command: The First Two Years.* Washington, DC: Center for Military History, United States Army, 1st printing 1962, reprint 1989.

Playfair, I. S. O., and C. J. C. Molony. *The Mediterranean and Middle East.* Vol. 1: *The Germans Come to the Help of Their Ally (1941).* London: Her Majesty's Stationery Office, 1956.

Playfair, I. S. O. *The Mediterranean and Middle East (September 1941 to September 1942).* Vol. 3: *British Fortunes Reach Their Lowest Ebb.* London: Her Majesty's Stationery Office, 1960.

Roskill, Stephen W. *The War at Sea 1939–1945.* Vol. 1: *The Defensive.* London: Her Majesty's Stationery Office, 1954.

Roskill, Stephen W. *The War at Sea 1939–1945.* Vol. 2: *The Period of Balance.* London: Her Majesty's Stationery Office, 1956.

Roskill, Stephen W. *The War at Sea 1939–1945.* Vol. 3: *The Offensive, Part 2, 1st June 1944–14th August 1945.* London: Her Majesty's Stationery Office, 1961.

Smith, Robert Ross. *United States Army in World War II: The War in the Pacific. The Approach to the Philippines.* Washington, DC: Center of Military History, United States Army, 1948.

Ziemke, Earl F., and Magna E. Bauer. *Stalingrad to Berlin: The German Defeat in the East.* Washington, DC: Center of Military History, United States Army, (1968) 1987.

Ziemke, Earl F., and Magna E. Bauer. *Moscow to Stalingrad: Decision in the East.* Washington, DC: Center of Military History, United States Army, 1987.

General Histories

Addington, Larry H. *The Patterns of War Since the Eighteenth Century.* Bloomington: Indiana University Press, 1984.

Beck, Ludwig. *Studien.* Edited by Hans Speidel. Stuttgart: K. F. Koehlers Verlag, 1955.
Beevor, Antony. *Stalingrad.* London: Penguin Books, 1999.
Bellamy, Christopher. *The Evolution of Modern Land Warfare: Theory and Practice.* London/New York: Routledge, 1990.
Belot, Raymond de. *The Struggle for the Mediterranean 1939–1945.* Translated by James A. Field, Jr. Princeton, NJ: Princeton University Press, 1951.
Bernhardi, Friedrich Adam Julius von. *Vom Heutigen Kriege.* 2 Vols. Berlin: E. S. Mittler & Sohn, 1912.
Bernhardi, Friedrich Adam Julius von. *On War of To-Day.* 2 Vols. Translated by Karl von Donat. New York, NY: Dodd, Mead & Company, 1914.
Bernhardi, Friedrich Adam Julius von. *The War of the Future in the Light of the Lessons of the World War.* 2nd ed. Translated by F. A. Holt. London: Hutchinson & Co., 1920.
Bitzel, Uwe. *Die Konzeption des Blitzkrieges bei der deutschen Wehrmacht.* Frankfurt, a.M: Peter Lang, 1991.
Bleck, Otto, W. Müller Loebnitz, and Ernst Kabisch, eds. *Das deutsche Wehrwesen in Vergangenheit und Gegenwart.* Stuttgart/Berlin: Konradin Verlag, 1936.
Blume, Wilhelm von. *Strategie. Ihre Aufgaben und Mittel.* Berlin: E. S. Mittler & Sohn, 1912.
Blumenson, Martin. *Sicily: Whose Victory?* New York, NY: Ballantine Books, 1969.
Boguslawski, Albrecht von. *Heerwesen und Kriegführung.* Berlin: Verlag von R. Eisenschmidt, 1897.
Boguslawski, Albrecht von. *Strategische Erörterungen betreffend die vom General von Schlichting vertretenen Grundsätze.* Berlin: Verlag von R. Eisenstadt, 1901.
Boog, Horst, ed. *Die Deutsche Luftwaffenführung 1935–1945. Führungsprobleme, Spitzengliederung, Generalstabsausbildung.* Stuttgart: Deutsche Verlags-Anstalt, 1982.
Boog, Horst, ed. *Luftkriegführung im Zweiten Weltkrieg. Ein internationaler Vergleich.* Herford/Bonn: Verlag E. S. Mittler & Sohn, 1993.
Bülow, Heinrich von. *Geist des neuern Kriegssystems hergeleitet aus dem Grundsätze einer Basis der Operationen.* Hamburg, 1799.
Burden, Raymond A. *Falklands: The Air War.* London/New York: Arms and Armour Press, 1986.
Caemmerer, Rudolph von. *The Development of Strategical Science During the 19th Century.* Translated by Karl von Donat. London: Hugh Rees, 1905.
Castex, Raoul. *Strategic Theories.* Selections Translated and Edited with an Introduction by Eugenia C. Kiesling. Annapolis, MD: Naval Institute Press, 1993.
Chaliand, Gérard. *The Art of War in World History: From Antiquity to the Nuclear Age.* Berkeley/Los Angeles/London: University of California Press, 1994.
Chandler, David G. *The Campaigns of Napoleon: The Mind and Method of History's Greatest Soldier.* New York: Scribner, 1966.
Citino, Robert M. *The Path to Blitzkrieg: Doctrine and Training in the German Army, 1920–1939.* Boulder, CO/London: Lynne Rienner, 1999.
Clausewitz, Carl von. *Vom Kriege.* Commented by Werner Hahlweg. Bonn: Ferdinand Duemmlers Verlag, 1952.
Clausewitz, Carl von. *On War.* Edited and translated by Michael Howard and Peter Paret. New York, NY: Alfred A. Knopf, 1993.
Cohen, Eliot A. *Revolution in Warfare? Air Power in the Persian Gulf.* Annapolis, MD: Naval Institute Press, 1995.
Condell, Bruce, and David T. Zabecki, eds. *On the German Art of War: Truppenfuehrung.* Boulder, CO: Lynne Rienner, 2001.
Connelly, Owen. *The Wars of the French Revolution and Napoleon, 1792–1815.* London/New York: Routledge, Taylor & Francis, 2006.
Cooper, Matthew. *The German Air Force 1933–1945: An Anatomy of Failure.* London: Jane's Publishing, 1981.
Corbett, Julian S. *Some Principles of Maritime Strategy.* London: Longmans, Green, 1918.
Cordesman, Anthony H. *The Iraq War: Strategy, Tactics, and Military Lessons.* Washington, DC: Center for Strategic and International Studies, 2003.

Corum, James A. *The Luftwaffe: Creating the Operational Air War, 1918–1940.* Lawrence, KS: University Press of Kansas, 1997.

Craig, Gordon A., and Felix Gilbert. *Makers of Modern Strategy: From Machiavelli to the Nuclear Age.* Princeton, NJ: Princeton University Press, 1986.

Cruttwell, Charles R. M. F. *A History of the Great War 1914–1918.* Oxford: Clarendon, 1934.

Däniker, Gustav. *Raum, Kraft und Zeit in der Militärischen Kriegführung.* Frauenfeld: Verlag von Huber & Co. Aktiengesellschaft, 1944.

D'Este, Carlo. *Fatal Decision: Anzio and the Battle for Rome.* 1st ed. New York, NY: Harper Perennial, 1992.

Douhet, Giulio. *The Command of the Air.* 2nd ed. Translated by Dino Ferrari. New York, NY: Coward-McCann, (1927) 1942.

Dűppler, Jòrg, ed. *Seemacht und Seestrategie im 19. und 20. Jahrhundert.* Hamburg: Verlag E. S. Mittler & Sohn, 1999.

Dupuy, Trevor N. *Elusive Victory: The Arab-Israeli Wars: 1947–1974.* Garden City, NY: Military Book, 1978.

Egger, Carl von. *Die Strategie mit Berűcksichtigung der neuen Kriegsmittel.* Basel: Schweig-hauserische Verlagsbuchhandlung, 1870.

Eglin, Hans-Herbert. *Einfűhrung in die operative Kunst der Landstreitkräfte.* Dresden: Military Academy Friedrich Engels, 1974.

Erfurth, Waldemar. *Der Vernichtungssieg. Eine Studie über das Zusammenwirken getrennter Heeresteile.* Berlin: Verlag von E. S. Mittler & Sohn, 1939.

Evans, David C., ed. *The Japanese Navy in World War II: In the Words of Former Japanese Naval Officers.* 2nd ed. Annapolis, MD: Naval Institute Press, 1986.

Evans, David C., and Mark R. Peattie. *Kaigun: Strategy, Tactics, and Technology in the Imperial Japanese Navy, 1887–1941.* Annapolis, MD: Naval Institute Press, 1997.

Evolution of Soviet Operational Art 1927–1991. 2 Vols. 1st ed. Translated by Harold S. Orenstein. London/Portland, OR: Frank Cass Publishers, 1995.

Fiebeger, G. J. *Elements of Strategy.* West Point, NY: United States Military Academy Press, 1910.

Foertsch, Hermann. *Kriegskunst Heute und Morgen.* Berlin: Zeitgeschichte-Verlag Wilhelm Andermann, 1939.

Freedman, Lawrence, and Virginia Gamba-Stonehouse. *Signals of War: The Anglo-Argentine Conflict of 1982.* Princeton, NJ: Princeton University Press, 1991.

Freytag-Loringhoven, Hugo von. *Die Macht der Persönlichkeit im Krieg. Studien nach Clausewitz.* Berlin: E. S. Mittler und Sohn, 1911.

Frieser, Karl-Heinz. *Blitzkrieg-Legende. Der Westfeldzug 1940.* Munich: R. Oldenbourg Verlag, 1995.

Fugate, Bryan I. *Operation Barbarossa, Strategy and Tactics on the Eastern Front, 1941.* Novato, CA: Presidio Press, 1984.

Fuller, J. F. C. *The Reformation of War.* London: Hutchinson, 1923.

Fuller, J. F. C. *The Foundations of the Science of War.* London: Curtis Brown, 1926.

Fuller, J. F. C. *The Conduct of War 1789–1961.* New York, NY: Da-Capo Press, reprint of 1961 edition.

Gat, Azar. *A History of Military Thought: From the Enlightenment to the Cold War.* Oxford: Oxford University Press, 2001.

Glantz, David M. *Soviet Military Operational Art: In Pursuit of Deep Battle.* London: Frank Cass, 1990.

Goltz, Colmar von der. *Kriegführung. Kurze Lehre ihrer wichtigsten Grundsätze und Formen.* Berlin: R. v. Decker's Verlag, 1895.

Gooch, John, ed. *Decisive Campaigns of the Second World War.* London: Frank Cass, 1990.

Gordon, Michael R., and Bernard E. Trainor. *The General's War: The Inside Story of the Conflict in the Gulf.* Boston: Little, Brown and Company, 1995.

Grenfell, Russell. *The Art of the Admiral.* London: Faber and Faber, 1937.

Groos, Otto. *Seekriegslehren Im Lichte Des Weltkrieges. Ein Buch für den Seemann, Soldaten und Staatsmann.* Berlin: Verlag von E. S. Mittler & Sohn, 1929.
Gundelach, Karl. *Die deutsche Luftwaffe im Mittelmeer 1940–1945.* 2 Vols. Frankfurt a. M: Peter D. Lang, 1981.
Hagerman, Edward. *The American Civil War and the Origins of Modern Warfare: Ideas, Organization, and Field Command.* Bloomington, IN: Indiana University Press, 1992.
Halder, Franz. *Hitler als Feldherr.* Munich: Münchener-Dom Verlag, 1949.
Handel, Michael I. *Clausewitz and Modern Strategy.* Ilford, Essex: Frank Cass, 1986.
Handel, Michael I. *Strategic and Operational Deception in the Second World War.* London: Frank Cass, 1987.
Handel, Michael I. *Intelligence and Military Operations.* London: Frank Cass, 1990.
Handel, Michael I. *Masters of War: Classical Strategic Thought.* 3rd and expanded ed. London: Frank Cass, 2001.
Harrison, Richard W. *The Russian Way of War: Operational Art, 1904–1940.* Lawrence, KS: University Press of Kansas, 2001.
Headerson, William D. *Cohesion: The Human Element in Combat: Leadership and Societal Influence in the Armies of the Soviet Union, the United States, North Vietnam, and Israel.* Washington, DC: National Defense University, 1985.
Hough, Richard, and Denis Richards. *The Battle of Britain: The Greatest Air Battle of World War II.* New York/London: W. W. Norton, 1989.
Howard, Michael, ed. *The Theory and Practice of War: Essays Presented to Captain B. H. Liddell Hart on his Seventieth Birthday.* New York, NY: Fredrick A. Praeger, 1966.
Howard, Michael. *Strategic Deception in the Second World War.* New York, NY/London: W. W. Norton, 1995.
Hubatsch, Walther. *Hitlers Weisungen fuer die Kriegsführung 1939–1945. Dokumente des Oberkommandos der Wehrmacht.* Koblenz: Bernard & Graefe Verlag, 1983.
Hughes, Daniel J., ed. *Moltke on the Art of War: Selected Writings.* Translated by Harry Bell and Daniel J. Hughes. Novato, CA: Presidio, 1993.
Jacobsen, Hans-Adolf, and Jürgen Rohwer, eds. *Entscheidigungsschlachten des zweiten Weltkrieges.* Frankfurt, a.M: Verlag für Wehrwesen Bernard & Graefe, 1960.
Johnstone, H. M. *The Foundations of Strategy.* London: George Allen & Unwin, 1914.
Jomini, Antoine-Henri de. *The Art of War.* Translated by G. H. Mendel and W. P. Craighill. Westport, CT: Greenwood, 1971. Originally published, Philadelphia, PA: J. P. Lippincott & Co., 1862.
Jones, Archer. *Civil War: Command and Strategy: The Process of Victory and Defeat.* New York, NY: The Free Press, a Division of Macmillan, 1992.
Krauss, Alfred. *Theorie und Praxis in der Kriegskunst.* Munich: J. F. Lehmanns Verlag, 1936.
Latimer, Jon. *Deception in War.* Woodstock/New York, NY: Overlook Press, 2001.
Lawrence, James. *Imperial Rearguard: Wars of Empire, 1919–1985.* London: Brassey's, 1988.
Leeb, Wilhelm Ritter von. *Die Abwehr.* Berlin: E. S. Mittler & Sohn, 1938.
Lehman, John, and Harvey Sicherman, eds. *America the Vulnerable: Our Military Problems and How to Fix Them.* Philadelphia, PA: Foreign Policy Research Institute, 2002.
Liddell Hart, B. H. *The Real War 1914–1918.* Boston/New York: Little, Brown and Company, (1930) 1964.
Liddell Hart, B. H. *Strategy.* New York, NY: Frederick A. Praeger, 1954.
Luvaas, Jay, ed. and trans. *Fredrick the Great on the Art of War.* New York, NY: De Capo Press, 1999.
Luvaas, Jay, ed. and trans. *Napoleon on the Art of War.* New York, NY: The Free Press, 1999.
Macksey, Kenneth. *Why the Germans Lose at War: The Myth of German Military Superiority.* London: Greenhill/Harrisburg, PA: Stackpole, 1996.
Magenheimer, Heinz. *Kriegswenden in Europa 1939–1945: Führungsentschlüsse, Hintergründe, Alternativen.* Munich, Landsberg am Lech: Guenter Olzog Verlag, GmbH, 1995.

Mahan, Alfred T. *Naval Strategy: Compared and Contrasted with the Principles and Practice of Military Operations on Land.* Boston: Little, Brown and Company, 1911.

Marshall, S. L. *The Officer as a Leader*. 1st ed. Harrisburg, PA: Stackpole Books, April 1966.

McKercher, B. J. C., and Michael A. Hennessy, eds. *The Operational Art: Developments in the Theories of War.* Westport, CT: Praeger, 1996.

McNeilly, Mark. *Sun Tzu and the Art of Modern Warfare.* Oxford: Oxford University Press, 2001.

Menning, Bruce W. *Bayonets before Bullets: The Imperial Russian Army, 1861–1914.* Bloomington/Indianapolis, IN: Indiana University Press, 1992, paperback reprint, 2000.

Merkens, Heinrich. *Ausgewählte Kriegswissenschaftliche Schriften Friedrichs des Grossen.* Jena: Hermann Costenoble, 1876.

Middlebrook, Martin. *The Fight for the 'Malvinas': The Argentine Forces in the Falklands War.* London: Viking, 1989.

Militärgeschichtlichen Forschungsamt (MGFA), ed. *Operatives Denken und Handeln in deutschen Streitkräften im 19. und 20. Jahrhundert.* Herford/Bonn: Verlag E. S. Mittler & Sohn, 1988.

Militärgeschichtlichen Forschungsamt (MGFA), ed. *Development, Planning and Realization of Operational Conceptions in World War I and II.* Herford/Bonn: E. S. Mittler & Sohn, 1989.

Militärgeschichtlichen Forschungsamt (MGFA), ed. *Operational Thinking of Clausewitz, Moltke, Schlieffen, and Manstein.* 2nd ed. Freiburg im Breisgau: Militärgeschichtliches Forschungsamt, 1989.

Millotat, Christian E. O. *Das preussisch-deutsche Generalstabssystem: Wurzeln-Entwicklung-Fortwirken.* Zurich: VDF Hoschschulverlag AG, 2000.

Moltke, Helmuth von, Sr. *Ausgewählte Werke. Vol. 1: Feldherr und Kriegslehrmeister.* Berlin: Verlag von Reimar Hobbing, 1925.

Momyer, William. *Airpower in Three Wars.* Washington, DC: Department of the Air Force, 1978.

Muller, Richard. *The German Air War in Russia.* Baltimore, MD: Nautical & Aviation Publishing Company of America, 1993.

Müller, Rolf-Dieter, and Hans-Erich Volkmann. *Die Wehrmacht. Mythos und Realität* Munich: R. Oldenbourg Verlag, 1999.

Murray, Williamson, and Robert H. Scales, Jr. *The Iraq War: A Military History.* Cambridge, MA/London: Belknap Press of Harvard University Press, 2003.

Naveh, Shimon. *In Pursuit of Military Excellence: The Evolution of Operational Theory.* London: Frank Cass, 1997.

Neumann, J. *Grundzüge der Strategie. Ein Leitfaden für das Studium der Kriegsgeschichte.* Vienna, 1870.

Neznamov, A. *Sovremennaya voyna. Deystviya polevoy armii.* Sankt Petersburg: Gr. Skachkov, 1911.

Ninković, Jovo. *Nastanak i Razvoj Ratne Operacije.* Belgrade: Vojno-Izdavački Zavod, 1984.

Ninković, Jovo. *Teorijske osnove operatike.* Belgrade: Vojno Izdavački Zavod, 1985.

O'Connell, D. P. *The Influence of Law on Sea Power.* Manchester: Manchester University Press, 1975.

Oetting, Dirk W. *Motivation und Gefechtswert. Vom Verhalten des Soldaten im Kriege.* Frankfurt, a. M/Bonn: Report Verlag, 1988.

Ose, Dieter. *Entscheidung im Westen 1944. Der Oberbefehlshaber West und die Abwehr der allierten Invasion.* Stuttgart: Deutsche Verlags-Anstalt, 1982.

Ottmer, Hans-Martin, and Heiger Ostertag, eds. *Ausgewählte Operationen und ihre militärhistorischen Grundlagen.* Herford/Bonn: Verlag E. S. Mittler & Sohn, 1993.

Paret, Peter, and Gordon A. Craig, eds. *Makers of Modern Strategy: From Machiavelli to the Nuclear Age.* Princeton, NJ: Princeton University Press, 1 March 1986.

Philippi, Alfred, and Ferdinand Heim. *Der Feldzug Gegen Sowjetrussland 1941 bis 1945. Ein operativer Überblick.* Stuttgart: W. Kohlhammer Verlag, 1962.

Phillips, Thomas R., ed. *Roots of Strategy: A Collection of Military Classics.* Harrisburg, PA: Military Service Publishing, 1940.

Post, Walter. *Unternehmen Barbarossa. Deutsche und Sowjetische Angriffspläne 1940/41* Hamburg/Berlin/Bonn: Verlag E. S. Mittler & Sohn, 1995.

Potter, Elmer B. *The Great Sea War: The Story of Naval Action in World War II.* Englewood Cliffs, NJ: Prentice Hall, 1960.

Rabenau, Friedrich von. *Operative Entschlűsse gegen einen an Zahl űberlegenen Gegner.* Berlin: E. S. Mittler & Sohn, 1935.

Rendulic, Lothar. *Grundlagen militärischer Fűhrung.* Herford/Bonn: Maximilian Verlag, 1967.

Richmond, Herbert. *National Policy and Naval Strength and Other Essays.* London/ New York/Toronto: Longmans, Green, 1934.

Robertson, Scot. *The Development of RAF Strategic Bombing Doctrine, 1919–1939.* Westport, CT: Praeger, 1995.

Roskill, Stephen W. *The Strategy of Sea Power: Its Development and Application.* London: Collins, 1962.

Rothenberg, Gunther E. *The Art of Warfare in the Age of Napoleon.* Bloomington: Indiana University Press, 1978.

Savinkin, A. E., et al. *Postizheniye Voyennogo Iskusstva. Ideynoe naslediye A. Svechina.* Moscow: Voyenniy Universitet; Ruskii put, 1999.

Schlichting, Sigismund von. *Taktische und strategische Grundsätze der Gegenwart,* 3 parts. Berlin: Ernst Sigfried Mittler und Sohn, 1898–1899.

Schűler, Klaus A. Friedrich. *Logistik im Russlandfeldzug. Die Rolle der Eisenbahn bei Planung, Vorbereitung und Durchfűhrung des deutschen Angriffs auf die Sowjetunion bis zur Krise vor Moskau im Winter 1941/42.* Frankfurt a.M: Peter Lang Verlag, 1987.

Seaton, Albert. *The Russo-German War 1941–45.* Novato, CA: Presidio Press, 1971.

Singh, Jasjit. *Air Power in Modern Warfare.* New Delhi: Lancer International, 1988.

Snyder, Jack. *The Ideology of the Offensive: Military Decision Making and the Disasters of 1914.* Ithaca, NY/London: Cornell University Press, 1984.

Sokolovsky, V. D. *Soviet Military Strategy.* 3rd ed. Moscow: Voyenizdat, 1968. Edited, with an analysis and Commentary by Harriett Fast Scott. New York, NY: Crane, Russak & Company, 1975.

Stanger, Wilhelm. *Grundzűge der Lehre von der Strategie. Studienbehelf fűr die K.K. Kriegschule. Vol. 1: Theorie mit Kűrzeren Bespielen.* Vienna: Verlag der K.K. Kriegschule, 1884.

Starkey, Armstrong. *War in the Age of Enlightenment 1700–1789.* Westport, CT: Praeger, 2003.

Stearley, Ralph F., Robert M. Lee, and James C. McGehee. *Air Power in the European Theater of Operations.* Study No. 56. General Board, United States Forces, European Theater, 1946.

Stephens, Alan, ed. *The War in the Air 1914–1994.* Maxwell AFB, AL: Air University Press, January 2001.

Summers, Harry G. *On Strategy: A Critical Analysis of the Vietnam War.* New York: Bantam Doubleday Dell, 1984.

Sun-Tzu. *The Art of War.* Translated with Introduction and Commentary by Ralph D. Sawyer, with the collaboration of Mei-chün Lee Sawyer. Boulder, CO: Westview Press, 1994.

Svechin, Aleksander A. *Strategiya.* Moscow: Voeynny Vestnik, 1927.

Tantum, W. H., IV, and E. J. Hoffschmidt, eds. *The Rise and Fall of the German Air Force: History of the Luftwaffe in WW2.* Old Greenwich, CT: WE, 1969.

Telp, Claus. *The Evolution of Operational Art, 1740–1813: From Frederick the Great to Napoleon.* London: Frank Cass, 2005.

Thoss, Bruno, and Hans-Erich Volkmann. *Erster Weltkrieg-Zweite Weltkrieg. Ein Vergleich.* Paderborn/Munich/Vienna/Zurich: Ferdinand Schöningh, 2002.

Till, Geoffrey, John Hattendorf, Richard Hill, Barry Hunt, Peter Nailor, Bryan Ranft, Stephen Roskill and Craig Symonds. *Maritime Strategy and the Nuclear Age.* 2nd ed. New York: St. Martin's Press, 1984.

Triandafillov, V. K. *Nature of the Operations of Modern Armies.* Translated by William Burhans, Russ-Eng Translations. Woodbridge, VA/Moscow: Leningrad, 1929.

Trotter, William R. *Frozen Hell: The Russo-Finnish Winter War of 1939–40.* Chapel Hill, NC: Algonquin Books, 2000.

Uhle-Wettler, Franz. *Höhe- und Wendepunkte Deutscher Militär Geschichte.* Mainz: v. Hase & Koehler, 1984.

Ungváry, Krisztián. *The Siege of Budapest: One Hundred Days in World War II.* Translated from the Hungarian by Ladislaus Löb. New Haven/London: Yale University Press, 2005.

Vallance, Andrew B. G. *Doctrine of Air Power Strategy and Operational Art.* New York, NY: St. Martin's Press, 1996.

Vauthier, Paul. *Die Kriegslehre des Generals Douhet.* Berlin: Rowohlt, 1935.

Waldstätten, Freiherr von, ed. *Erzherzog Karl. Ausgewählte militärische Schriften.* Berlin: Richard Wilhelmi, 1882.

Wallach, Jehuda L. *The Dogma of the Battle of Annihilation: The Theories of Clausewitz and Schlieffen and Their Impact on the German Conduct of Two World Wars.* Westport, CT: Greenwood Press, 1986.

Wegener, Wolfgang. *The Naval Strategy of the World War.* Translated by Herwig Holger. Annapolis, MD: Naval Institute Press, 1989.

Wilkinson, Spencer. *The Brain of an Army: A Popular Account of the German General Staff.* London: Macmillan, 1890.

Will, Thomas. *Operative Führung: Versuch einer begrifflichen Bestimmung im Rahmen von Clausewitz' Theorie 'Vom Kriege.'* Hamburg: Verlag Dr. Kovac, 1997.

Zehrer, Hartmut. *Der Golfkonflikt. Dokumentation, Analyse und Bewertung aus militärischer Sicht.* Herford/Bonn: Verlag E.S. Mittler & Sohn, 1992.

Zerboni, K. Von. *Die Elemente Militärwesen.* Vienna, 1870.

Ziemke, Earl F. *The Red Army 1918–1941: From Vanguard of World Revolution to U.S. Ally.* London/New York: Frank Cass, 2004.

Monographs

Abt, Frederic E. *The Operational End State: Cornerstone of the Operational Level of War.* Fort Leavenworth, KS: School of Advanced Military Studies, U.S. Army Command and General Staff College, 1988.

Agee, Collin A. *Peeling the Onion: The Iraqi Center of Gravity in Desert Storm.* Fort Leavenworth, KS: School of Advanced Military Studies, U.S. Army Command and General Staff College, May 1992.

Arbeitspapier. *Operative Führung.* Hamburg: Führungsakademie der Bundeswehr, August 1992.

Athens, Arthur J. *Unraveling the Mystery of Battlefield Coup d'Oeil.* Fort Leavenworth, KS: School of Advanced Military Studies, U.S. Army Command and General Staff College, February 1993.

Bahnemann, Jörg. *Der Begriff der Strategie bei Clausewitz, Moltke und Liddell Hart. Eine Untersuchung der Beziehungen zwischen politischer und militärischer Führung.* Hamburg: Führungsakademie der Bundeswehr, March 1966.

Barron, Michael J. *Operational Level Command: Who is in Charge?* Fort Leavenworth, KS: U.S. Army and General Staff College, School of Advanced Military Studies, 1988.

Barry, Robert F., II. *Who's Zooming Who? Operational Fires and the FSCL.* Fort Leavenworth, KS: U.S. Army and General Staff College, School of Advanced Military Studies, 1993.

Bartele, Manfred. *Beiträge des Generals von Scharnhorst zur Weisungsführung in der preussischen Armee.* Hamburg: Führungsakademie der Bundeswehr, September 1970.

Bassenge, Jürgen. *Zur Zielsetzung der Deutschen Angriffe in Frankreich, Frühjahr bis Sommer 1918. Überlegungen zum Problem des Kulminationspunktes.* Hamburg: Führungsakademie der Bundeswehr, 1964.

Batson, William R. *Centers of Gravity, Lines of Operations, and the Normandy Campaign.* Fort Leavenworth, KS: School of Advanced Military Studies, U.S. Army Command and General Staff College, May 1987.

Beck, J. *Schlagkraft-ein Schlagwort? Zur Problematik der Erfassung und Bewertung militärischer Leistungsfähigkeit.* Hamburg: Fűhrungsakademie der Bundeswehr, October 1978.

Behrendt, Harmut. *Die Handlungsfreiheit der militärischen Fűhrung-Möglichkeiten und Grenzen aufgrund des heutigen Kriegsbildes.* Hamburg: Fűhrungsakademie der Bundeswehr, January 1968.

Benjamin, David J. *Prerequisite for Victory: The Discovery of the Culminating Point.* Fort Leavenworth, KS: School of Advanced Military Studies, U.S. Army Command and General Staff College, May 1986.

Besterhorn, Jörg. *'Freie Operationen'-zur Bedeutung eines neuen Begriffes im Rahmen der operativen Schulung des Heeres.* Hamburg: Fűhrungsakademie der Bundeswehr, January 1999.

Betson, William R. *Centers of Gravity, Lines of Operations, and the Normandy Campaign.* Fort Leavenworth, KS: School of Advanced Military Studies, U.S. Army Command and General Staff College, June 1987.

Bloechl, Timothy D. *Operation Just Cause: An Application of Operational Art.* Fort Leavenworth, KS: U.S. Army and General Staff College, School of Advanced Military Studies, 1993.

Boatner, Michael E. *Operational Synchronization Methodology.* Fort Leavenworth, KS: School of Advanced Military Studies, U.S. Army Command and General Staff College, 1993.

Bolick, Joseph A. *The Influence and Reasons for Acceptance or Rejection of Operational Level Intelligence During the 1914 Marne and 1943 Kursk Campaigns.* Fort Leavenworth, KS: School of Advanced Military Studies, U.S. Army Command and General Staff College, 1988.

Boyce, Earl, J. *Cover and Deception in World War II – Its Lessons and Doctrine Implications.* Maxwell AFB: Air Command and Staff College, 1982.

Brant, Bruce A. *Retrograde at the Operational Level of War.* Fort Leavenworth, KS: School of Advanced Military Studies, U.S. Army and General Staff College, May 1987.

Brinkley, Phillip L. *The Operational Commander's Will: An Intangible Element in Victory.* Fort Leavenworth, KS: School of Advanced Military Studies, U.S. Army Command and General Staff College, May 1986.

Browne, Joseph Edward. *Deception and the Mediterranean Campaigns of 1943–1944.* Carlisle Barracks, PA: U.S. Army War College, 1986.

Bryan, William H. *Close Air Support Doctrine: Dynamic Feature or Dogmatic Past.* Maxwell AFB, AL: Air War College, April 1994.

Bullington, Terry W. *Considerations for the Organization and Employment of an Operational Reserve.* Fort Leavenworth, KS: School of Advanced Military Studies, U.S. Army Command and General Staff College, May 1986.

Burgdorf, Charles E. *An Appreciation for Vulnerability to Deception at the Operational Level.* Fort Leavenworth, KS: School of Advanced Military Studies, U.S. Army Command and General Staff College, May 1987.

Burgess, Ronald L. *Operational Intelligence: Is It a Panacea?* Fort Leavenworth, KS: School of Advanced Military Studies, U.S. Army Command and General Staff College, April 1987.

Carter, Gregory A. *Some Historical Notes on Air Interdiction in Korea.* Santa Monica, CA: RAND, September 1966.

Clark, Robert L., IV. *The Essential Elements of Operational Surprise.* Fort Leavenworth, KS: School of Advanced Military Studies, U.S. Army Command and General Staff College, May 1987.

Coomler, James D. *The Operational Culminating Point: Can You See It Coming?* Fort Leavenworth, KS: School of Advanced Military Studies, U.S. Army Command and General Staff College, May 1986.

Cordesman, Anthony H. *The "Instant Lessons" of the Iraq War, Third Working Draft.* Washington, DC: Center for Strategic and International Studies, April 14, 2003.

Cowan, David M. *The Utility of the Operational Pause in Sequencing Battles to Achieve an Operational Advantage.* Fort Leavenworth, KS: School of Advanced Military Studies, U.S. Army Command and General Staff College, 1987.

Craft, Douglas W. *An Operational Analysis of the Persian Gulf War.* Carlisle Barracks, PA: Strategic Studies Institute, U.S. Army War College, 1992.

Crain, William F. *Infantry, The Heart of Operational Art: Translating Strategic Objectives into Tactical Missions.* Fort Leavenworth, KS: School of Advanced Military Studies, U.S. Army Command and General Staff College, Second Term (1989) 1990.

Crenshaw, Charles T. *Distinctions Between Tactical and Operational Level of War – Are Some More Important Than Others?* Fort Leavenworth, KS: School of Advanced Military Studies, U.S. Army Command and General Staff College, May 1986.

Dail, Robert T. *Does the U.S. Army Really Understand Operational War? A Logistics Perspective.* Fort Leavenworth, KS: School of Advanced Military Studies, U.S. Army Command and General Staff College, May 1988.

Dehen, Herbert. *Der Balkanfeldzug 1941. Beispiel einer Offensive aus der Bewegung ohne vorbereitenden Aufmarsch.* Hamburg: Führungsakademie der Bundeswehr, January 1963.

Dixon, Richard J. *Operational Sequencing: The Tension between Simultaneous and Sequential Operations.* Fort Leavenworth, KS: School of Advanced Military Studies, U.S. Army Command and General Staff College, 1994.

Donohue, Kevin S. *Dynamic Synchronization Matrix for Campaign Planning.* Fort Leavenworth, KS: School of Advanced Military Studies, U.S. Army Command and General Staff College, 1993.

DuBois, Edmund L., Wayne P. Hughes, Jr., and Lawrence J. Low. *A Concise Theory of Combat.* Monterey, CA: Institute for Joint Warfare Analysis, 2nd printing, October 1998.

Duncan, Stephen C. *Intelligence and Defensive Culminating Point – Piercing the Fog.* Fort Leavenworth, KS: School of Advanced Military Studies, U.S. Army Command and General Staff College, 1995.

Ebel, Todd J. *Without Coup D'oeil, but with Responsibility: Determining Centers of Gravity and Decisive Points.* Fort Leavenworth, KS: School of Advanced Military Studies, U.S. Army Command and General Staff College, December 1993.

Eggington, Jack B. *Ground Maneuver and Air Interdiction, A Matter of Mutual Support at the Operational Level of War.* Maxwell, AFB, AL: School of Advanced Airpower Studies, Air University Press, August 1984.

Ellermann, Werner. *Die grundlegenden deutschen Überlegungen zur Abwehr einer Invasion in Frankreich.* Hamburg: Führungsakademie der Bundeswehr, 1962.

Ellison, Gregory W. *Operational Art: The Missing Link in the Iran-Iraq War.* Fort Leavenworth, KS: School of Advanced Military Studies, U.S. Army Command and General Staff College, April 1988.

Engelhaupt, Hans-Joachim. *Operatives Denken und Planen Mansteins 'Philosophie.' Eine Untersuchung mit Hinblick auf die Bedingungen eines modernen Kriegs-/Einsatzgebietes.* Hamburg: Führungsakademie der Bundeswehr, January 1993.

Enzenberger, Kurt. *Die allierte und deutsche Luftwaffenführung im 2. Weltkrieg – Gedanken zum Problem einheitlicher Führung.* Hamburg: Führungsakademie der Bundeswehr, 1964.

Estorf, Wolfgang. *Führungsvorstellungen zur Abwehr einer invasion in Frankreich 1944 – Darstellung und Vergleich der Auffassungen Hitlers, von Rundstedts und Rommels.* Hamburg: Führungsakademie der Bundeswehr, 1967.

Fastabend, David A. *A Theory of Conflict and Operational Art.* Fort Leavenworth, KS: School of Advanced Military Studies, U.S. Command and General Staff College, 1988.

Fix, Robert G. *Operational Encirclement: Quick Decisive Victory or a Bridge Too Far?* Fort Leavenworth, KS: School of Advanced Military Studies, U.S. Command and General Staff College, 1993.

Flanigan, David B. *The Culminating Point and U.S. Army Tactical Doctrine.* Fort Leavenworth, KS: School of Advanced Military Studies, U.S. Army Command and General Staff College, 1992.

Flynn, Michael T. *Climax or Conclusion: Culmination in Defense.* Fort Leavenworth, KS: School of Advanced Military Studies, 1994.

Fontaine, Yves J. *Operational Sustainment in an Immature Theater.* Fort Leavenworth, KS: School of Advanced Military Studies, U.S. Army Command and General Staff College, 1990.

Frandsen, Herbert L. *Counterblitz: Conditions for a Successful Counteroffensive.* Fort Leavenworth, KS: School of Advanced Military Studies, U.S. Army Command and General Staff College, 1990.

Franklin, Charles D. *Time, Space, and Mass at the Operational Level of War: The Dynamics of the Culminating Point.* Fort Leavenworth, KS: School of Advanced Military Studies, U.S. Army Command and General Staff College, April 1988.

Franz, Wallace P. *The Character of Modern War: Theory-Doctrine-Practice at the Operational Level.* Carlisle Barracks, PA: Strategic Studies Institute, U.S. Army War College, December 1981.

Frostic, Fred L. *Air Campaign Against the Iraqi Army in Kuwaiti Theater of Operations.* Santa Monica, CA: RAND, 1994.

Fulbright, Barry D. *The Effects of Airpower on the Offensive Culminating Point.* Fort Leavenworth, KS: School of Advanced Military Studies, U.S. Army Command and General Staff College, April 1988.

Gardner, Gregory C. *Generalship in War: The Principles of Operational Command.* Fort Leavenworth, KS: School of Advanced Military Studies, U.S. Army Command and Staff College, May 1987.

Garrett, Jerry D. *Strategic Airpower as Operational Fires: Integrating Long-Range Bombers into Campaign Design.* Fort Leavenworth, KS: School of Advanced Military Studies, U.S. Army Command and General Staff College, May 1991.

Giles, Phillip Kevin, and Thomas P. Galvin. *Center of Gravity: Determination, Analysis, and Application.* Carlisle Barracks, PA: Center for Strategic Leadership, U.S. Army War College, January 1996.

Glantz, David M. *From the Don to the Dnepr: A Study of Soviet Offensive Operations, December 1942–August 1943.* Fort Leavenworth, KS: U.S. Army Command and General Staff College, 1985.

Goehring, J. Russell. *Sequencing Operations: The Critical Path of Operational Art.* Fort Leavenworth, KS: School of Advanced Military Studies, U.S. Army Command and General Staff College, May 1987.

Göttelmann, Heribert. *Die deutschen Angriffe im Frűhjahr/Sommer 1918 in Frankreich und die Ardennen-Offensive 1944/45. Vergleich von Zielsetzung-Durchfűhrung-Ergebnis.* Hamburg: Fűhrungsakademie der Bundeswehr, December 1968.

Gregory, Charles R. *Operational Reserves – Renewing the Offensive Spirit.* Fort Leavenworth, KS: School of Advanced Military Studies, U.S. Army Command and General Staff College, April 1988.

Greiffenberg, Hans von. *Deception and Cover Plans.* Washington, DC: Department of the Army, Office of the Chief of Military History, MSS OP-044A, May 1952.

Gribling, Richard H. *The Yom Kippur War: Insights into Operational Theory.* Fort Leavenworth, KS: U.S. Army and General Staff College, School of Advanced Military Studies, 1988.

Grimsley, William F. *Commander's Preparation for the Future Operational Battlefield.* Fort Leavenworth, KS: U.S. Army and General Staff College, School of Advanced Military Studies, 1993.

Griswold, Myron J. *Considerations in Identifying and Attacking the Enemy's Center of Gravity.* Fort Leavenworth, KS: School of Advanced Military Studies, U.S. Army Command and General Staff College, May 1986.

Gross, David F. *Logistics Implications of the Operational-Level Offensive.* Fort Leavenworth, KS: School of Advanced Military Studies, U.S. Army Command and General Staff College, 1984.

Gutwald, Rick. *The Effects of Operational Encirclement.* Fort Leavenworth, KS: School of Advanced Military Studies, U.S. Command and General Staff College, 1987.

Hamilton, William W. *Operational Vision – An Essential Trait for Army Operational Commanders.* Fort Leavenworth, KS: School of Advanced Military Studies, U.S. Army Command and General Staff College, May 1991.

Hammond, Charles O. *Operational Fires and Unity of Command.* Fort Leavenworth, KS: School of Advanced Military Studies, U.S. Army Command and General Staff College, 1990.

Hanisch, Norbert. *Untersuchen Sie die operativen Ideen Manstein hinsichtlich Schwerpunktbildung, Überraschung, Initiative und Handlungsfreiheit an den Beispielen Westfeldzug 1940 (Sichelschnitt-Plan) und Operation Zitadelle.* Hamburg: Führungsakademie der Bundeswehr, 15 January 1988.

Hannak, Gerd. *Das Urteil Über Den Landkrieg In Der Luftkriegslehre Douhets.* Hamburg: Führungsakademie der Bundeswehr, September 1969.

Heinstein, Rudolf. *Zur Strategie des mehrfrontenkrieges. Das Problem der "inneren und äusseren Linie" dargestellt am Beispiel des Ersten Weltkrieges.* Hamburg: Führungsakademie der Bundeswehr, 10 November 1975.

Hettfleisch, Hans-Christian. *Operatives Denken in den sowjetischen Streitkräften-die Entwicklung von der Oktoberrevolution bis heute.* Hamburg: Führungsakademie der Bundeswehr, December 1988.

Hooper, Thomas A. *National Power, Strategic Goals, and Operational Art in the First World War: The Relationship.* Fort Leavenworth, KS: School of Advanced Military Studies, U.S. Army Command and General Staff College, May 1988.

House, John M. *Do Doctrinal Buzzwords Obscure the Meaning of Operational Art?* Fort Leavenworth, KS: School of Advanced Military Studies, U.S. Army Command and Staff College, April 1989.

Huber, Thomas M. *Pastel: Deception in the Invasion of Japan.* Fort Leavenworth, KS: School of Advanced Military Studies, U.S. Army Command and General Staff College, 1988.

Hufenbach, Gottfried. *Die Unterstützung des Heeres in der Auffassung der deutschen Luftwaffe bis zum Vorabend des Zweiten Weltkrieges. Ein Beitrag zum Problem der Führungskoordination von Teilstreitkräften.* Hamburg: Führungsakademie der Bundeswehr, 1975.

Irwin, Ash. *The Levels of War, Operational Art and Campaign Planning, Occasional Paper No. 5.* Camberley, Surrey, England: Strategic and Combat Studies Institute, 1993.

Ivey, William D. *Objectives and Success – Linking National Policy Objectives and Military Strategic Objectives to Achieve Success.* Carlisle Barracks, PA: U.S. Army War College, 1996.

Izzo, Lawrence L. *A Perspective of the Operational Level of War and Its Implications.* Fort Leavenworth, KS: School of Advanced Military Studies, U.S. Army Command and General Staff College, 1986.

Johnson, Darfus L. *Center of Gravity: The Source of Operational Ambiguity and Linear Thinking in the Age of Complexity.* Fort Leavenworth, KS: School of Advanced Military Studies, U.S. Army Command and General Staff College, December 1998.

Johnson, John D. *Sequencing Operations: Considerations for the Operational Planner.* Fort Leavenworth, KS: School of Advanced Military Studies, May 1991.

Jordan, Stephen T. *Operational Art: Practical Utility or Defunct Doctrinal Concept.* Fort Leavenworth, KS: School of Advanced Military Studies, U.S. Army Command and General Staff College, December 1991.

Kalb, John F. *A Foundation for Operational Planning: The Concept of Center of Gravity, Decisive Point and the Culminating Point.* Fort Leavenworth, KS: School of Advanced Military Studies, U.S. Army Command and General Staff College, April 1987.

Kellett, Anthony. *Combat Motivation, ORAE Report No. R77*. Ottawa, Ontario, Canada: Department of Defence, November 1980.

Kidder, Bruce L. *Center of Gravity: Dispelling the Myths*. Carlisle Barracks, PA: U.S. Army War College, 1996.

Kievit, James O. *Operational Art in the 1944 Ardennes Campaign*. Fort Leavenworth, KS: School of Advanced Military Studies, U.S. Army Command and General Staff College, May 1987.

Kipp, Jacob W. *Mass Mobility, and the Red Army's Road to Operational Art, 1918–1936*. Fort Leavenworth, KS: Soviet Army Studies Office, U.S. Army Combined Arms Center, 1987.

Kirchbach, Hans-Peter von. *Clausewitz and the Culmination Point of Victory*. Carlisle Barracks, PA: U.S. Army War College, March 1989.

Kirkland, Donald E. *Rommel's Desert Campaigns, February 1941–September 1942: A Study in Operational Level Weakness*. Fort Leavenworth, KS: School of Advanced Military Studies, U.S. Army Command and General Staff College, May 1986.

Klink, Ernst. *Die Begriffe 'Operation' und 'operativ' in ihrer militärischen Verwendung in Deutschland*. Freiburg, i.Br.: Militärgeschichtliches Forschungsamt, December 1958.

Kolditz, Thomas A. *Exploring the Conditions for Decisive Operational Fires*. Fort Leavenworth, KS: School of Advanced Military Studies, U.S. Command and General Staff College, June 1993.

Könen, Erich H. *Die operativen Ideen Mansteins hinsichtlich Nuetzung des Raumes, Gewinnen der Initiative, Schaffen von Handlungsfreiheit und Wahl zwischen offensivem and defensivem Vorgehen. Eine Untersuchung anhand der Beispiele 'Rochade' und Schlacht bei Kharkow des Winterfeldzuges 1942/43*. Hamburg: Fűhrungsakademie der Bundeswehr, November 1988.

Kriesel, Wolf-Dietrich. *Weltkriegserfahrung und Kriegsvorbereitung: Die militärischen Lehren des 1. Weltkrieges und deren Umsetzung bei der Vorbereitung des 2. Weltkrieges*. Hamburg: Fűhrungsakademie der Bundeswehr, October 1983.

Kriwanek, Thomas M. *The Operational Center of Gravity*. Fort Leavenworth, KS: School of Advanced Military Studies, U.S. Army Command and General Staff College, May 1986.

Krueger, Daniel W. *Calculated Risk? – Military Theory and the Allies' Campaign in Italy, 1943–1944*. Fort Leavenworth, KS: School of Advanced Military Studies, U.S. Army Command and General Staff College, May 1988.

Kuykendall, Anthony N. *Operational Level Campaign Planning*. Fort Leavenworth, KS: School of Advanced Military Studies, U.S. Army Command and General Staff College, April 1988.

Landry, Alan D. *Time Estimation at the Operational Level of War*. Fort Leavenworth, KS: School of Advanced Military Studies, U.S. Army Command and General Staff College, June 1990.

Lange, Werner. *Raum, Zeit und Mittel als Faktoren strategischer Zielsetzung. Überlegungen zu Napoleons Feldzug in Russland 1812*. Hamburg: Fűhrungsakademie der Bundeswehr, November 1964.

Leaf, Daniel P. *Unity of Command and Interdiction*. Maxwell Air Force Base, AL: College of Aerospace Doctrine, Research, and Education, Airpower Research Institute, July 1994.

Leggemann, Christian. *Die taktisch/operativen Ideen Mansteins hinsichtlich Schwerpunktbildung, Überraschung, Initiative und Handlungsfreiheit*. Hamburg: Fűhrungsakademie der Bundeswehr, January 1989.

Lidsba, Achim. *"Fűhren durch Auftrag" als Norm und Form der Fűhrung im Heer, im Frieden wie im Krieg – Tendenzen des "Fűhren mit Auftrag" in Gegenwart und Zukunft*. Hamburg: Fűhrungsakademie der Bundeswehr, October 1979.

Liedtke, Gert. *Der Balkanfeldzug 1941: Beispiel Einer Offensive Aus Der Bewegung Ohne Vorbereitender Aufmarsch*. Hamburg, Fűhrungsakademie der Bundeswehr, January 1963.

Lovatt, Brian A. *The Essentials of Operational Momentum*. Fort Leavenworth, KS: School of Advanced Military Studies, U.S. Army Command and General Staff College, May 1987.

Lucchese, Donna. *The Relationship of Center of Gravity Analysis, Targeting for Effect, and Measuring Success.* Carlisle Barracks, PA: Army War College, 1998.

Lynch, Timothy D. *Operational Synchronization: A Revised Doctrinal Perspective.* Fort Leavenworth, KS: School of Advanced Military Studies, U.S. Army Command and General Staff College, May 18, 1990.

Madden, Robert W. *A Thousand Points of Light: Integrating Operational Fires into Campaign Design.* Fort Leavenworth, KS: School of Advanced Military Studies, U.S. Army Command and General Staff College, May 1990.

Mahaney, Michael. *Operational Durability: The Marines and Operational Maneuver from the Sea.* Fort Leavenworth, KS: School of Advanced Military Studies, U.S. Army Command and General Staff College, May 2001.

Marr, John E. *War in the Falklands: Perspectives British Strategy and Use of Air Power.* Maxwell AFB, AL: Air War College Research Report, Air University, 1988.

Martin, Jerome V. *Victory from Above: Air Power Theory and the Conduct of Operations Desert Shield and Desert Storm, Research Report No. AU-ARI-92-8.* Maxwell AFB, AL: Air University Press, June 1994.

Marx, Harry. *Raum-Zeit-Mittel als Faktoren strategischer Zielsetzung. Überlegungen zu Napoleons Feldzug in Russland 1812.* Hamburg: Führungsakademie der Bundeswehr, 1964.

May, Günter. *Zum Probleme der Auftragstaktik in schwierigen Lagen: Der "Haltebefehl" Hitlers, die deutsche Heeresführung und die Entschlussfreiheit des Truppenführers während der Winterkrise des Russlandfeldzuges 1941/42.* Hamburg: Führungsakademie der Bundeswehr, 1983.

McGinnis, Thomas M. *Jomini and the Ardennes: An Analysis of Lines of Communications and Decisive Points.* Fort Leavenworth, KS: School of Advanced Military Studies, U.S. Army Command and General Staff College, May 1988.

McGraw, Donald C., Jr. *Operational Pauses: Minimizing the Effects of Culminating in Joint Operations.* Fort Leavenworth, KS: School of Advanced Military Studies, U.S. Army Command and General Staff College, May 1991.

McRobb, Nick. *Luftlandeoperationen: Möglichkeiten, Grenzen und Gültigkeit in Zukunft.* Hamburg: Führungsakademie der Bundeswehr, February 1999.

Meisner, Bruce L. *The Culminating Point – A Viable Operational Concept or Some Theoretical Nonsense?* Fort Leavenworth, KS: School of Advanced Military Studies, U.S. Army Command and General Staff College, March 1986.

Metz, Steven. *Eisenhower as Strategist: The Coherent Use of Military Power in War and Peace.* Carlisle Barracks, PA: Strategic Studies Institute, U.S. Army War College, February 1993.

Mixon, Robert W., Jr. *A Dramatic Challenge to Operational Theory: The Sinai Campaign, October 1973.* Fort Leavenworth, KS: School of Advanced Military Studies, U.S. Army Command and General Staff College, May 1987.

Moore, T. D. *Logistics Intelligence: The First Step in Operational Sustainment.* Fort Leavenworth, KS: School of Advanced Military Studies, U.S. Army Command and General Staff College, 1990.

Mueller, D. Holger. *The Character of Manoeuvre at the Operational Level.* Toronto: Advanced Military Studies Course, Canadian Forces College, 1999.

Nichols, Howard V. *Operational Level Logistics: An Examination of U.S. Army Logistical Doctrine for the Operational Level of War.* Fort Leavenworth, KS: School of Advanced Military Studies, U.S. Army Command and General Staff College, May 1986.

O'Brien, John. *Coup d'Oeil: Military Geography and the Operational Level of War.* Fort Leavenworth, KS: School of Advanced Military Studies, U.S. Army Command and General Staff College, December 1991.

Orlov, Morton, II. *Definitions and Doctrine: Operational Language and Understanding in Combined Arms Warfare.* Fort Leavenworth, KS: School of Advanced Military Studies, U.S. Army Command and Staff College, December 1993.

Pagonis, William G., and Michael D. Krause. *Operational Logistics and the Gulf War, Land Warfare Papers No. 13*. Arlington, VA: Association of the United States Army, Institute of Land Warfare, October 1992.

Palmer, Peter J. *Operational Main Effort and Campaign Planning.* Fort Leavenworth, KS: School of Advanced Military Studies, U.S. Army Command and General Staff College, 1991.

Parks, Wayne A. *Operational-Level Deep Operations: A Key Component of Operational Art and Future Warfare.* Fort Leavenworth, KS: School of Advanced Military Studies, U.S. Army Command and General Staff College, May 1998.

Partridge, Mark S. *Avoiding Operational Surprise: Lessons from World War II.* Fort Leavenworth, KS: School of Advanced Military Studies, U.S. Army Command and General Staff College, April 1987.

Pellegrini, Robert. *The Links between Science, Philosophy, and Military Science: Understanding the Past, Implications for the Future.* Maxwell AFB, AL: School of Advanced Airpower Studies, Air University Press, August 1997.

Pennypacker, William S. *Sequels: Thinking About the Future.* Fort Leavenworth, KS: School of Advanced Military Studies, U.S. Army Command and General Staff College, May 1988.

Peterson, Stephen P. *Operational Sustainment: The Impact of Critical Decisions upon Operational Design.* Fort Leavenworth, KS: School of Advanced Military Studies, U.S. Army Command and General Staff College, May 1989.

Petrole, Gary P. *Understanding the Operational Effect.* Fort Leavenworth, KS: School of Advanced Military Studies, U.S. Army Command and General Staff College, September 1991.

Pickar, Charles K. *Blitzkrieg: Operational Art or Tactical Craft?* Fort Leavenworth, KS: School of Advanced Military Studies, U.S. Army Command and General Staff College, May 1992.

Plocher, Hermann. *The German Air Force Versus Russia, 1941.* Maxwell AFB, AL: USAF Historical Division, Aerospace Studies Institute, Air University, July 1965.

Porter, Lanning M. *Preconceptions, Predilections, and Experience: Problems for Operational Level Intelligence and Decisionmaking.* Fort Leavenworth, KS: School of Advanced Military Studies, U.S. Army Command and General Staff College, 1989.

Powers, George W. *The Civil War Campaigns of 1864 – Operational and Tactical Defeat Leading to Strategic Victory.* Fort Leavenworth, KS: School of Advanced Military Studies, U.S. Army Command and General Staff College, February 1988.

Quinlan, Kenneth J. *Initiative in the Operational Defense – Is It Possible?* Fort Leavenworth, KS: School of Advanced Military Studies, U.S. Army Command and General Staff College, May 1987.

Read, Steven N. *Planning for the Unplannable: Branches, Sequels, and Reserves.* Fort Leavenworth, KS: School of Advanced Military Studies, U.S. Army Command and General Staff College, June 1990.

Redlinger, Mark J. *Hans Delbrueck and Clausewitz's Culminating Points.* Fort Leavenworth, KS: School of Advanced Military Studies, U.S. Army Command and General Staff College, May 1988.

Reece, Ralph G. *Operational Fires.* Maxwell AFB, AL: Air War College, 1989.

Rember, William B. *Air Power in the Great War: Operational Lessons Learned or Lost?* Fort Leavenworth, KS: U.S. Army and General Staff College, School of Advanced Military Studies, 1993.

Rhoades, George H. *Decentralized Command and Control – Reality or Myth?* Carlisle Barracks, PA: U.S. Army War College, 1990.

Rice, Edward A., Jr. *The Operational Commander and War Termination on Favorable Terms.* Newport, RI: Naval War College, February 1989.

Rice, William J. *Operational Fires: What's in a Name?* Fort Leavenworth, KS: School of Advanced Military Studies, U.S. Army Command and General Staff College, June 1990.

Richter, Harry. *Die logistische Sicherstellung der Operationen im Falle "Barbarossa." Die Planung und Vorbereitungen der Versorgungsführung bei der Erstellung des Feldzugplanes und die Gründe für die Entwicklung zur krisenhaften Versorgungslage während der Offensive auf Moskau im Jahre 1941 bei der Heeresgruppe Mitte.* Hamburg: Führungsakademie der Bundeswehr, November 1986.

Riley, Don T. *The Evolution of Operational Art – The Reconquest of Burma, 1943–1945.* Fort Leavenworth, KS: School of Advanced Military Studies, U.S. Army Command and General Staff College, September 1987.

Rios, Leon H. *The Linkage of the Strategic and Operational Levels of War.* Fort Leavenworth, KS: School of Advanced Military Studies, U.S. Army Command and General Staff College, May 1986.

Rios, Leon H. *Seeking a Final Victory: Creating Conditions for Conflict Resolution.* Carlisle Barracks, PA: U.S. Army War College, 1993.

Ritter, Herbert. *Bei allen Überlegungen zur Gefechtsführung gilt das Dogma, dass der Angreifer dem Verteidiger 3:1 an Kräften überlegen sein muss, um mit höher Wahrscheinlichkeit Erfolg zu haben. Beurteilen Sie diese Behauptung vor dem Hintergrund kriegsgeschichtliche Beispiele für die Divisionebene.* Hamburg: Führungsakademie der Bundeswehr, October 1983.

Robertson, William G., and Lawrence A. Yates. *Block by Block: The Challenges of Urban Operations.* Fort Leavenworth, KS: U.S. Army Command and General Staff College Press, 2003.

Rockwell, Christopher A. *Operational Sustainment, Lines of Communication and the Conduct of Operations.* Fort Leavenworth, KS: School of Advanced Military Studies, U.S. Army Command and General Staff College, May 1987.

Rodriguez, Joseph O., Jr. *How to Maximize the Advantages of Interior Lines at the Operational Level.* Ft. Leavenworth, KS: School of Advanced Military Studies, May 1987.

Roh, Daniel J. *Operational Pause vs. Offensive Culmination: Lessons in Eisenhower's Broad Front Strategy.* Fort Leavenworth, KS: School of Advanced Military Studies, U.S. Army Command and General Staff College, April 1988.

Santala, Russel D. *Operational Art in the Spanish-American War: An Analysis of the American Way of War in Major Regional Contingency.* Fort Leavenworth, KS: U.S. Army and General Staff College, School of Advanced Military Studies, 1993.

Savoia, Thomas A. *Deception at the Operational Level of War.* Fort Leavenworth, KS: School of Advanced Military Studies, U.S. Army Command and General Staff College, May 1986.

Schlott, John E. *Operational Vision: The Way Means Reach the End.* Fort Leavenworth, KS: School of Advanced Military Studies, United States Army Command and General Staff College, May 1992.

Schmitt, John F. *A Practical Guide for Developing and Writing Military Concepts, DART Working Paper # 02-43.* McLean, VA: Hicks & Associates, Defense Adaptive Red Team, December 2002.

Schroedel, Joseph. *The Art and Science of Operational Maneuver.* Fort Leavenworth, KS: School of Advanced Military Studies, U.S. Command and General Staff College, May 1988.

Scudder, John V. *Synchronization and the Joint Task Force Planning Process.* Fort Leavenworth, KS: School of Advanced Military Studies, U.S. Command and General Staff College, 1993.

Seebens, Dieter. *Grundlagen, Auffassungen und Pläne für eine Kriegführung in der Ostsee 1935–1939.* Hamburg: Führungsakademie der Bundeswehr, August 1971.

Seifert, Martin. *Der Balkanfeldzug 1941: Beispiel einer Offensive aus der Bewegung ohne vorbereitenden Aufmarsch.* Hamburg: Führungsakademie der Bundeswehr, March 1963.

Sittnick, Stephen M. *Are the Operational Operating Systems an Adequate Framework for Operational Synchronization?* Fort Leavenworth, KS: U.S. Army and General Staff College, School of Advanced Military Studies, 1993.

Smith, Carleton M. *Logistics Principles in Third Wave Warfare.* Carlisle, PA: Army War College, 1999.

Smith, Philip A. *Bombing to Surrender: The Contribution of Airpower to the Collapse of Italy 1943.* Maxwell AFB, AL: Air University Press, August 1998.

Snodgrass, James G. *Operational Maneuver – From the American Civil War to the OMG: What Are Its Origins and will It Work Today?* Fort Leavenworth, KS: School of Advanced Military Studies, U.S. Army Command and General Staff College, February 1988.

Springman, Jeffrey A. *The Relationship Among Tasks, Centers of Gravity, and Decisive Points.* Fort Leavenworth, KS: School of Advanced Military Studies, U.S. Army Command and General Staff, May 1998.

Starry, Michael D. *Deception and the Operational Level of War.* Fort Leavenworth, KS: School of Advanced Military Studies, U.S. Army Command and General Staff College, 1987.

Steffan, D. Grant. *Changing the Campaign Plan in Midstream: Deciding Whether to Cancel an Operation.* Fort Leavenworth, KS: School of Advanced Military Studies, U.S. Army Command and General Staff College, 1993.

Stewart, D. *Operation Anvil/Dragoon: The Invasion of Southern France.* Fort Leavenworth, KS: School of Advanced Military Studies, U.S. Army Command and General Staff College, 1984.

Stohry, Kimble D. *Airpower in Your Hip Pocket: Under What Conditions Should an Operational Commander Constitute an Air Reserve?* Fort Leavenworth, KS: School of Advanced Military Studies, U.S. Army Command and General Staff College, May 1990.

Strategic Studies Institute. *The Malvinas/Falklands War of 1982: Lessons for the United States and Latin America.* Carlisle Barracks, PA: Strategic Studies Institute, U.S. Army War College, August 1983.

Stuart, Paul H. *Operational Maneuver: Creator of the Decisive Point.* Fort Leavenworth, KS: School of Advanced Military Studies, U.S. Army Command and General Staff College, May 1987.

Sweeney, Patrick C. *Risk – The Operational Edge in the Peninsula Campaign of 1862.* Fort Leavenworth, KS: School of Advanced Military Studies, U.S. Army Command and General Staff College, April 1988.

Taddonio, Frank T. *The Concept of the Culminating Point of Attack.* Fort Leavenworth, KS: School of Advanced Military Studies, U.S. Army Command and General Staff College, May 1986.

Tallant, Shawn R. *Fleet Admiral Ernest J. King: Strategist, Leader & Clausewitzian.* Maxwell AFB, AL: Air War College, Spring 1994.

Thomas, Kent. *Second Manassas: An Operational Dynamics Perspective.* Fort Leavenworth, KS: School of Advanced Military Studies, U.S. Army Command and General Staff College, April 1988.

Tokar, Leonard G. *U.S. Doctrine for Command and Control of Operational Fires.* Fort Leavenworth, KS: School of Advanced Military Studies, U.S. Army Command and General Staff College, April 1996.

Turner, Albert F., Jr. *The Operational Reserve: What Should It Be Used For?* Fort Leavenworth, KS: School of Advanced Military Studies, U.S. Army Command and General Staff College, April 1988.

Vanderbeek, Walter A. *The Decisive Point: The Key to Victory.* Fort Leavenworth, KS: School of Advanced Military Studies, U.S. Army Command and General Staff College, April 1988.

Wasson, James N. *Innovator or Imitator: Napoleon's Operational Concepts and the Legacies of Bourcet and Guibert.* Fort Leavenworth, KS: School of Advanced Military Studies, United States Army Command and General Staff College, May 1998.

Watts, Barry D. *Clausewitzian Friction and Future War.* Washington, DC: National Defense University, October 1996.

Wehner, Randolph B. *Sustainment Improvisation – Expanding the Realm of the Possible.* Fort Leavenworth, KS: School of Advanced Military Studies, U.S. Army Command and General Staff College, May 1987.

Weimer, Michael B. *Preparing to Be Deceived: Counterdeception at the Operational Level of War.* Fort Leavenworth, KS: School of Advanced Military Studies, U.S. Army Command and General Staff College, May 1988.

Wells, Gordon M. *The Center of Gravity Fad: Consequence of the Absence of an Overarching American Theory of War.* Arlington, VA: AUSA's Institute of Land Warfare, March 2001.

Wenning, Peter. *Aufmarschplanung als strategische Führungsproblem. Eine vergleichende Untersuchung am Beispiel des Schlieffenplans 1914, der NATO Aufmarschplanung im Rahmen der "Flexible Response" un dem neuen strategischen Konzept der NATO.* Hamburg: Führungsakademie der Bundeswehr, December 1997.

Werb, Franz. *Raum – Zeit – Mittel als Faktoren Strategischer Zielsetzung.* Hamburg: Führungsakademie der Bundeswehr, February 1965.

Wieker, Volker. *Operation, Operative Führung zwischen Taktik und Strategie – Bedeutung und Entwicklung des Begriffes seit Moltke d. Ae. – Bearbeitung aus der Sicht der Landstreitkräfteführung.* Hamburg: Führungsakademie der Bundeswehr, 1987.

Williams, Brett T. *Effects-Based Operations: Theory, Application and the Role of Airpower.* Carlisle Barracks, PA: U.S. Army War College, April 2002.

Zajac, Daniel L. *The German Invasion of Yugoslavia: Insights from Crisis Action Planning and Operational Art in a Combined Environment.* Fort Leavenworth, KS: School of Advanced Military Studies, U.S. Army Command and General Staff College, May 1993.

Articles

Adams, Dwight L., and Clayton R. Newell. "Operational Art in the Joint and Combined Arenas." *Parameters* (June 1988).

Allen, Ralph L. "Piercing the Veil of Operational Art." *Parameters* 4 (Winter 1986).

Anderson, Jeffrey W. "Operational Art on the Eastern Front." *Military Review* 6 (June 1988).

Arthur, Stanley R., and Marvin Pokrant. "Desert Storm at Sea." *Proceedings/Naval Review* 117, no. 5 (May 1991).

Atkeson, Edward B. "The Operational Level of War." *Military Review* 3 (March 1987).

Bahnemann, Jörg. "Der Begriff der Strategie bei Clausewitz, Moltke und Liddell Hart." *Wehrwissenschaftliche Rundschau* 1 (January 1968).

Baker, Robert G. "Friedrich von Bernhardi: Clausewitz Disciple and Prussian Militarist." *Marine Corps Gazette* 5 (May 1991).

Baldwin, Hanson W. "Okinawa: Victory at the Threshold." *Marine Corps Gazette* 12 (December 1950); 1 (January 1951).

Banisch, Werner W. "Leadership at the Operational Level." *Army* 8 (August 1987).

Beck-Broichsitter, Helmut. "Über die Beharrlichkeit im Angriff." *Militär Wehrwissenschaftliche Rundschau* 1 (January 1944).

Blumenson, Martin. "Sicily and Italy: Why and What For?" *Military Review* 2 (February 1966).

Blumenson, Martin. "A Deaf Ear to Clausewitz: Allied Operational Objectives WWII." *Parameters* (Summer 1993), 16.

Böckmann, Herbert von. "Der moderne Feldherr." *Europäische Sicherheit* 1 (January 1951).

Bolle, Hermann. "Wesen und Merkmale der Oberen Führung unter besonderer Berücksichtigung der Abgrenzung zur Mittleren Führung." *Wehrkunde* 11 (November 1967).

Bolt, William J., and David Jablonsky. "Tactics and the Operational Level of War." *Military Review* 2 (February 1987).

Bötticher, Friedrich von. "Feldherrntum und Generalstab: Grundsätze und Geist." *Wehrkunde* 9 (September 1964).

Brand, Dieter. "Freie Operationen (1)." *Truppenpraxis/Wehrausbildung* 10 (October 1997).
Brand, Dieter. "Freie Operationen (2)." *Truppenpraxis/Wehrausbildung* 11 (November 1997).
Brand, Dieter. "Plädoyer für freie Operationen." *Österreichische Militärische Zeitschrift* 2 (March–April 1998).
Brodie, Bernard. "Strategy as an Art and Science." *Naval War College Review* 1 (Winter 1998) (reprint of the article published in February 1959).
Büntemeyer, Fredi. "Information als entscheidende Dimension der Auseinandersetzung im 21. Jahr-hundert. Sicherheits- und militärpolitische Auswirkungen moderner Informationstechnologies." *Soldat und Technik* 9 (September 1996).
Burnhauser, Anton. "Grundbegriffe der Strategie. Ein Versuch zu ihrer Erläuterung und Abgrenzung." *Wehrkunde* 10 (October 1966).
Cochenhausen, Friedrich von. "Wie betreibt der Anfänger am zweckmässigsten Kriegsgeschichte?" *Militärwissenschaftliche Mitteilungen* 9–10 (September–October 1931).
Cochenhausen, Friedrich von. "Schlieffen. Zu seinem 100-jährigen Geburtstage 28." *Militärwissenschaftliche Mitteilungen (Vienna)* 2 (February 1933).
Collins, Lawton J. "Leadership at Higher Echelons." *Military Review* 5 (May 1990).
Connolly, Richard L. "Exercise of Command." *Naval War College Review* 11 (November 1955).
D'Amura, Ronald M. "Campaigns: The Essence of Operational Warfare." *Parameters* 17, no. 1 (1987).
DeVries, Paul T. "Maneuver and the Operational Level of War." *Military Review* 2 (February 1983).
Duffner, Robert W. "Conflict in the South Atlantic: The Impact of Air Power." *Air University Review* (March–April 1984).
Eder, Philip, and Bruno Günter Hofbauer. "Operation Enduring Freedom."*Österreichische Militärische Zeitschrift* 1 (January–February 2002)
Fallon, Willard G. "Combating the Ballistic Missile Threat." *Proceedings* 7 (July 1994).
Farndale, Martin. "The Operational Level of Command." *R.U.S.I. Journal* 3 (Autumn 1988).
Fastabend, David A. "That Elusive Operational Concept." *Army Magazine* (June 2001).
Franz, Wallace P. "Grand Tactics." *Military Review* 12 (December 1981).
Ganz, Harding A. "Questionable Objective: The Brittany Ports, 1944." *Journal of Military History* 59 (January 1995).
Gerner, Mark H. "Leadership at the Operational Level." *Military Review* 6 (June 1987).
Gibson, Edmund A. "Leaders and Leadership." *Proceedings* 3 (March 1954).
Goldschmidt, Klaus. "Grundlagen der Strategie." *Wehrwissenschaftliche Rundschau* 1 (January 1969).
Greaves, Fielding L. "Intelligence, Key to Victory." *Military Review* 9 (September 1965).
Greiner, Gottfried. "Militärische Operationen heute (Ein Beitrag zur Klärung)." *Wehrkunde* 8 (August 1965).
Guderian, Heinz. "Raum und Zeit in der modernen Kriegführung." *Geopolitik* 1 (January 1951).
Hall, George M. "Culminating Points." *Military Review* 7 (July 1989).
Handel, Michael. "Intelligence and Deception." *Journal of Strategic Studies* (March 1982), 123.
Harley, Jeffrey. "Information, Technology, and the Center of Gravity." *Naval War College Review* 50 (Winter 1997).
Harned, Glenn M. "Quality Versus Quantity: Limits of Operational Art." *Army* 6 (June 1986).
Heine, Wilhelm. "Die Bedeutung der Verkehrswege für Planung und Ablauf militärischer Operationen (dargestellt an Beispielen der Kriegsgeschichte)." *Wehrkunde* 8 (August 1965).
Heuser, Beatrice. "Was ist Strategie? Definitionen zur Kunst des Feldherrn." *Militärgeschichte.-Zeitschrift für Historische Bildung* 3 (2005).
Hitz, Hans. "Taktik und Strategie." Zur Entwicklung kriegwissenschaftliclie. Begriffe, *Wehr-wissenschaftiche Rundschau* 11 (November 1956).

Holder, L. D. "A New Day for Operational Art." *Army* 3 (March 1985).
Hooker, Richard D., Jr. "The Mythology Surrounding Maneuver Warfare." *Parameters* 23 (Spring 1993).
Hugh, Deri A. "British Strategic Air Operations, May to November 1918." *Cross & Cockade International* 37, no. 1 (2006).
Hughes, Daniel J. "Schlichting, Schlieffen, and the Prussian Theory of War in 1914." *Journal of Military History* 59 (April 1995).
Jablonski, Walter. "Die Seekriegführung im Vierten Nahostkrieg aus ägyptischer Sicht." *Marine Rundschau* 7 (July 1978).
Jablonsky, David. "Part I: Strategy and the Operational Level of War." *Parameters* (Spring 1987), 73.
Jacobs, Walter Darnell. "The Art of Operations: Soviet Theoreticians Have Wedged What They Call the Operational Art Between Tactics and Strategy but to Little Apparent Practical Purpose." *Army* 12 (November 1961).
Jelden, Helmut. "Soldat und Raum." *Zeitschrift für Geopolitik* 7 (July 1941).
Johnston, Paul. "Doctrine Is Not Enough: The Effect of Doctrine on the Behavior of Armies." *Parameters* 30, no. 3 (Autumn 2000).
Kettgen, Hans. "Gedanken über Logistik in Theorie und Praxis (I)." *Wehrkunde* 8 (August 1964).
Kiliani, Emanuel. "Die Operationslehre des Grafen Schlieffen und ihre deutschen Gegnern (II. Teil)." *Wehrkunde* 3 (March 1961).
Kindsvatter, Peter S. "Operational Maneuver: The Lessons of WWII." *Army* 8 (August 1986).
Kirin, Stephen J. "Synchronization." *Naval War College Review* 4 (Autumn 1996).
Knox, Dudley W. "The Role of Doctrine in Naval Warfare." *Proceedings* 41, no. 2 (March–April 1915).
Kraska, Peter. "Operative Leitlinie 1994." *Trupenpraxis* 5 (May 1994).
Krause, Michael D. "Moltke and the Origins of Operational Art." *Military Review* 9 (September 1990).
Krumpelt, Ihno. "Die Bedeutung des Transportwesen für den Schlachterfolg." *Wehrkunde* 5 (May 1965).
Liddell Hart, B. H. "The Objective in War: National Object and Military Aim, A Lecture Delivered at the Naval War College on 24 September 1952." *Naval War College Review* 5, no. 4 (1952).
Liddell Hart, B. H. "The Ratio of Troops to Space." *Armor* 3 (May–June 1960).
Lind, William S. "The Operational Art." *Marine Corps Gazette* 4 (April 1988).
Lukić, Božidar. "Suština Faktora Vremena i Prostora kao Vojnih Kategorija." *Vojno Delo* 6 (November–December 1974).
Luttwak, Edward N. "The Operational Level of War." *International Security* 5, no. 3 (Winter 1980/81).
Luvaas, Jay. "Thinking at the Operational Level." *Parameters* 1 (Spring 1986), 34.
Mantey, Friedrich von. "Umfassung, Umgehung und Durchbruch. Eine Schlieffen Studie." *Wehr und Wissen* 10 (October 1931).
Marcy, Scott A. "Operational Art: Getting Started." *Military Review* 9 (September 1990).
Marolz, Josef. "Die Entwicklung der operativen Führung, 1. Angriff – Erster Weltkrieg."*Österreichische Militärische Zeitschrift* 2 (March–April 1973).
Massel, Philip. "Speed vs. Haste." *The Fighting Forces* (August 1947). Reprint, *Military Review* 9 (September 1948).
McKenzie-Smith, Robert H. "Crisis Decision-Making in Israel: The Case of the October 1973 Middle East War." *Naval War College Review* 29, no. 3 (Summer 1976).
McMillan, Joseph. "Talking to the Enemy: Negotiations in Wartime." *Comparative Strategy* 2, no. 4 (October–December 1992).
Meehan, John F., III. "The Operational Trilogy." *Parameters* 16, no. 1 (Autumn 1986).

Mendel, William W., and Lamar Tooke. "Operation Logic: Selecting the Center of Gravity." *Military Review* 6 (June 1993).
Milano, James M. "How Rommel Applied Lessons to His Afrika Korps Operations." *Armor* 5 (September–October 1991).
Moser, Hans-Wilhelm. "Waterloo 1815/Ostpreussen 1914. Operieren auf der Inneren Linie." *Truppenpraxis* 4 (April 1987).
Müller, Christian. "Anmerkungen zur Entwicklung von Kriegsbild und operativ-strategischem Szenario im preussisch-deutschen Heer vor dem Ersten Weltkrieg." *Militärgeschichtliche Mitteilungen* 57 (1998).
Mussel, Walter. "Strategie als Aufgabe der Politik." *Wehrkunde* 4 (April 1966).
Newell, Clayton R. "What is Operational Art?" *Military Review* 9 (September 1990).
Pergent, J. "Neue Begriffe – Operationsgebiet und Kriegstheater." *Wehrkunde* 2 (February 1959).
Pierce, Terry C. "The Tactical-Strategic Link." *Proceedings* 9 (September 1990).
Podlesny, Robert E. "Infrastructure Networks are Key Vulnerabilities." *Proceedings* 125, no. 2 (February 1999).
Record, Jeffrey. "Operational Brilliance, Strategic Incompetence: The Military Reformers and the German Model." *Parameters* 16, no. 1 (Autumn 1986).
Regele, Oskar. "Strategie-Operation-Taktik." *Militär-Wochenblatt* 9 (4 September 1926).
Rehm, Walter. "Die Zahl im militärischen Denken der Gegenwart." *Wehrkunde* 9 (September 1960).
Reid, Brian Hold. "J. F. C. Fuller and B. H. Liddell Hart: A Comparison." *Military Review* (1990), 71.
Rendulic, Lothar. "Operative Beherrschung des Raumes." *Wehrwissenschaftlichen Rundschau* 1 (January 1964).
Runals, Stephen E. "A Different Approach." *Military Review* 10 (October 1987).
Samarđic, Milorad. "Vreme kao faktor oružane borbe." *Vojno Delo* 5 (September–October 1980).
Sas, Anthony. "Invasion of Russia." *Military Review* 6 (June 1971).
Sass, Hubertus. "Über das Verhältniss von Politik und Kriegführung-ganzheitliches Denken bei Clausewitz." *Wehrwissenschaftliche Rundschau* 6 (June 1970).
Schenke, Wolf. "Der Raum als Waffe." *Zeitschrift für Geopolitik* 9 (September 1938).
Schering, Walther. "Clausewitz' Lehre von Zweck und Mittel." *Wissen und Wehr* 9 (September 1936).
Scherz, Reimar, and Theodor Hodouschek. "Kampf um Zeit und Information. Die Digitalisierung des Gefechtsraumes-Der Ansatz des deutschen Heeres." *Truppenpraxis/Wehrausbildung* 11 (November 1999).
Schinzer, Dankmut. "Vom Niedergang der operativen Führung: Die Mahnung Stalingrads." *Truppenpraxis* 4 (April 1987).
Schmitt, John F. "The Use of the Reserve in Combat." *Marine Corps Gazette* (March 1990), 64.
Schneider, James L. "Theoretical Implications of Operational Art." *Military Review* 9 (September 1990).
Schoch, Bruce. "Logistics of the Falklands War." *Army Logistican* 3 (May–June 1986).
Schubert, Hans-Joachim. "Mehr Raum, weniger Zeit und Kräfte." *Truppenpraxis* 5 (May 1995).
Semenchikov, V., and G. Bryuchovski. "Zur Entwicklung der Theorie der Luftoperationen." *Militärwesen* 3 (March 1983).
Sexton, Donald J. "Phantoms of the North: British Deceptions in Scandinavia, 1941–1944." *Military Affairs* (October 1983), 112.
Snow, Joel J. "Airland Battle Doctrine Tenets in Opposition." *Military Review* 9 (October 1987).
Sodenstern, Georg von. "Operationen." *Wehrwissenschaftliche Rundschau* 1 (January 1953).
Stewart, Charles T. "Time as a Concept in Military Strategy." *Military Review* 4 (April 1959).
Stockfisch, Dieter. "Die moralischen Grössen – in Vergessenheit geraten." *Truppenpraxis/Wehrausbildung* 11 (November 1997).

Strachwitz, Karl Ernst. "Wissen ist Macht." *Truppenpraxis/Wehraubildung* 9 (September 1996).

Sutten, Charles G. "Command and Control at the Operational Level." *Parameters* (Winter 1986), 19.

Tschischwitz, Erich von. "Part 1: Der Kulminationspunkt des Angriffes im Landkrieg." *Militär-wissenschaftliche Rundschau (Berlin)* 7, no. 4 (1942).

Tschischwitz, Erich von. "Der Kulminationspunkt des Angriffes im Landkrieg. 2nd Part." *Militärwissenschaftliche Rundschau* 8, no. 1 (1943).

Turlington, John E. "Truly Learning the Operational Art." *Parameters* 17, no. 1 (Spring 1987).

Vad, Erich. "Operative Führung: Grundlagen, Merkmale und Perspektiven."*Ősterreichische Militärische Zeitschrift* 2 (March–April 1998).

Wahlde, Pete von. "A Pioneer of Russian Strategic Thought: G. A. Leer, 1829–1904." *Military Affairs* 4 (December 1971).

Weth, Wolf. "Fűhrungsgrundlagen und Ihre Bedeutung. Betrachtung űber einige historische Ereignisse und die mὸgliche Nutzanwendung der dabei gewonnen Erkentnisses." *Wehrkunde* 2 (February 1967).

Whaley, Barton. "Toward a General Theory of Deception." *The Journal of Strategic Studies* 5 (1982).

Wien, Otto. "Wandel der Kriegfűhrung." *Allgemeine Schweizerische Zeitschrift* 1 (January 1957).

Zais, Mitchell M. "Strategic Vision and Strength of Will: Initiatives for Theater Command." *Parameters* 15, no. 1 (Winter 1985).

Zeitzler, Kurt. "Withdrawals of the German Army on the Eastern Front." *Military Review* 8 (August 1960).

Zeitzler, Kurt. "Mensch und Raum im Kriege (Ein deutsches Problem im zweiten Weltkrieg)." *Wehrkunde* 11 (November 1961).

Ziemke, Earl F. "The Soviet Theory of Deep Operations." *Parameters* 13, no. 2 (June 1983).

Zimmer, Winfried. "Clausewitz and the Human Dimension of War." *Military Review* 74, no. 2 (March 1994).

Index

Note: Page numbers in *italic* indicate a figure on the corresponding page.

For Product Safety Concerns and Information,
please contact our EU representative GPSR@taylorandfrancis.com
Taylor & Francis Verlag GmbH, Kaufingerstraße 24,
80331 München, Germany

Printed by Integrated Books International,
United States of America